SOURCES OF ALCHEMY AND CHEMISTRY
SIR ROBERT MOND STUDIES IN THE HISTORY OF EARLY CHEMISTRY

AMBIX VOLUME 60 SUPPLEMENT 1 2013

The *Four Books* of Pseudo-Democritus

Matteo Martelli

GENERAL EDITORS

Lawrence M. Principe (Johns Hopkins University)
Jennifer M. Rampling (Princeton University)

Editorial Advisory Board

Charles Burnett (Warburg Institute)
Michèle Mertens (Université de Liège)
Cristina Viano (CNRS, Paris)

Routledge
Taylor & Francis Group
LONDON AND NEW YORK

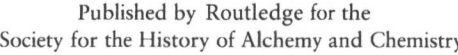
Published by Routledge for the
Society for the History of Alchemy and Chemistry

www.ambix.org

First published 2013 by Maney Publishing Ltd.

Published 2017 by Routledge
2 Park Square, Milton Park, Abingdon, Oxon OX14 4RN
711 Third Avenue, New York, NY 10017, USA

Routledge is an imprint of the Taylor & Francis Group, an informa business

Front cover: Byzantine ourobouros, fol. 196 of Codex Parisinus graecus 2327, copied by Theodoros Pelekanos in Crete in 1478 from a lost manuscript of an early medieval tract attributed to Synosius (Synesius) of Cyrene (d. 412).

ISBN 13: 978-1-909662-28-5 (pbk)
ISSN 0002-6980 (print); ISSN 1745-8234

SOURCES OF ALCHEMY AND CHEMISTRY:
SIR ROBERT MOND STUDIES IN THE HISTORY OF EARLY CHEMISTRY

THE *FOUR BOOKS* OF PSEUDO-DEMOCRITUS
MATTEO MARTELLI

AMBIX   VOLUME 60   SUPPLEMENT 1   2013

CONTENTS

General Editors' Foreword

LAWRENCE M. PRINCIPE and JENNIFER M. RAMPLING

General Editors, Sources of Alchemy and Chemistry

This volume marks the inauguration of *Sources of Alchemy and Chemistry: Sir Robert Mond Studies in the History of Early Chemistry*: a new series of publications that presents critical editions of core texts in the history of alchemy and early chemistry, together with English translations, commentary and scholarly apparatus. The series is produced and distributed under the aegis of *Ambix* and the Society for the History of Alchemy and Chemistry, and named in honour of the Society's first and only President, Sir Robert Mond (1867–1938).

It has often been the case that misunderstandings about alchemy have arisen and persisted because of the inaccessibility of key texts, or faulty readings thereof. Now, with the field attracting an unprecedented level of interest, it is more crucial than ever for scholars to have ready access to dependable editions of primary sources. This need for reliable texts and translations is perhaps most pressing in regard to foundational materials written in Greek, Syriac, Arabic, Hebrew, and Medieval Latin. It is therefore fitting that the inaugural volume of this series should be devoted to one of the very earliest alchemical texts to have come down to us, albeit in fragmentary form: the *Four Books* of the pseudo-Democritus, dating from the first century AD. In this volume, Matteo Martelli presents not only a fresh edition and translation of the surviving Greek fragments, but also includes, for the first time, additional materials that have come down to us in Syriac. To these are added an edition of the most influential Byzantine commentary on the *Four Books* – the dialogue of Synesius and Dioscorus – and Matthaeus Zuber's previously unpublished Latin translations of 1606. Together, these materials offer significant insights into an ancient alchemical source and its medieval and early modern reception: an end that will be pursued further in planned future volumes devoted to the works of Zosimus of Panopolis, Muhammad ibn Zakariyā al-Rāzī, and many others.

The publication of this series has been made possible by the exceptional generosity of Robert Temple. His sponsorship means that all subscribers to *Ambix* and SHAC members, both individual and institutional, will receive *Sources* as part of their normal annual subscription. Given the present difficulty of acquiring support for the painstaking task of preparing critical editions and translations, such sponsorship is particularly precious and far-sighted. The present issue has also received valuable support from the Alexander von Humboldt Stiftung, through Professor Philip van der Eijk's project "Medicine of the Mind, Philosophy of the Body." It is our hope that the new series will in fact bring more serious attention to bear upon textual scholarship, while also providing opportunity, support, and encouragement for the talented scholars who devote themselves to such crucially important labours.

Dedication: for Giulia

Acknowledgements

When I was asked to translate my Italian book on ps.-Democritus into English, I envisaged producing just an English version of my study without either introducing new material or changing the structure of the book. However, when I started working on this project, I changed my original plans. I realised, in fact, that the project opened the possibility of both making available the results of further investigations I had carried on during the last two years (especially on the Oriental tradition) and correcting or making clearer some parts of my earlier research. Consequently, this book represents an English revised version of my Italian *Pseudo-Democrito* that has been expanded in some parts and reduced in others. I decided to leave aside the philological questions related to the Byzantine tradition and to the *constitutio textus* (Chapter 1 of my *Pseudo-Democrito* and the philological notes in the commentary), since I have reprinted the same Greek text as that established in the Italian version (although rectified in some respects). Next to the Greek text, I have presented a new English translation, with a selection of notes focusing in particular on the technical problems arising from ps.-Democritus' *Four Books* and Synesius' *Commentary*. Moreover, I have added two unpublished early translations of these writings: (a) the Syriac translation handed down by the Cambridge manuscript Mm. 6.29, that preserves some parts of the *Four Books* lost in their original version; and (b) the handwritten Latin translation by Matthaeus Zuber (1606), which is preserved in the manuscript *Vindobonensis* lat. 11427. This textual material is introduced by a general survey—largely based on Chapters 2–3 of the Italian version—on ps.-Democritus' *Four Books* and their legacy in the Byzantine, Oriental, and Western traditions.

I have incurred many obligations in preparing this manuscript. Robert Temple contacted me many years ago, when my Italian book had not yet been published. With intense curiosity, he encouraged me to carry on my investigation, which would have never reached its final stage without his passionate and generous support. His personal and scholarly generosity is greatly reflected in this project. Moreover, I had the privilege of discussing many parts of the book with Lawrence Principe and Jennifer Rampling, who carefully read the manuscript. I owe to them a special debt of gratitude for their deep comments on several technical and historical questions. I could rely on their great proficiency in the history of chemistry, and benefit from their fresh insights that allowed me to rethink many aspects of my investigation. I cannot thank Jennifer and Lawrence enough for their great help and for the care they have given to this project. In addition, I am grateful to the reviewers of my Italian *Pseudo-Democrito*, whose comments and suggestions have been taken into account in this new version of the book.

Particular thanks must be given to Philip van der Eijk, who has been a constant source of inspiration for my research. I am currently working under his supervision within the project 'Medicine of the Mind, Philosophy of the Body,' based at the Humboldt Universität zu Berlin and generously supported by the Von Humboldt Stiftung. During the last years I had the good fortune to present several aspects of my research during the weekly seminars organised by Philip, and to benefit from the feedback received from a large group of both young and established scholars.

Finally, my greatest and warmest thanks go to my family and to Giulia, without whose love and encouragement this work would never have taken shape.

Abbreviations

CAAG = Marcellin Berthelot and Charles-Émile Ruelle, *Collection des anciens alchimistes grecs*, 3 vols. (Paris: Georges Steinheil, 1887–88)

CGL = Georg Goetz et al., *Corpus Glossariorum Latinorum*, 7 vols. (Leipzig: Teubner, 1888–1903)

CMA II = Marcellin Berthelot and Rubens Duval, *La chimie au Moyen-Âge*, vol. 2: *L'alchimie syriaque* (Paris: Imprimerie Nationale, 1893)

CMA III = Marcellin Berthelot and Octave Victor Houdas, *La chimie au Moyen-Âge*, vol. 3: *L'alchimie arabe* (Paris: Imprimerie Nationale, 1893)

CMAG = *Catalogues de manuscrits alchimiques grecs*, 8 vols. (Bruxelles: Union Académique Internationale, 1924–32)

CMG = *Corpus medicorum graecorum* (Leipzig and Berlin)

CPF = *Corpus dei papiri filosofici greci* (Firenze: Olschki, 1989–)

DELG = Pierre Chantraine, *Dictionnaire étymologique de la langue grecque*, 4 vols. (Paris: Klincksieck, 1968–80)

DPhA = *Dictionnaire des philosophes antiques*, ed. Richard Goulet, (Paris: éd. Du CNRS, 1989–)

FGrH = Felix Jacoby, *Die Fragmente der Griechischen Historiker*, third edition, 3 vols., 9 tomes (Berlin and Leiden: Brill, 1958)

FHG = Karl Müller, *Fragmenta historicorum graecorum*, 5 vols. (Paris: A.F. Didot, 1841–70)

*LSJ*⁹ = Henry George Liddell, Robert Scott, Henry S. Jones, *A Greek–English Lexicon*, ninth edition (Oxford: Clarendon Press, 1996)

OLD = *Oxford Latin Dictionary* (Oxford: Clarendon Press, 1982)

PG = Jacques Paul Migne, *Patrologiae cursus completus. Series Graeca*, 161 vols. (Paris: 1857–96)

PGM = Karl Preisendanz, *Papyri Graecae Magicae. Die Griechischen Zauberpapyri*, 2 vols. (Leipzig and Berlin: Teubner, 1928–31)

PL = Jacques Paul Migne, *Patrologiae cursus completus. Series Latina*, 221 vols. (Paris: 1844–64)

RE = *Paulys Realencyclopädie der classischen Altertumswissenschaft* (Stuttgart: Metzler, 1893–1980)

SyrLex. Suppl. = Jesse Payne Margoliouth, *Supplement to the Thesaurus Syriacus* (Oxford: Clarendon Press, 1927)

ThGL = Henricus Stephanus, *Thesaurus Linguae Graecae*, 9 vols. (Paris: A.F. Didot, 1831–1865)

ThLL = *Thesaurus Linguae Latinae* (Berlin 1900–)

ThSyr = Robert Payne Smith, *Thesaurus Syriacus*, 2 vols. (Oxford: Clarendon Press, 1879–1901)

INTRODUCTION

§ 1. The *Four Books* of pseudo-Democritus[1]

The four alchemical books ascribed to the Greek atomist Democritus rank among the most ancient examples of Western alchemical writing. The *Four Books* cover a range of technical questions and recipes, similar to those handled in the earliest surviving chemical manuscripts: the Leiden and Stockholm papyri (third century AD). The *Books* also played a central role in the development of alchemy as a discipline.[2] Democritus is frequently cited by the alchemists whose treatises make up the Greek *Corpus alchemicum*, including Zosimus of Panopolis and Synesius.[3] The preoccupation with Democritus continued in the work of later Byzantine writers, including Olympiodorus, Stephanus, and Christianus,[4] as well as the alchemist 'Anonymous,' whose short history of alchemy sets Democritus among the founders of the art.[5]

The *Four Books* no longer exist in their original form, although fragments survive in two treatises: *Physika kai mystika* (*Natural and Secret Questions*) and *Peri asēmou poiēseōs* (*On the Making of Silver*). These texts are actually epitomized

---

[1] This introduction is based upon material previously published in Italian, in Matteo Martelli, *Pseudo-Democrito, scritti alchemici con il commentario di Sinesio* (Paris-Milan: S.É.H.A.-Archè, 2011), Chs. 1–3.

[2] Besides the studies of Kopp, Berthelot, and Lippmann (see *infra*, pp. 5–7), several scholars count the four books by ps.-Democritus among the earliest examples of alchemical writing: F. Sherwood Taylor, "The Origins of Greek Alchemy," *Ambix*, 1 (1937): 37–8; Eric John Holmyard, *Alchemy* (Harmondsworth: Penguin Books, 1957; reprint, New York, 1990), 25–6 |who did not, however, consider them in his previous work, *The Makers of Chemistry* (Oxford: Oxford University Press, 1931)|; James R. Partington, *A Short History of Chemistry* (New York: St. Martin's Press, 1957), 20–1; Robert P. Multhauf, *The Origins of Chemistry* (London: Oldbourne, 1966), 94–101.

[3] The Greek *Corpus alchemicum* represents the collection of Greek writings that have been included in alchemical Byzantine anthologies handed down by several manuscripts (see *infra*, p. 7). These anthologies, however, preserve just a selection of an originally wider corpus of treatises composed between the first and the fourteenth century AD, which have been selected, often epitomized, and reworked by anonymous collectors.

[4] I list all passages in which Zosimus or later alchemists refer to the *Four Books* (at least in the form preserved by the Byzantine tradition) in the apparatus under the translation of the relevant ps.-Democritean excerpts.

[5] *CAAG* II 424,6–425,9. The *pinax* at the start of *Marcianus gr.* 299 |see Joseph Bidez et al., *Catalogue des manuscrits alchimiques grecs*, 8 vols (Bruxelles: Union Académique Internationale, 1924–1932), vol. 2, 20–3 (hereafter *CMAG*)| lists three works ascribed to the φιλόσοφος Ἀνεπίγραφος. At fol. 2ᵛ, nn. 3–4, we have the titles Ἀνεπιγράφου φιλοσόφου πε(ρὶ) θείου ὕδατος and Τοῦ αὐτοῦ περὶ χρυσοποιίας ("the philosopher 'Anonymous', *On Divine Water*" and "*On the Making of Gold* by the same author"). These treatises are preserved in the codex at fols. 78ʳ5–79ʳ10 and fols. 79ʳ11–92ᵛ24. The same list at fol. 2ᵛ, n. 13, mentions another work with the title of Ἀνεπιγράφου φιλοσόφου περὶ χρυσοποιίας ("the philosopher 'Anonymous,' *On the Making of Gold*"): this is found at fols. 181ʳ1–184ᵛ4, under the simple specification Ἀνεπιγράφου φιλοσόφου ("by the philosopher Anonymous"). According to Jean Letrouit, this name conceals two different authors, whom he calls 'Anépigraphe 1' (author of the third treatise listed in the *pinax*) and 'Anépigraphe 2' (author of the first two treatises), both dating to the eighth/ninth century: Letrouit, "Chronologie des alchimistes grecs," in *Alchimie: art, histoire et mythes. Actes du Iᵉʳ Colloque international de la Société d'Étude de l'Histoire de l'Alchimie*, ed. Didier Kahn and Sylvain Matton (Paris-Milan: S.É.H.A-Arché, 1995), 63–5.

---

versions of what were originally four books dealing with dyeing or "colouring" processes: namely, the making of gold, silver, precious stones, and purple dye.[6] Of these, the sections on gold and purple are now found within the *Physika kai mystika*,[7] while the section on silver is now in the *Peri asēmou poiēseōs*. To this material can be added three lists of substances, usually grouped under the title of *Katalogoi* (*Catalogues*), which can also be ascribed to ps.-Democritus.[8]

The aim of this volume is to reconstruct, as far as possible, the original *Four Books* of ps.-Democritus. Through analysis of the *Physika kai mystika* and *Peri asēmou poiēseōs*, and comparison with other traditions – both indirect references in later Greek writings, and direct transmission via Syriac – we can reconstruct a single organized treatise, offering a rational presentation of four different technical fields related to dyeing techniques. The fragmentary and epitomized version in which these books have come down to us makes exhaustive scrutiny of their original contents extremely difficult. Nevertheless, the surviving evidence still allows us to identify the books as fundamental documents, from which we can attempt to reconstruct the principal features of alchemy during its infancy, the kinds of craftsmanship this art involved, and its possible relationship to an Egyptian or, more broadly, Near Eastern tradition.

As I shall argue, the technical content of the *Four Books*, coupled with their importance to later alchemical writing, requires us to reconsider how early 'alchemy' is to be defined. In light of an ancient tradition that unanimously placed Democritus among the first and most important fathers of alchemy, we must accept that the topics covered in his books were considered to be key aspects of the discipline. The crafts of changing base metals into gold and silver, making precious stones, and dyeing wool purple, therefore represent the technical background against which alchemy developed. Indeed, the four books on these subjects under the name of Democritus correspond in substance to the topics covered by the Leiden and Stockholm papyri. This wide range of interests contrasts with a more restricted and perhaps later definition of alchemy, which usually focuses on the transformation of base metals into gold and silver. Indeed, it is this narrower notion of alchemy that seems to have determined the selection of pseudo-Democritus' writings for inclusion in Byzantine anthologies: the best preserved sections of the four original books are precisely those dealing with gold-making (*chrysopoiea*) and silver-making (*argyropoeia*).

The *Four Books* also offer insight into an ancient tradition that linked the Greek atomist, Democritus, to the wisdom of Egypt and Persia. According to this tradition, Democritus was taught the secrets of alchemy by his master, the Persian magus

---

[6]   Syn. Alch. § 1, ll. 12–3; see *infra*, pp. 13–8.

[7]   Ps.-Dem. Alch. *PM* §§ 1–2 corresponds to the section on purple dyeing, which must be augmented by considering some recipes preserved by the Syriac tradition (see *infra*, pp. 8–11); Ps.-Dem. Alch. *PM* §§ 4–20 corresponds to the section on gold-making.

[8]   This ascription is made on the basis of comparison between a commentary on ps.-Democritus by the author Synesius and some sections of the so-called *Chemistry of Moses*. See *infra*, pp. 26–9

Ostanes. He is also described as travelling to Egypt to learn geometry and astrology; to Persia to be educated by the *magi*; and even to have reached India.[9] This close connection with Middle Eastern traditions left its mark on the pseudepigraphic works circulating under Democritus' name, including the *Four Books*.

A pseudepigraphic work is, by its nature, particularly difficult to date: the attribution of any writing to a mythical or ancient philosopher constitutes an attempt to remove it from the historical context in which it was produced, projecting it back to a remote (and not necessarily historical) past, closer to the mythical sources of human knowledge. The legend of the Greek philosopher Democritus, initiated by the Persian magus Ostanes in Egypt, unites three influential cultural traditions that were brought together during the Hellenistic and Late Antique period. Against this strongly syncretic background the author of the *Four Books* has lost his historical identity, which is very difficult to reconstruct today. Already in the second century AD, Aulus Gellius (X 12,8 = 68 [55] B 300,7 D-K) warned his readers against the many pseudepigrapha circulating under the name of Democritus when he wrote that "many fictions of this kind[10] seem to have been attached to the name of Democritus by ignorant men, who sheltered themselves under his reputation and authority."[11]

The difficulties of reconstructing the *Four Books* are examined below, in five relatively technical sections. § 2 introduces the evidence for direct transmission of the *Books*: a group of manuscripts which preserve epitomes of the original Greek text and its Syriac translation. In § 3 I use these documents, as well as indirect evidence from other sources, to reconstruct the likely order and structure of the original *Books*. The books on gold and purple are examined in § 4, while § 5 introduces the books on silver and precious stones. In § 6, I compare two indirect sources for the *Catalogues*: a commentary on the *Four Books* by Synesius, and the so-called *Chemistry of Moses*.

The next four sections address the difficulties associated with dating and authorship. The date of the *Four Books* is considered in § 7, and likely reasons for their attribution to Democritus in § 8. In § 9, I review the claim that the *Books* were written by the philosopher Bolos of Mendes (third/second century BC), and argue against Bolos's authorship. Finally, § 10 examines an influential commentary on the *Books*: the dialogue between Synesius and Dioscorus, which I include in my edition.

---

[9] Regarding Egypt, see Diog. Laert. IX 35, depending on Demetrius of Magnesia's Περὶ ὁμωνύμων ποιητῶν καὶ συγγραφέων |*On Poets and Authors of the Same Name*; fr. 29 in Jirgen Mejer, "Demetrius of Magnesia: On Poets and Authors of the Same Name," *Hermes*, 109 (1981): 469| and Antisthenes of Rhodes' Φιλοσόφων διαδοχαί (*Successions of Philosophers*; FGrH 508 F 12). Diodorus Siculus (I 98,2 = 68 |55| A 1 D-K) claims that Democritus spent five years in Egypt; see also Cic. *De fin.* V 19,50 (= 68 |55| A 13 D-K): "quid de Pythagora? quid de Platone aut de Democrito loquar? a quibus propter discendi cupiditatem videmus ultimas terras esse peragratas." Regarding Persian magi, see *infra*, pp. 69–73. Regarding India, see Strab. XV 1,38 (= 68 |55| A 12 D-K), depending on Megasthenes.

[10] Aulus Gellius here criticizes the spurious information provided by Pliny the Elder (*NH* XVIII and X) about Democritus and his works (especially his supposed work *On the Power and Nature of the Chameleon*).

[11] Translation by John C. Rolfe, *The Attic Nights of Aulus Gellius*, 3 vols. (London: Loeb, 1927), vol. 2, 245. See also Diog. Laert. IX 49,12–4 (= 68 |55| A 33 D-K): τὰ δ' ἄλλα, ὅσα τινὲς ἀναφέρουσι εἰς αὐτόν, τὰ μὲν εκ τῶν αὐτοῦ διεσκεύασται, τὰ δ' ὁμολογουμένως ἐστὶν ἀλλότρια ("With regard to the other books attributed to him, some of them are compilations from his own writings, others are uniformly spurious").

In the last three sections, I engage with some of the substantive issues raised by the
Four Books, and their implications for the history of alchemy more generally. In particular, I show that the definition of early alchemy should be reconsidered in light of
the colouring techniques discussed in the *Four Books*, and tease out some of the
Egyptian and Persian elements that have contributed to the legend of Democritus
the alchemist.

1.1. Early modern editions

Unlike some other early alchemical works, the surviving excerpts of the *Four Books*
did not attract the attention of Renaissance humanist scholars.[12] As a result, no
complete edition of the Greek text was prepared until the work of Berthelot and
Ruelle in the late nineteenth century, while three Latin translations appeared
between the middle of the sixteenth and the middle of the seventeenth century. Of
these Latin versions, *Natural and Secret Questions* and *On the Making of Silver*
were edited around 1573 in Padua by the Calabrian scholar Domenico Pizzimenti,
under the title of *Democritus Abderita De arte magna. Sive de rebus naturalibus.
Nec non Synesii, et Pelagii, et Stephani Alexandrini, et Michaelis Pselli in eundem
commentaria* (*Democritus of Abdera, On the Great Art or On Natural things,
along with the commentaries on this work by Synesius, Pelagius, Stephanus of Alexandria, and Michael Psellus*).[13]

Analysis of the various alchemical manuscripts owned by Pizzimenti – in particular, the *Neapolitanus* III D 17 and III D 18 kept today in the Biblioteca Nazionale in
Naples – shows that the scholar examined these texts closely in order to create his
translation.[14] The first of these was transcribed by the copyist Cornelius Murmuris
from one of the most influential collections of Greek alchemical writings, *Marcianus
gr. 299*, in 1565.[15] The second is a copy of *Vaticanus gr. 1174*, a manuscript that
Pizzimenti may already have examined before he bought *Neapolitanus* III D 18.[16]
Both of Pizzimenti's Neapolitan codices include several marginal notes that testify

[12] See Sylvain Matton, "L'influence de l'humanisme sur la tradition alchimique," *Micrologus*, 3 (1995): 309–41.

[13] On the dating of this edition see Maria R. Formentin, "Domenico Pizzimenti Vibonense: maestro, interprete, copista
del sec. XVI," in *Testi medici latini antichi. Le parole della medicina: Lessico e storia*, ed. Maurizio Baldin, Marialuisa Cecere and Daria Crismani (Bologna: Pàtron, 2004), 692 n. 7. However, John Ferguson, "On the First Editions
of the Chemical Writings of Democritus and Synesius," *Proceedings of the Philosophical Society of Glasgow*, 16
(1884–85): 36–8, in his discussion of the various editions of Pizzimenti's works, proposed pushing back the date
of the first edition to 1572 (see also Matton, "L'influence de l'humanisme," 319 n. 5), the year in which Pizzimenti
also published his translation of Psellus' letter on the making of gold: *Pselli tractatus De auri conficiendi ratione ad
Michaelem Cerularium Dom. Pizimentio Veron. Interprete*, (Padua: Simon Galignanus, 1572); see Formentin,
"Domenico Pizzimenti," 692 n. 6. Pizzimenti's translation of Synesius' dialogue was also reprinted by Fabricius,
alongside the Greek text in the eighth volume of his *Bibliotheca Graeca* (233–48); see Matton, "L'influence de l'humanisme," 318 n. 7. On Pizzimenti's interest in alchemy, see François Secret, "Notes sur quelques alchimistes italiens
de la Renaissance," *Rinascimento*, 23 (1973): 211–7.

[14] On *Neapolitanus* III D 17, see *CMAG* II 217–24; we read the ownership note at fol. II: δομενικοὺ (*sic*) τοῦ πιζιμεντίου. On III D 18, see *CMAG* II 225–30; we read at fol. I: τούτο (*sic*) τὸ βιβλίον ἐστὶ Δομινικοὺ (*sic*) τοῦ πιζιμεντίου,
and at fol. I: Δομινίκου τοῦ πιζιμεντίου.

[15] See *Neapolitanus* III D 17, fol. 189ʳ. The same copyist made three other copies of the *Marcianus*: the *Vind. Med. Gr. 2*
and 3, and the *Vratislav. R 46*; see Andrè-Jean Festugière, "Alchymica," in *Hermétisme et mystique païenne* (Paris:
Aubier-Montaigne, 1967), 218.

[16] On *Vaticanus gr. 1174*, see Festugière, "Alchymica," 226.

to a deep and critical comparison between the two manuscripts and other sources. According to Formentin's analysis, these notes are in the hand of Pizzimenti, who conducted a philological inquiry — often suggesting possible corrections to the Greek texts – before publishing his Latin translation.[17]

After Pizzimenti's version, a handwritten Latin translation of the whole *Corpus alchemicum* was prepared in 1606 by Matthaeus Zuber (1570–1623), a poet and professor in Nuremberg, and friend of the hermetic philosopher and alchemist Heinrich Khunrath (ca. 1560–1605).[18] Zuber's translation, preserved in MS. *Vindobonensis lat.* 11427, is edited in this volume for the first time.[19] It was very probably based on the Greek text preserved in a copy of the *Marcianus gr.* 299.[20]

Finally, a few decades after Zuber's work, a third handwritten Latin translation of the *Corpus alchemicum* was composed by an anonymous scholar and dedicated to the emperor Ferdinand II (1637–1657). This still unedited translation is preserved in the manuscript *Vindobonensis V* 11453.[21] Two copies made by the same copyist, Thomas Perdiller, at the end of the seventeeth century, are handed down by the manuscripts *Vindobonensis V* 11456 (1677) and *Gothanus A* 147.[22]

1.2. Modern scholarship

After these early attempts to translate and interpret the works of ps.-Democritus, it was not until the end of the nineteenth century that scholars began to conduct deeper investigations into this literature. Among them, Hermann Kopp[23] and Marcellin Berthelot (who, with Charles-Émile Ruelle, published the first critical edition of the ps.-Democritean treatises)[24] highlighted the primacy of these writings and

[17] Formentin, "Domenico Pizzimenti," 695–7. Prof. Jean-Marc Mandosio has also drawn my attention to another translation into Czech, preserved in Ms. *Voss. Chym.* F. 3 |see Petrus C. Boeren, *Codices Vossiani chymici* (Leiden: Bibliotheca Universitatis Leidensis, 1975), 7–13|. The codex was checked by Mandosio in November 2006, and I warmly thank him for the following information: "Si tratta di una poderosa raccolta di testi alchemici in lingua ceca, intitolata 'Knila dokonalého vmieni chymiczého' ('Libro perfetto dell'arte chimica'), copiata negli anni 1582–1585 da Bavor Rodovsky z Hustiran, collaboratore dell'alchimista Pietro Vok, detto Rosenberg, che acquistò il volume nel 1589. Rodovsky e Vok facevano parte del circolo alchemico dell'imperatore Rodolfo II di Asburgo. Dopo la morte di Vok nel 1611, il codice passò tra varie mani e fu preso con molti altri dalle truppe svedesi durante la guerra dei Trent'anni. Si ritrovò così nella biblioteca della regina Cristina di Svezia, che consegnò tutti i suoi manoscritti alchemici allo studioso leidense Isaac Vossius nel 1654, quando rinunciò al trono. Il codice contiene (fols. 115ᵛ–118ᵛ) un 'Traktat Synesi o kamenu filozoffskem' ('Trattato di Sinesio sulla pietra filosofica'), indicato da Boeren, *Codices Vossiani*, 9, come *Pseudo-Synesius, Tractatus de lapide philosophica*. L'*incipit* (da me copiato) recita: 'Dioscorowi Kniezy welite bolynie ... Alexandru z bozi molosti Synesius mudzecz.'"

[18] See Matton, "L'influence de l'humanisme," 320. Matthaeus Zuber composed a dedicatory epigram to Khunrath, which was printed in Khunrath's work *Amphitheatrum sapientiae aeternae* (Hanover, 1609), 14.

[19] *CMAG* IV 68–85.

[20] This could be *Monacensis gr.* 112 (*CMAG* IV 247–72), which is copied from a lost manuscript derived from *Marcianus gr.* 299 (see *CMAG* IV 13–7 and Festugière, "Alchymica," 218).

[21] See *CMAG* IV 47–58.

[22] See *CMAG* IIV 59–67 and 140–4 respectively; See also Matton, "L'influence de l'humanisme," 321 n. 2.

[23] See in particular Hermann F.M. Kopp, *Beiträge zur Geschichte der Chemie*, 3 vols. (Braunschweig: F. Vieweg und Sohn, 1869–75), vol. 1, 108–37; at pp. 137–43. He edited the Latin translation of the *Four Books* published by Pizzimenti, adding to the footnotes some passages of the Greek texts taken from the volumes of the *Notices et extraits des manuscrits de la bibliothèque du roi et autres bibliothèques* (see. e.g., Kopp, *Geschichte der Chemie*, vol. 1, 137 n. 62; 141 nn. 64–8).

[24] *Collection des anciens alchimistes grecs*, 3 vols. (Paris: Georges Steinheil, 1887–88), vol. 2, 41–53 (hereafter *CAAG*). The *Physika kai mystika* and the *Peri asēmou poïēseōs* were published at the beginning of the second section ("Traités démocritains"), which effectively opens the edition of Greek alchemical works; the first section ("Indications

drew attention to the problem of their authenticity.[25] It was immediately realised that the *Four Books* were pseudepigraphic – that is, they were not produced by the historical Democritus, the fifth-century BC Greek atomist. By analysing several sources dating from the first centuries AD (especially Pliny the Elder, Columella, Seneca, and Diogenes Laertius), these scholars inferred the existence of a wider pseudo-Democritean production that was probably already circulating during the third-second centuries BC.[26]

In particular, their attention was attracted by the controversial figure of Bolos of Mendes. Bolos is mentioned in ancient sources as the author of naturalistic treatises that circulated under the name of Democritus, and which seemed similar to the alchemical writings of ps.-Democritus.[27] The works of ps.-Democritus were also found to have points of similarity with the recipe book preserved in the so-called 'Leiden Papyrus' (*PLeid.X.*), which had been published by Leemans in 1885[28] and translated by Berthelot in the first volume of his *Collection des anciens alchimistes grecs*.[29]

Twentieth-century historical-philological surveys improved and deepened scholarly knowledge about the origins of alchemy. The edition of the 'Stockholm Papyrus' (*PHolm.*), published by Lagercrantz,[30] enabled Lippmann to open his impressive study on the history of alchemy by examining both papyrological sources and the works of ps.-Democritus.[31] This approach was reinforced by subsequent studies, which continued to insist on thematic similarities between the abovementioned sources. Indeed, both the Leiden and Stockholm papyri and the *Four Books* include recipes explaining how to process metals, dye fabric purple, and make artificial precious stones. The Stockholm papyrus also explicitly attributes one of its recipes to Democritus.[32]

[24] *Continued*
    générales"), includes in particular excerpts or treatises which were considered as propaedeutical to the reading of the collection. Prior to Berthelot, some passages of the *Physika kai mystika* (taken from the codex *Parisinus gr.* 2325) were translated into French by Ferdinand Hoefer, *Histoire de la chimie*, 2nd ed., 2 vols. (Paris: Firmin Didot frères, fils et Cie, 1866–69), vol. 1, 276–9.

[25] See, for instance, Marcellin Berthelot, *Les origines de l'alchimie* (Paris: Georges Steinheil, 1885), 145–6 : "Démocrite et les traditions qui s'y rattachent jouent un rôle capital dans l'histoire des origines de l'alchimie. En effet, parmi les livres venus jusqu'à nous et qui contiennent des recettes et des formules pratiques, l'ouvrage le plus ancien de tous, celui que les auteurs ayant quelque autorité historique citent, et qui n'en cite aucun, c'est celui de Démocrite, intitulé *Physica et Mystica*." Prior to these investigations, Hoefer (*Histoire de la chimie*, vol. 1, 277), for instance, dated ps.-Democritus to the time of Zosimus and Olympiodorus.

[26] Kopp, *Geschichte der Chemie*, vol. 1, 109; Berthelot, *Origines de l'alchimie*, 148–59; see also Marcellin Berthelot, "Des origines de l'alchimie et des œuvres attribuées à Démocrite d'Abdère," *Journal des Savants*, 49 (1884): 517–27.

[27] See Berthelot, *Origines de l'alchimie*, 156–8; Berthelot, "Alchimie et Démocrite d'Abdère," 522–3.

[28] Conrad Leemans, *Papyri Graeci Musei Antiquarii Publici Lugduni Batavi*, 2 vols. (Leiden: Brill, 1843–85), vol. 2, 199–259.

[29] *CAAG* I 3–70 (this section precedes the explanation of ps.-Democritus' recipes, pp. 70–3) and Marcellin Berthelot, "Papyrus de Leyde," *Mémoires de l'Académie des sciences de l'Institut de France*, 49 (1906): 266–307.

[30] Otto Lagercrantz, *Papyrus Graecus Holmiensis. Recepte für Silber, Steine und Purpure* (Uppsala: A.B. Akademiscka Bockhandeln, 1913). A new critical edition of the two papyri was published by Robert Halleux, *Papyrus de Leyde. Papyrus de Stockholm. Fragments de recettes* (Paris: Les Belles Lettres, 1981).

[31] Edmund O. von Lippmann, *Entstehung und Ausbreitung der Alchemie*, 3 vols. (Berlin: Julius Springer, 1919–54), vol. 1, at 1–27 and 27–46 respectively.

[32] *PHolm.* 2; see *infra*, pp. 35–6.

This new information raised a real ps.-Democritean *quaestio* concerning the possible role played by the Egyptian polygrapher Bolos of Mendes in the early history of alchemical literature. Following the influential studies of Wellmann, who tried to attribute the greater part of the ps.-Democritean production to Bolos (including both the *Physika kai mystika* and *Peri asēmou poiēseōs*), scholars continued to consider this question.[33] While Kroll, and more recently Letrouit, clearly distinguish between ps.-Democritus the alchemist and Bolos, the most common trend has been to consider the two authors as somehow related.[34] As Halleux notes, "it is generally admitted that [the *Four Books*] lead back to Bolos in one way or one another."[35]

However, as discussed in section 9, there are several reasons why we cannot attribute the *Four Books* to Bolos, the extent of whose work is difficult to evaluate given the present state of our knowledge.[36] The surviving extracts from the *Four Books* therefore represent the earliest example of a treatise unanimously accepted as a fundamental work in the subsequent alchemical tradition. These books, although in some way related to the wider ps.-Democritean production concerning different arts (*technai*), have specific features that are characteristic of the new field of alchemy.

§ 2. Alchemical works by ps.-Democritus: the direct manuscript tradition

The major source for the *Four Books* is a group of medieval and early modern manuscripts containing epitomes of the original works, both in Greek and in translation into Syriac. This edition makes use of four Byzantine manuscripts:

Marcianus gr. 299 (tenth-eleventh century), signified as **M**
Parisinus gr. 2325 (thirteenth century), signified as **B**
Parisinus gr. 2327 (1478), signified as **A**
Vaticanus gr. 1174 (fourteenth-fifteenth century), signified as **V**[37]

This edition further compares this Greek tradition with the more extensive material preserved in three Syriac manuscripts:

Cambridge University Library, Mm. 6.29 (fifteenth century), signified as **SyrC**
British Library, *Oriental* 1593 (fifteenth-sixteenth century) and *Egerton* 709 (sixteenth century), signified as **SyrL**

---

[33] Max Wellmann, *Die Georgika des Demokritos* (Berlin: "Abhandlungen der Preußischen Akademie der Wissenschaften, Phil.-hist. Klasse" 4, 1921) and *Die Physika des Bolos Demokritos und der Magier Anaxilaos aus Larissa, Teil I* (Berlin: "Abhandlungen der Preußischen Akademie der Wissenschaften. Phil.-hist. Klasse" 7, 1928).

[34] Wihlelm Kroll, "Bolos und Democritos," *Hermes*, 69 (1934): 230–1; see also Jackson P. Hershbell, "Democritus and the Beginnings of Greek Alchemy," *Ambix*, 34 (1987): 5–8. Letrouit, "Chronologie," 17.

[35] Halleux, *Papyrus de Leyde*, 73–4. My translation from the original French.

[36] See *infra*, pp. 36–44.

[37] For more detailed descriptions of these manuscripts, with particular focus on their possible relationships, see Martelli, *Pseudo-Democrito*, 3–54.

These Syriac texts preserve several books (not always corresponding to the Greek ones) under the name of the same author. Unfortunately, we cannot give an exact date for the translation of ps.-Democritus' writings into Syriac: thus the *Corpus Syriacum* might have been compiled *after* the original *Four Books* had been epitomised and, for that reason, cannot reliably help us reconstruct their original structure.[38] Nevertheless, as we shall see, the presence of several different readings, together with some passages preserved only in **SyrC**, makes the *Corpus Syriacum* a valuable source that probably stems from a different branch of transmission than the one preserved in the Byzantine manuscripts.[39]

The index of manuscript **M**, which lists the titles of almost all the works preserved in the manuscript, records two Democritean texts: "Democritus, *On the Making of Purple and Gold: Natural and Secret Questions*" and "*On the Making of Silver* by the same author."[40] This manuscript also preserves the text of these works: the first under the simplified title of "Democritus, *Natural and Secret Questions (Physika kai mystika)*," and the second entitled *On the Making of Silver (Peri asēmou poiēseōs)*.[41] Manuscripts **B** and **A** also preserve texts with the same titles, "Democritus, *Natural and Secret Questions*" and *On the Making of Silver (Peri poiēseōs asēmou)*.[42]

A singular situation is attested, however, by manuscript **V**. The excerpt handed down in **M**, **B**, and **A** under the title of *Natural and Secret Questions* has here been divided into two sections: *Excerpts by Democritus: On Natural Purple* and *Excerpts from Democritus' Natural and Secret Questions*.[43] **V** also includes a third work entitled *On Silver (Peri argyrou)* which corresponds to the *On the Making of Silver*.[44] This situation is schematized in Table 1 of the Appendix.

The Syriac manuscripts preserve long passages which overlap with this Byzantine tradition.[45] A fairly complete translation of the abovementioned Greek excerpts appears in **SyrC**, which includes six books ascribed to Democritus.[46] These books are divided into two sections within the manuscript.

---

[38] See Martelli, *Pseudo-Democrito*, 55–6.

[39] In this edition, I did not consider the Arabic tradition, which surely deserves deeper investigation to help us better understand the role it played in the transmission of the *Four Books*. A few preliminary studies on this topic have been published so far: see Fuat Sezgin, *Geschichte des arabischen Schrifttums*, vol. 4: *Alchimie-Chemie, Botanik-Agrikultur* (Leiden: Brill, 1971), 49–50; Manfred Ullmann, *Die Natur- und Geheimwissenschaften im Islam* (Leiden: Brill, 1972), 156–60; Hans Daiber, "Democritus in Arabic and Syriac Tradition," in *Proceedings of the First International Congress on Democritus* (Xanthi: International Democritean Foundation, 1984), 251–65.

[40] M, fol. 2ʳᵛ: Δημοκρίτου περὶ πορφύρας καὶ χρυσοῦ ποιήσεως· φυσικὰ καὶ μυστικά and τοῦ αὐτοῦ περὶ ἀσήμου ποιήσεως. Index edited in *CMAG* II 20–2; French translation by Berthelot *CAAG* I 174–6 and Michèle Mertens, *Zosime de Panopolis, Mémoires authentiques* (Paris: Les Belles Lettres, 1995), XXIII-XXV.

[41] At M, fols. 66ᵛ27–71ʳ6, and M, fols. 71ʳ7–72ᵛ8, respectively.

[42] At (B, fols. 8ᵛ10–17ʳ16; A, fols. 24ᵛ5–29ᵛ4) and (B, fols. 17ʳ16–20ʳ18; A, fols. 29ʳ4–31ʳ22) respectively. A (fols. 258ʳ17–259ᵛ26) also preserves a third work, titled Δημοκρίτου βίβλος πέντη προσφωνηθεῖσα Λευκίππῳ (*Democritus' Fifth Book Addressed to Leucippus*) which probably post-dates the composition of the abovementioned works, and cannot be ascribed to the same author (see *infra*, p. 17). It will be not taken into account in this section.

[43] The two sections are at V, fols. 33ᵛ13–35ᵛ16 (Ἐκ τῶν Δημοκρίτου περὶ πορφύρας φυσικῆς) and fols. 1ʳ1–7ʳ16 (Ἐκ τῶν Δημοκρίτου φυσικῶν καὶ μυστικῶν), respectively.

[44] V, fols. 7ʳ17–10ᵛ8.

[45] See also Martelli, *Pseudo-Democrito*, 54–9.

[46] Partial French translation by Marcellin Berthelot and Rubens Duval, *La chimie au Moyen-Âge*, vol. 2: *L'alchimie syriaque* (Paris: Imprimerie Nationale, 1893), 267–93 (hereafter *CMA* II).

The first section – the Syriac text of which is edited here for the first time[47] – comprises books 1–3, which are transmitted under the following titles:

1. **SyrC**, fols. 90ᵛ1–94ʳ3: *Book by Democritus: On the Making of Shiny Gold* (= *1SyrC*). This text is an almost complete translation of *Natural and Secret Questions* (*Physika kai mystika*), §§ 5–20.
2. **SyrC**, fols. 94ʳ4–96ᵛ2: *Second Book by the Philosopher Democritus* (= *2SyrC*). This part matches the Greek *On the Making of Silver* (*Peri asēmou poiēseōs*), and also preserves one additional section (§ 5) that was probably lost in the Byzantine tradition.
3. **SyrC**, fols. 96ᵛ3–98ʳ2: *Again by Democritus: I greet you wise men* (= *3SyrC*). Several recipes concerning both the making of precious stones (§§ 1–4) and the dyeing of wool purple (§§ 5–7) are collected in this text.

This third section does not match any texts found in the Byzantine manuscripts under the name of Democritus. However, some of its recipes may be ascribed to him on the basis of information preserved by the Greek indirect tradition. In particular it is possible to find several similarities.

For instance, in this third section, § 2 corresponds to a long quotation that a Byzantine alchemist – called simply "the philosopher Anonymous" – has taken from ps.-Democritus.[48] The Greek text is incorrectly edited as a work of the alchemist Zosimus[49] by Berthelot-Ruelle, who followed only manuscript **A**, without taking **M** into consideration.[50] The third recipe (*3SyrC* § 3) describes a technique in which an ingredient called *kōmaris* (ܟܘܡܪܝܣ = κώμαρις/κόμαρις) seems to play an important role. Although this recipe is not preserved in the Byzantine tradition, some manuscripts[51] include an interesting section on the making of precious stones, entitled *Deep Tincture of Stones, Emeralds, Rubies and Jacinths from the Book Taken out from the Sancta Sanctorum of Temples*,[52] which cites Democritus several times regarding the use of this ingredient.[53]

Finally, the last three recipes (*3SyrC* §§ 5–7) concern dyeing wool purple. Even if these texts are not preserved in the Byzantine tradition, they deal with the same topic covered in the first two paragraphs of *Physika kai mystika*. In addition, the last

---

47 See *infra*, pp. 152–87. In this section, I refer to my edition of the Syriac text.

48 See also *CMA* II 25 n. 2 and 273 n. 1.

49 Letrouit, "Chronologie," 64.

50 *CAAG* II 122,4–17; see also *CAAG* II 202,16–7. This section is preserved in **M**, fols. 84ʳ11ff.; see *3SyrC* § 2 n. 3.

51 In particular **B**, fols. 160ᵛ1–173ᵛ8 and **A**, fols. 147ʳ1–159ʳ5.

52 Καταβαφὴ λίθων καὶ σμαράγδων καὶ λυχνίτων καὶ ὑακίνθων ἐκ τοῦ ἀδύτου τῶν ἱερῶν ἐκδοθέντος βιβλίου; edited in *CAAG* II 350–64. This section comprises various recipes on the making of precious stones in addition to some more theoretical paragraphs, where the compiler discusses the identification of several ingredients, their use, and some technical details on the basis of ancient authors (including Maria, Democritus, Ostanes, and Zosimus) who are often quoted.

53 See, in particular, *CAAG* II 356–7, which reports a sentence by Democritus (357, ll. 11f.: ἐπίχριε ὅσον βούλει, λειώσας αὐτόν, καὶ ἔσται μαργαρίτης) that matches ll. 3–4 of the *3SyrC* § 3 (ܐܝܟ ܕܨܒܐ ܐܢܬ ܟܠ ܡܐ ܫܘܦ ܩܡ ܗܘܝܐ ܘܗܘ ܡܪܓܢܝܬܐ). See also *CMA* II 26 n. 3.

recipe concerns a cold process of dyeing (called ܦܘܪܦܘܪܐ ܡܩܪܪܬܐ, "cold purple"), a technique that seems to have been ascribed to Democritus.[54]

The second section of **SyrC** – which has not been included in the present edition – is composed of three additional books that seem to comprise a "collection of recipes from various eras after the time of Zosimus."[55] The first book is not introduced by any title and closes with the sentence: "End of the first book by the wise Democritus."[56] The second book is simply called *Democritus' Book*.[57] Lastly, the third book begins: "Under the guidance and the supervision of God we are going to copy another book by Democritus."[58]

Of the alchemical books included in the London manuscripts (**SyrL**), some of which are explicitly attributed to Democritus,[59] the first and second deserve particular attention:

(1) *From the Teaching of Democritus: First Part of the First Treatise on the Making of Gold.*[60] This is a partial translation – less complete than the text handed down in *1SyrC* – of *Natural and Secret Questions* (*Physika kai mystika*) §§ 5–20, concerning the making of gold.

(2) *Again by the same author.*[61] This text is a partial translation of *On the Making of Silver* (again, less complete than *2SyrC*). This book is followed by an appendix introduced by the sentence: "Democritus: greetings to the wise men."[62] Several recipes have been collected after this incipit, most of which are also preserved – quite often in a different and more complete form – in the second section of **SyrC**, ascribed to Democritus.[63] The first recipes are particularly interesting (*CMA* II

[54] *CAAG* II 355,2f.: ὃν τρόπον καὶ ἐπὶ τῆς ψυχροβαφῆς ἐδικαίωσε πορφύρας ("he |i.e possibly Democritus| also considered this way suitable for the cold dyeing of purple").

[55] *CMA* II 275 n. 1.

[56] **SyrC**, fols. 98ʳ14–101ʳ17. Partial French translation in *CMA* II 275–7.

[57] **SyrC**, fols. 102ʳ1–111ᵛ14. Partial French translation in *CMA* II 277–91.

[58] **SyrC**, fols. 112ʳ1–116ᵛ5.

[59] Ten books, with some appendixes, are marked as "Doctrine de Démocrite" by Berthelot-Duval, who edited the Syriac text (*CMA* II 1–60) and provided a French translation (1–106). However, not all of these books are preserved under the name of Democritus. If we omit the first two books, more extensively described in this chapter, the other treatises include the following titles:

(3) 15,20–18,24: ܡܐܡܪܐ ܕܓ. ܕܕܝܡܩܪܝܛܘܣ, *Third Treatise by Democritus* (French transl. 31–6). This book seems to show many similarities with the third book of **SyrC**, second section (see *CMA* II 291–3).

(4) 19,1–21,15: ܡܐܡܪܐ ܕܕ ܕܐܝܦܘܩܪܛܝܣ, *Fourth Treatise by Hippocrates* (Fr. tr. 37–41). The name of Hippocrates has been interpreted by Berlthelot-Duval (*CMA* II 37 n. 1) as a misreading of the original name of Democritus.

(5) 21,16–27,6: ܡܢ ܡܐܡܪܐ ܕܗ. ܕܦܝܠܘܣܘܦܘܣ, *From the Fifth Treatise of the Philosopher* (Fr. tr. 42–50).

(6) 27,6–40,13: ܡܢ ܟܬܒܐ ܕܫܬܝܬܝܐ, *From the Sixth Book* (Fr. tr. 51–70); the second part comprises a sort of appendix, introduced by the title: ܕܕܝܡܩܪܝܛܘܣ ܡܛܠ ܚܝܘܬܐ ܠܘܩܒܠ ܐܦܐ, *From Democritus, About Animals with Two Faces.*

(7) 40,13–42,7: *sine titulo* (Fr. tr. 70–74); the first recipe is titled: ܦܘܪܫܐ ܕܥܒܕ ܣܐܡܐ, *Chapter on the Making of Silver.*

(8) 42,8–45,21: *sine titulo* (Fr. tr. 75–81); the first recipe is titled: ܡܛܠ ܗܢ ܕܐܝܠܝܢ ܐܢܘܢ ܡܣܡܩܢܐ, *On Which Are the |Substances| that Turn Red.*

(9) 45,22–50,13: *sine titulo* (Fr. tr. 82–90); the first recipe is titled: ܡܛܠ ܙܝܘܐ ܠܬܡܗܐ ܛܠ, *On the Amazing Mercury.* This book ends with the mark (50,13) ܫܠܡ, "the end."

(10) 50,14–60: ܕܥܣܪ̈ܐ, *Tenth* (*Book/Treatise*) (Fr. tr. 82–106). The book seems to end at 59,5 with the explicit ܫܠܡ ܡܐܡܪܐ ܕܥܣܪܐ ("End of the tenth treatise"); however, some recipes have been added after this point.

[60] See *CMA* II 10,3–12,4.

[61] See *CMA* II 12,5–13,8.

[62] See *CMA* II 13,9–15,19.

[63] See in particular *CMA* II 282–9.

25, §§ 1,2, and 4), since they correspond to the first paragraphs of the third book by ps.-Democritus, preserved in the first section of **SyrC** (*3SyrC* §§ 1–3).

2.1. The Physika kai mystika — Natural and Secret Questions (PM)

The excerpt handed down by **M**, **B**, and **A** under the title of *Natural and Secret Questions* begins with a technical section on the purple dyeing of wool. This section, which is not preserved (at least in the same form) in the Syriac tradition, is divided into two parts. The first is a long recipe (*PM* § 1) explaining how to dye wool purple by means of two natural substances, named *bryon thalassion* (βρύον θαλάσσιον) and *lakcha* (λακχά).[64] The second part is a catalogue of pigments employed in such processes (*PM* § 2). These fall into two distinct groups: substances that, although valued by the author's predecessors, should not be considered long-lasting dyes; and substances that, despite their efficacy, were not appreciated at the time when the work was composed. This section, as we shall see, bears some similarity to analogous lists of substances which, according to Synesius' commentary and to the so-called *Chemistry of Moses*,[65] probably belonged to ps.-Democritus' books *On the Making of Gold* and *On the Making of Silver* in their original form.[66]

A more narrative section (*PM* § 3) then follows, in which the author gives a first person account of his initiation into the alchemical art.[67] After the unexpected death of his master (almost certainly to be identified with the Persian magus Ostanes, although this name is not mentioned in the passage), the author tried to conjure up his master's spirit from Hades in order to secure the arcane teaching that had been kept in his books. None but the master's son was permitted to find these books, which had been hidden. The account ends with the discovery of the treatises during a festival in an Egyptian temple, when a column collapsed, making the precious books available. The author and his friends thus uncovered the secret of secrets hidden inside the column. This was a fundamental teaching, thought to encapsulate the rules underlying every natural combination: "Nature delights in nature, nature conquers nature, nature masters nature" (ἡ φύσις τῇ φύσει τέρπεται, καὶ ἡ φύσις τὴν φύσιν νικᾷ, καὶ ἡ φύσις τὴν φύσιν κρατεῖ). The assembled company marvelled at the short but powerful formula, into which the master had condensed all his knowledge.

At this point, the author briefly relates how he came to Egypt in order to spread his own teaching about the natures (*PM* § 4). Such a section, however, seems to contradict the previous one, in which our author is already in Egypt. The passage – probably

---

[64] I refer to the subdivision of paragraphs adopted in the present edition.

[65] The so-called *Chemistry of Moses* is a collection of recipes, mostly dealing with the treatment of metallic or mineral substances, preserved in Byzantine manuscripts (**A**, fols. 268ᵛ–278ᵛ is the earliest witness) under the name of the biblical patriarch. The recipe book (*CAAG* II 300–5) opens with a brief introduction (300, 3–6) based on two passages of *Exodus* (31,1–5 and 35,30–5), according to which Moses relied on the expertise of Bezalel, the chief architect of the Tabernacle and a craftsman skilled in working and engraving metals (gold, silver, and iron), stone, and wood. Although Zosimus already attributed some alchemical recipes to Moses (*CAAG* II 182,16–7; 183,5–7; 216,12–21), the date of the recipe book remains uncertain.

[66] See *infra* pp. 26-9

[67] For a more extensive analysis of this section, with particular attention to the identification of the personalities involved, see *infra*, p. 20 and Ps.-Dem. Alch. *PM* § 3 n. 21.

no longer in its original position[68] — may have been intended as a kind of introduction to the following section (*PM* §§ 5–20), which contains several recipes on the making of gold. From this point, the Byzantine and the Syriac tradition run in parallel. This section's focus on gold-making is emphasised by the titles which introduce it in both **SyrL** (*From the Teaching of Democritus: First Part of the First Treatise on the Making of Gold*) and **SyrC** (*Book by Democritus: On the Making of Shiny Gold*). The contents of the Greek and Syriac manuscripts are compared in Table 2 in the Appendix.

These gold-making recipes include two paragraphs (*PM* §§ 15–16) in which the author criticizes young students (οἱ νέοι) who do not understand the relevance of his treatise and carry out only perfunctory and superficial investigations. The author first stresses the importance of gaining solid experience, which must be based on deep knowledge of substances and their natural properties, before he starts to focus in more detail on liquid substances (*PM* § 16, ll. 183–5): "Let this suffice about dry substances [lit. dry powders] and about how people must approach this writing. Let us deal with washes [i.e. dyeing liquids] in the following part."

An analogous section – although with some differences – is also preserved in the Syriac tradition. The London manuscripts preserve only the second part of this section, positioned at the beginning of the collection, before ps.-Democritus' books.[69] The Syriac text is shorter than the Greek version, from which it seems to be derived. The Cambridge manuscript, however, runs parallel with the Byzantine tradition, preserving a similar section in the same position, immediately after the malachite (*chrysokolla*) recipe (*1SyrC* §§ 11–12).

The following part contains three further recipes, which cover different dyeing processes based on the use of liquid ingredients and plant juices. Finally, the treatise closes with a brief, theoretical paragraph in which the author draws attention to the unity of the principles for dyeing. At the very beginning of the text, the Greek version introduces the alchemist Pammenes, who is presented as the teacher of the whole Egyptian clergy.[70] In the Syriac translation, however, Democritus himself addresses his own teaching to the Egyptian priests.[71]

2.2. The *Peri asēmou poiēseōs* – On the Making of Silver (AP).

After the section on gold-making, all the manuscripts include a book on making silver, which is preserved under slightly different titles: *Peri asēmou poiēseōs* in **M**, *Peri poiēseōs asēmou* in **B** and **A**, and simply *Peri argyrou* in **V**. In the Syriac tradition, the treatment of the precious metal is not mentioned in the titles, although it is emphasised in the subtitle that introduces this section: "Consider the power of the drugs [lit. herbs] that lead [us] to silver; these are as follows."[72] Several recipes are

---

[68] See *infra* pp. 18–9.
[69] Edited in *CMA* II 1–2; it more or less corresponds to *PM* § 16, 170ff.
[70] See Ps.-Dem. Alch. *PM* § 20.
[71] See in particular *1SyrC*, § 16, which is more complete than the text handed down by **SyrL** and edited by Berthelot-Duval in *CMA* II 12, 2–5.
[72] See also Ps.-Dem. Alch. *PM* § 20, ll. 228f.

collected in the following folia, with a specific focus on the treatment of ingredients employed for the whitening (λεύκωσις) of base metals. (See Appendix, Table 3.)

After the first four recipes (2SyrC §§ 1–4 ≈ AP §§ 1–5), **SyrC** alone preserves a more theoretical passage (2SyrC § 5), similar to the two paragraphs (PM §§ 15–16) that divide the section on gold-making. Again, the author insists on the hidden natural properties of ingredients that alchemists are expected to know and use. As in the analogous passage in PM § 16, this section ends by indicating the new topic to be covered in the following recipes: "Let us move on to the waters in which both the white and the yellow drugs [lit. herbs] are boiled down." Accordingly, the second part of the book includes four recipes with a specific focus on liquid ingredients and washes.

Finally, the last paragraph (AP § 10) makes explicit reference to the books on gold- and silver-making: "You have received everything useful for gold and silver." Similarly, the Syriac translation reads (2SyrC § 10): "Now we receive all the methods for preparing gold and silver." The author then claims to have left nothing out, apart from distilling and subliming processes; topics which, he asserts, have been covered in his other treatises. Unfortunately, these other treatises have not been preserved in the *Corpus alchemicum*.

§ 3. The structure of the lost *Four Books* of ps.-Democritus

It is possible to advance some hypotheses about the original structure of ps.-Democritus' alchemical works by comparing the direct manuscript tradition with information preserved by later alchemists (the indirect tradition). The works ascribed to Democritus became quite popular among Late Antique and Byzantine alchemical authors, who packed their own treatises with quotations from the earlier alchemist. Although an extensive examination of this rich tradition would certainly yield precious information about the ps.-Democritean writings, it would also go far beyond the scope of this edition. I shall therefore focus only on those passages that refer explicitly to the *structure* of the *Four Books*, in order to better understand their original form and extent.

All modern scholars agree that the excerpts handed down by the Byzantine manuscripts should be regarded as an epitome of a more extended and organic work. This appears from passages in two later works, Synesius' commentary and Syncellus' *Chronicle*, both of which present Democritus as the author of four distinct alchemical books:

> He [i.e. Democritus] took his basic principles from him and composed four books on dyeing, on gold, silver, [precious] stones and purple. I stress this point: he wrote by taking his basic principles from the great Ostanes.[73]

> Democritus of Abdera wrote about gold, silver, stones and purple in an ambiguous way.[74]

---

[73] Syn. Alch. § 1, ll. 11–4: ἐκ τούτου λαβὼν ἀφορμὰς συνεγράψατο βίβλους τέσσαρας βαφικάς, περὶ χρυσοῦ καὶ ἀργύρου καὶ λίθων καὶ πορφύρας· Λέγω δή· τὰς ἀφορμὰς λαβών, συνεγράψατο παρὰ τοῦ μεγάλου Ὀστάνου. This passage has been edited on the basis of **M** by Letrouit, "Chronologie," 76 (t. L).

In addition to these testimonies, the *Corpus alchemicum* preserves a passage by the alchemist Olympiodorus that bears a striking similarity to Synesius' words:

> Democritus, by taking [his knowledge] from these [aphorisms],[75] composed four books under the title of *Principle*.[76]

Although Olympiodorus does not specify the topics of these books, he seems to indicate the title under which all four books were collected: the *Principle*. The reliability of such information is, however, questionable. Olympiodorus closely follows the passage by Synesius, which is probably his source. The same expressions are recognizable in both the texts: ἐκ τούτων λαβών in Olympiodorus is close to Synesius' words ἐκ τούτου λαβὼν ἀφορμάς, while both authors employ the verb συγγράφω ("to compose"). Given that Olympiodorus also quotes Synesius just a few lines before the abovementioned passage, it seems quite likely that he was misreading his source when he gave the title of *Principle* (Ἀφορμή) to the books of ps.-Democritus.[77] Specifically, Synesius' first line may have sounded ambiguous, leading Olympiodorus to take "principles" (ἀφορμάς) to refer to the title of the four books mentioned immediately after. Synesius himself seems to have been aware of the risk of being misread, hence his reassertion in the following line, that ps.-Democritus wrote his own books by following the principles that he had learned from his master Ostanes.

The *Corpus alchemicum* preserves three further passages in which the *Four Books* are explicitly mentioned. The first comes from an excerpt by the alchemist Zosimus (ca. 300 AD) and represents the earliest reference to these books:

> In his four books he [i.e. Democritus] went through the different points concerning the water of untouched sulphur. In the book on silver he treats: 'Chian earth, *asteritēs*, and silver spume according to its own application (ἐπὶ τῆς ἰδίας αὐτοῦ ἐπιβολῆς)';[78] in [the book on] yellow [i.e. gold]: 'Sinope's earth, Attic ochre, and Phrygian stone'; in [the book on] stones: 'Goat's blood and Physalis juice'; and afterwards (ὕστερον δέ): 'I'm saying something useful: sulphurous substances are mastered by sulphurous substances and liquid substances by the corresponding liquid substances; for sulphurous substances are held by sulphurous substances.'[79]

[74] Syncell. 297, ll. 24–8 Mosshammer: Δημόκριτος Ἀβδηρίτης [...] συνέγραψε περὶ χρυσοῦ καὶ ἀργύρου καὶ λίθων καὶ πορφύρας λοξῶς. See Letrouit, "Chronologie," 79 (t. Z).

[75] I.e. the aphorisms on the φύσις. According to this interpretation of the syntagma ἐκ τούτων λαβών (see Letrouit, "Chronologie," 78), the pronoun τούτων would refer to the aphorisms on natures (ἡ φύσις τῇ φύσει τέρπεται κτλ.), which Olympiodorus quotes a few lines earlier. Berthelot-Ruelle suggest adding ἀφορμάς after λαβών.

[76] *CAAG* II 102,17–8: ὁ δὲ Δημόκριτος ἐκ τούτων λαβὼν συνεγράψατο βιβλία τέσσαρα τῷ τῆς Ἀφορμῆς ὀνόματι. This passage has been edited on the basis of the M reading by Letrouit, "Chronologie," 76 (t. V). Olympiodorus refers again to the *Four Books* in *CAAG* II 78,11f. (= Letrouit's t. T).

[77] See Matteo Martelli, "L'opera alchemica pseudo-democritea: un riesame del testo," *Eikasmos*, 14 (2003): 161–84, on 166f. For a different interpretation, see Letrouit, "Chronologie," 78, who considered the title Ἀφορμή authentic.

[78] The text is unclear at this point. Manuscripts (see the following note) have: γῆν Χίαν, ἀστερίτην καὶ ἀφροσέληνον καὶ τῆσ ἰδίας αὐτοῦ ἐπιβολῆς. I have considered it necessary to change καί to ἐπί. Zosimus may here refer to the application (to metallic bodies) either of sulphur water or of the compound whose preparation was described in the recipe he is quoting (Ps.-Dem. Alch. *AP* § 2).

[79] Translation based on the revised Greek text published with complete scholarly apparatus in Martelli, *Pseudo-Democrito*, 68 (= *CAAG* II 186,3–9, reedited on the basis of M by Letrouit, "Chronologie," 76 t. G).This

Zosimus here presents the *Four Books* in a different order than either Synesius or Syncellus, placing the book on silver before the one on gold.[80] The first quotation may have been taken from *AP* § 2, ll. 14ff., where the same three ingredients are listed in the same order, and where ps.-Democritus use the technical term "applications" (ἐπιβολαί) that seems to have been kept by Zosimus in his quotation.[81] The second quotation, on the other hand, is likely to come from the book on gold-making, explicitly cited by Zosimus with the words "in the yellow" (ἐν δὲ τῷ ξανθῷ). However, since what survives of that book does not preserve any passage similar to the quotation, it probably comes from a section that was not included in the Byzantine epitome. A similar interpretation can be offered for the third citation, taken from ps.-Democritus' lost book on stones, which was not preserved in the Byzantine tradition.[82]

This third quotation by Zosimus deserves particular attention, since it is not easy to reconstruct its original position within the *Four Books*, at least in the form they had before being epitomized. Zosimus introduces this third quotation with the phrase "and afterwards" (ὕστερον δέ), resulting in a certain ambiguity. Berthelot, for example, translated it as: "and further on, that which is useful: 'sulfurous substances are mastered by sulfurous substances.'" Similarly, Letrouit's translation reads: "and following, so that I should say something useful to you: 'sulfurous substances are mastered by the sulfurous substances.'"[83] Both interpretations are somewhat unclear, since they do not clarify whether Zosimus is quoting a later passage from the book on stones, or a different ps.-Democritean book that followed the one on stones – perhaps the one on purple – whose title Zosimus does not mention explicitly.

The same quotation recurs in a later treatise by the Byzantine alchemist Christianus, who refers extensively to the *Four Books* in a way that recalls the abovementioned passage from Zosimus:

> 'Since I have just now finished my account related to my second treatise and I am going to present a complete exposition of the methods [for the making] of stones, I come to my third treatise by premising something useful for my writing, that is: sulphurous substances are mastered by sulphurous substances, and liquid substances by the corresponding liquid substances.' The wise man from Abdera [i.e. Democritus] opened his fourth

---

[79] *Continued*
passage is part of Zosimus' excerpt (*CAAG* II 181–6) that is handed down in M, fols. 156ʳ7–157ʳ10 and V, fols. 112ʳ1–113ᵛ13 under the title Περὶ ϻ, and in B, fols. 143ʳ2–144ᵛ10 and A, fols. 129ᵛ25–131ʳ7 under the title Περὶ θείου ἀθίκτου ὕδατος.

[80] The correct order of the alchemical processes concerning gold and silver is also discussed by Synesius (§ 8).

[81] Zosimus quoted this passage several times. In the same excerpt Περὶ θείου ὕδατος, just before the abovementioned reference to the *Four Books* (*CAAG* II 185,17–8), he writes: ἐν μὲν τῷ θείῳ λευκῷ· γῆ χεία καὶ ἀστερίτης καὶ ἀφροσέληνον. In addition, Zos. Alch. IV 68–70 Mertens (= *CAAG* II 226,23–5) reads: καὶ ἐὰν μὲν λευκοῦ θείου χρεία, συλλείου τῷ ὕδατι γῆν Χίαν, ἀστερίτην, ἀφροσέληνον ὀπτόν, <στῖμι> Κοπτικόν, Σαμίαν, Καρικήν, Κιμωλίαν ἢ στιλβάδα κτλ. In each case, the three ingredients listed by Zosimus seem to be somehow related to white sulphur, a substance that plays only a marginal role in *AP* § 2 (see l. 11, where the ingredient is said to be employed for purifying tin).

[82] See *infra*, pp. 25–6.

[83] *CAAG* III 183: "et plus loin ce qui est utile … Les sulfureux sont dominés par les sulfureux"; Letrouit, "Chronologie," 76 (t. F): "et ensuite: que je vous dise quelque chose d'utile: les sulfureux sont dominés par les sulfureux etc."

treatise in this way, by showing that the liquid substance, the corresponding liquid substance, and the sulphurous substance represent the key points of the [alchemical] treatment and that sulphurous substances are mastered by sulphurous substances, and liquid substances by the corresponding liquid substances. For nature delights in nature: and in the same way nature conquers nature and nature masters nature, according to what has been claimed by the philosopher himself and by his master Ostanes.[84]

The similarity between Christianus' and Zosimus' quotations from ps.-Democritus is obvious – both authors even use the same expressions. Christianus' quotation, however, seems more complete, for it presents ps.-Democritus (ll. 1–5) as introducing a new topic in his third book, the making of stones.[85] It is therefore likely that the above-mentioned quotations stem from the lost book on stones. On the basis of this supplementary information, we should therefore understand Zosimos' phrase "and afterwards" to refer to a position further on in the same original book, *On Stones*.

One element does not, however, fit this interpretation. Christianus, having reported the long section from ps.-Democritus, actually claims that it comes from the beginning of his *fourth* book. This seems to contradict the words he cites from ps.-Democritus himself, who clearly mentions his *third* book. Letrouit tried to solve this contradiction by assuming a textual lacuna after "the corresponding liquid substances" (l. 5), whereby the text of a second quotation, taken from the fourth book, has been omitted.[86] Alternatively, we could speculate that Christianus was reading the books of ps.-Democritus in a sequence different from that of the original.

Further doubts concerning the correct order of the *Four Books* arise when we consider the work of another Byzantine alchemist, the philosopher Anonymous, who wrote:[87]

The subdivisions of the principal [alchemical] procedures, then, are of 135 different kinds altogether, and it is not possible to recognize a larger or smaller number of procedures that are admitted on the basis of the only authentic constituent of the substances, according to their species and genres: this is the kind of knowledge about silver, gold, pearls, [precious] stones and purple that spreads through the four or five very famous books [by Democritus].[88]

---

[84] Translation based on the revised Greek text published with complete scholarly apparatus in Martelli, *Pseudo-Democrito*, 69–70 (*CAAG* II 395,2–396,2, reedited on the basis of M by Letrouit, "Chronologie," 79 t. W).

[85] This interpretation implies that the two opening particles πεποιημένος and ἐκθέμενος (see ll. 1–2: Τῆς δευτέρας πραγματείας ἄρτι τὸν λόγον πεποιημένος καὶ τῶν λίθων τὰς μεθόδους ἀφθόνως ἐκθέμενος, ἐπὶ τὴν τρίτην ἥκω πραγματείαν) refer to two different moments, as the different verbal tenses seem to indicate: the first one, which the adverb ἄρτι refers to, is a perfect; the second is an aorist, related to the verb ἥκω. The passage has been read in a similar way by Letrouit, "Chronologie," 79: "Après avoir exposé tout à l'heure le discours du second traité, développant ensuite amplement les procédés relatifs aux pierres, j'en viens au troisième traité etc." Conversely, Paul Tannery thought that the making of stones constituted the topic of the second book of ps.-Democritus, suggesting that the third book (concerning which the passage gives no information) could have dealt with the gold: Tannery, "Études sur les alchimistes grecs. Synésius à Dioscore," *REG*, 3 (1890): 283.

[86] Letrouit, "Chronologie," 79 n. 251.

[87] According to Letrouit's interpretation (see *supra*, n. 5), the passage here taken into account would belong to the treatise *On the Making of Gold* by the so-called "philosophe Anépigraphe 1."

[88] Translation based on Martelli, *Pseudo-Democrito*, 71 (= *CAAG* II 433,13–7). Even if ps.-Democritus is not explicitly mentioned in this passage, both Berthelot (*CAAG* II 409 n. 5) and Tannery ("Études sur les alchimistes grecs," 283) agree that the commentator refers to his books.

This commentator does not seem sure of the exact number of books by ps.-Democritus, and he adds a treatise on pearls that is never mentioned in other sources. This treatise is placed before the book on stones, which is accordingly shifted from third to fourth place in the list. We might suppose a similar situation in the case of Christianus, who seems to have considered the book on the making of stones – originally supposed to be the *third* treatise by ps.-Democritus – to be the *fourth* book by the ancient alchemist. However, Christianus lists only four books of Democritus, without any reference to a supposed fifth treatise on the making of pearls.[89] Such divergence may be explained by the fluctuating status of the corpus of alchemical writings ascribed to Democritus, which was doubtless enriched and expanded in the course of its transmission.[90]

Furthermore, we can guess that several alchemical treatises or passages were attributed to the philosopher as a result of the fortunes of the *Four Books*. The *Corpus alchemicum*, for instance, preserves a book under the title of *Fifth Book by Democritus Addressed to Leucippus*.[91] After a brief introduction, this book presents five elaborate recipes describing how to dye base metals (especially copper) white or yellow. The origin and date of this treatise are not clear at present, although the explicit designation "fifth book" implies that its author was aware of the authentic *Four Books* of ps.-Democritus. No later alchemist ever cites it. After analysing its opening section – which insists on the priority of the Egyptian tradition over the Persian – Bidez-Cumont concludes that the unknown author had "as his goal, to attribute priority in this science to the Egyptians, who were the initiators of the Phoenicians and for whom Leucippus was the intermediary."[92] All scholars agree that this excerpt cannot be attributed to ps.-Democritus, and consider it a later composition – a characteristic feature of the alchemical tradition, within which apocryphal texts flourished.[93]

In light of the sources discussed above, we see that Democritus was recognized by later alchemists as the author of four books on dyeing, thought to explain how to tinge base metals yellow and white (in order to make gold and silver), make precious stones, and dye fabrics, especially wool, using substitutes for the expensive Tyrian purple. The same topics recur in the recipes of the Leiden and the Stockholm papyri.[94] The original order of the four books remains uncertain; some sources state that the book on gold-making came before the one on silver-making, while the opposite order is indicated by others. The book on the making of stones probably came after these first two, even if some sources list this book in fourth position,

---

[89] *CAAG* II 396,3–6; see Letrouit, "Chronologie," 79 (t. X).

[90] See, for instance, the anthology of alchemical works preserved in the London Syriac manuscrips, which includes several books ascribed to Democritus: *supra*, n. 59 and *CMA* II, IX-XII.

[91] Δημοκρίτου βίβλος πέμπτη προσφωνηθεῖσα Λευκίππῳ. Its most ancient witness is manuscript A (fols. 258ʳ17–259ᵛ26), edited by Berthelot-Ruelle in *CAAG* II 53–6.

[92] Joseph Bidez and Franz Cumont, *Les mages hellénisés*, 2 vols. (Paris: Les Belles Lettres, 1938), vol. 1, 211: "pour but d'attribuer aux Égyptiens – initiateurs des Phéniciens et par leur intermédiaire de Leucippe lui-même – la priorité de la science."

[93] Letrouit, "Chronologie," 80 n. 253 spoke of a "processus de prolifération des faux alchimiques par amalgame."

[94] See, for instance, Halleux, *Papyrus de Leyde*, 35–52.

following an additional treatise on the making of pearls. Finally, the original ensemble concluded with a book on purple.

§ 4. *Physika kai mystika* and the original books on gold and purple

When we compare the information preserved in the indirect tradition with the excerpts handed down by the direct tradition, several common elements appear. The *pinax* (roughly, the table of contents) of manuscript **M** (fol. 2^r) explicitly states that the first excerpt, *Physika kai mystika*, covers the making of gold and purple by giving its title as "Democritus, *On the Making of Purple and Gold: Natural and Secret Questions.*"[95] The content confirms that the excerpt was composed from two distinct parts, the first on purple dyeing (*PM* §§ 1–2), and the second on treating base metals to produce gold (*PM* §§ 5–20).[96] The two sections are not amalgamated seamlessly, for *PM* § 4 does not continue the account of the previous paragraph where the author describes his initiation into the alchemical art after the collapse of the column containing the secret books of his master Ostanes. Indeed, the reader may have some trouble following the correct sequence of events, since in § 4 the author claims to have come to Egypt, while according to § 3 he should have been there already, since the collapsing column was part of an Egyptian temple.[97] In light of this seeming contradiction, Hershbell proposed a different interpretation of § 3, namely, that it describes the initiation of *Ostanes* by an unnamed master.[98] Consequently, the statement "I too have come to Egypt" (ἥκω δὲ κἀγὼ ἐν Αἰγύπτῳ) in § 4 would have the role of introducing Democritus to the story by having him claim to have visited Egypt in order to follow in his master's footsteps and reveal the art of alchemy. However, such an interpretation seems a little hasty: it would be curious, in fact, for an alchemical text attributed to Democritus to give such ample treatment to the initiation of Ostanes, whom later alchemists unanimously considered to be the master of Democritus. Even in ancient, non-alchemical sources, Ostanes is presented as a wise man who knew all natural secrets:[99] the same role he plays within the *Corpus alchemicum*. Indeed, according to Syncellus, Ostanes did not travel to Egypt in order to be initiated by Egyptian priests, but to become their master and to administer their temples.[100]

The interpretation given by Berthelot, Bidez-Cumont, and Festugière — namely that this break in the text should be understood as an artefact of the epitomised

---

[95] See *CMAG* II 21.
[96] The copist of **V** (or its model) was probably aware of this situation, since he copied down the two sections separately: see *supra*, p. 8.
[97] This interpretation is based on references to the temple of Memphis by Synesius (Syn. Alch. § 1, ll. 9–17) and Syncellus (297,24–298,1 Mosshammer).
[98] See Hershbell, "Democritus," 11–2 and Ingolf Vereno, *Studien zum ältesten alchemistischen Schrifttum. Auf der Grundlage zweier erstmals edierter Hermetica* (Berlin: Klaus Schwarz Verlag, 1992), 91–4.
[99] See Bidez-Cumont, *Mages*, vol. 2, 267–70.
[100] Syncell. 297,94–298,1 Mosshammer.

form in which ps.-Democritus' work has been handed down — therefore sounds more persuasive.[101] In addition, the manuscripts themselves seem to indicate a break between *PM* § 3 and § 4. In **B** and **A**, the last phrase of § 3 ends with a *dikolon* (:) and the first letter of § 4 is capitalised by rubrication.[102] In **M**, there is a simple horizontal dash in the left margin, just next to the beginning of § 4.[103]

Taken together, these factors suggest that the break must be a kind of a bridge between two original ps.-Democritean books which have been epitomised and juxtaposed by a later compiler. This hypothesis is partly confirmed by a passage by Synesius (§ 5, ll. 60ff.): "Turn your attention to what he said in the introduction of his book: 'I too came to Egypt to deal with natural substances, so that you may turn your mind away from the plurality of matter.'" Although Synesius does not specify which book he was quoting, comparison with the excerpts preserved in the Byzantine manuscripts allows us to recognise the same words in *PM* § 4, which may consequently be considered as the *introduction* to the book on gold-making, implying that § 1–3 come from another book, namely, the one on purple dyes. *PM* § 4 is in fact followed by several recipes that deal with gold-making, and the same recipes are explicitly connected to chrysopoeia in the Syriac tradition.

To sum up, several clues allow us to identify two different sections within the excerpt known as *Physika kai mystika*. These sections represent the extant portions of the two original books on gold and purple. However, it difficult to gauge what fraction of the originals has been preserved in the epitome, and how extensive the revision has been. In order to approach these problems, I will consider the two sections separately.

4.1. The Epitome of the Book on Purple

The *Physika kai mystika* preserves only short excerpts of the original book *On Purple*, whose content is consequently difficult to reconstruct. The book very probably included a technical section made up of several recipes (τάξεις), similar to *PM* § 1. Analogous recipes have been preserved in the Syriac tradition, which also mentions a *Book on Purples* (ܦܘܪ̈ܦܝܪܐ ܟܬܒܐ) that probably coincides with, or stems from, the work by ps.-Democritus.[104] In particular, SyrC preserves three recipes under Democritus' name that explain how to make purple (ܘܥܒܕ ܗܘ ܚܘܘܪܝ), while the same procedures and ingredients (especially ܣܘܣܘ

---

[101] See, respectively, *CAAG* III 45 n. 1; Bidez-Cumont, *Mages*, vol. 2, 311 n. 1; Andrè-Jean Festugière, *La révélation d'Hermès Trismégiste*, 4 vols. (Paris: Les Belles Lettres, 1944–54), vol. 1, 228.

[102] See Bidez-Cumont, *Mages*, vol. 2, 320 n. 12. See **A**, fol. 26ʳ2; the ms. **B** does not have the ἤ, but just the aspirate and accent, suggesting that the rubricator did not complete the word.

[103] Manuscript **V**, as we have already noted (*supra*, p. 8), closes this section – preserved in fols. 33ᵛ13–35ᵛ16 under the title of Ἐκ τῶν Δημοκρίτου περὶ πορφύρας φυσικῆς – with the aphorism on nature. The same aphorism is copied again at the beginning of the section Ἐκ τῶν Δημοκρίτου φυσικῶν καὶ μυστικῶν (**V**, fol. 1ʳ2–8): Ἡ φύσις τῇ φύσει τέρπεται, καὶ ἡ φύσις τὴν φύσιν νικᾷ, καὶ ἡ φύσις τὴν φύσιν κρατεῖ. Ἐθαυμάσαμεν πάνυ ὅτι ἐν ὀλίγῳ λόγῳ πᾶσαν συνήγαγε τὴν γραφήν. Ἥκω δὲ κἀγὼ ἐν Αἰγύπτῳ φέρων τὰ φυσικά, ὅπως τῆς πολλῆς περιεργείας καὶ συγκεχυμένης ὕλης καταφρονήσητε. There is no break between τὴν γραφήν and the following ἥκω.

[104] See *CMA* II 90 (see also 274, §§ 3–5).

= φῦκος) are mentioned in both *PM* §§ 1–2 and the three Syriac recipes (*3SyrC* §§ 5–7), which seem to be translations of part of the ps.-Democritean book on purple.

Additionally, *PM* § 2 consists of a list of dyeing substances in some way analogous to the catalogues (κατάλογοι) of ingredients employed for making gold and silver that were extensively quoted and commented upon by the alchemist Synesius and that have been partially preserved within the recipe book edited by Berthelot-Ruelle (*CAAG* II 300–315) under the title of *The Chemistry of Moses* (see Ps.-Dem. Alch. *Cat.* §§ 1–3).[105] As I shall demonstrate in section 6, these catalogues were actually part of the original work of ps.-Democritus, and perhaps opened the books on gold and silver, preceding the collections of recipes. A similar order is not attested, however, for the book *On Purple*, in which (at least according to the manuscript tradition) the catalogue of dyeing substances appears *after* the only recipe preserved in its original language.[106]

Lastly, many doubts remain concerning the position of *PM* § 3 within the original *Four Books*, and even regarding its authenticity. The paragraph seems to describe the initiation of Democritus by Ostanes; however, although both Synesius (§ 1, ll. 3–17) and Syncellus explicitly present Democritus as a pupil of the Persian magus in the Egyptian temple of Memphis,[107] neither mentions any of the details upon which the story of *PM* § 3 seems to be based, such as the master's early death or the collapse of the column. Such dissimilarities have led modern interpreters to puzzle over the authenticity of the paragraph, at least in its Byzantine form. Some have suggested that a similar account, explaining how Democritus learnt the saying about natures from Ostanes, might have fitted well into the introduction of the work.[108] On the one hand, no passages are preserved by the indirect tradition that can confirm this hypothesis; on the other, the so-called *Chemistry of Moses* preserves a glancing reference to *PM* § 3: "Having learned these things from the abovementioned master I was striving to combine natures. Nature in fact conquers nature, and nature masters nature."[109] This recalls the incipit and explicit of *PM* § 3, and also comes after a catalogue of dyeing substances which tallies with the last lines of *PM* § 2. The two sections in *The Chemistry of Moses* are therefore preserved in the same order as that attested by the *Physika kai mystika*. Consequently, we cannot rule out a similar account being included in the book on purple, after the list of substances employed in the dyeing procedures. A similar position could also explain the words "from the abovementioned master" (*PM* § 3, l. 35) that open the section, in which Democritus refers to his master Ostanes: evidently, the latter had already been mentioned in the previous books.

[105] See *supra*, n. 65.
[106] See Ps.-Dem. Alch. *PM* § 2 n. 6.
[107] See *infra*, pp. 69–70.
[108] See Bidez-Cumont, *Mages*, vol. 2, 314 n. 4 and 318 n. 1; Letrouit, "Chronologie," 79–80; in Festugière's opinion right after the sentence ἥκω κἀγώ κτλ. See Festugière, *Révélation d'Hermès*, vol. 1, 228.
[109] *CAAG* II 307,15-7: Ταῦτα παρὰ τοῦ εἰρημένου διδασκάλου μεμαθηκὼς ἠσκούμην ὅπως ἀκούσω τὰς φύσεις. Ἡ φύσις γὰρ τὴν φύσιν νικᾷ, καὶ ἡ φύσις τὴν φύσιν κρατεῖ. See also *infra*, Appendix, table 4 (p. 266).

4.2. The Epitome of the Book on Gold

The section on gold-making, covered by *PM* §§ 4–20, presumably derived from the original book *On Gold*, and is much more extended. *PM* § 4 provides the extant remains of the introduction, while the following recipes (§§ 5–20) describe how to process the solid and liquid substances used to treat base metals and change their colours. To an extent we can also reconstruct the original structure of the book, relying on information preserved by the indirect tradition. When citing individual passages from this book, several alchemists actually indicate the recipe from which their quotation comes. Usually, they refer to these recipes by using the term *taxis* (τάξις, 'recipe'),[110] followed by the name of the main ingredient – often the first ingredient mentioned in the recipe – on which the procedure is based, although sometimes they only give the name of the substance. I shall here focus only on those quotations which preserve valuable information concerning the original sequence of paragraphs in the book *On Gold*.

The incipit of the recipe of *PM* § 5 is quoted by both Synesius[111] and the commentator Christianus, who wrote:

> But at the right moment we will recall the words of his [i.e. Democritus'] first recipe […]. He says: 'Take mercury and make it solid with the body of *magnēsia*, or with the body of Italian stibnite, or with unburnt sulphur, or with silver spume, or with burnt lime, or with alum from Milos, or according to your knowledge.'[112]

Both quotations – which perfectly match *PM* § 5, ll. 67–70 – confirm the opening position of the recipe: a position that is reasserted in the manuscripts by the marginal note "⳨" (i.e. *chrysopoeia*) preserved in all codices next to the first line of this text. *PM* § 5 was in all probability the first recipe (τάξις) of the original book *On Gold*, as may also be inferred from the Syriac tradition. Both the London and the Cambridge manuscripts, in fact, open the book on gold-making with a translation of this recipe.

The three recipes of *PM* §§ 7, 9, and 11 are quoted one after the other by Zosimus within a single paragraph:

> Therefore he says in the pyrite [recipe]: 'take pyrite and process or crush it with vinegar and brine and so on,' words that hint at the white water of untouched sulphur. Afterwards in the cinnabar [recipe]: 'Make cinnabar white by means of oil or vinegar or honey or so on'; in the *androdamas* [recipe] again in a similar way: 'by means of brine or vinegar and brine'; then he adds: 'boil with water of untouched sulphur,' so that you understand that seawater and urine and vinegar and the oil mentioned in the cinnabar [recipe] and honey are the water of untouched sulphur.[113]

---

[110] See Ingeborg Hammer Jensen, *Die älteste Alchymie* (Copenhagen: Hovedkom-missionær, A.F. Høst & søn, 1921), 84; Letrouit, "Chronologie," 78.

[111] Syn. Alch. § 11, ll. 181–4.

[112] M, fol. 110ᵛ8–16; B, fol. 92ʳ4–11; A, fol. 93ʳ3–10 (= *CAAG* II 396,16–397,5): Ἀλλ' ἐπὶ καιροῦ τῶν λόγων τῆς πρώτης αὐτοῦ τάξεως μνησθησόμεθα [...]. Λαβών, φησίν, ὑδράργυρον πῆξον τῷ τῆς μαγνησίας σώματι, ἢ τῷ τοῦ Ἰταλικοῦ στίμεως σώματι, ἢ θείῳ ἀπύρῳ, ἢ ἀφροσελήνῳ, ἢ τιτάνῳ ὀπτῷ, ἢ στυπτηρίᾳ τῇ ἀπὸ Μήλου, ἢ ὡς ἐπινοεῖς (scholarly apparatus in Martelli, *Pseudo-Democrito*, 76).

[113] M, fol. 156ᵛ7–14; B, fols. 143ᵛ13–144ʳ2; V, fols. 112ᵛ15–113ʳ2; A, fol. 130ᵛ2–9 (= *CAAG* II 185,6–12): Φησὶ οὖν ἐπὶ τοῦ πυρίτου· λαβὼν πυρίτην, οἰκονόμει ἢ λείου ὀξάλμῃ καὶ τοῖς ἑξῆς, ὃ αἰνίττεται ὕδωρ θείου ἀθίκτου λευκόν. Εἶτα ἐπὶ

The first quotation is taken from *PM* § 7, a recipe that explains how to process pyrites using several ingredients, including the abovementioned brine with vinegar. However, the recipe does not include the verb λειόω ("to crush") used by Zosimus: the particle ἤ ("or") could suggest that Zosimus himself added this verb, perhaps in order to clarify the procedure, or by following a different version of the recipe.[114] Zosimus then moves on to a recipe for treating cinnabar, corresponding to *PM* § 9, whose incipit is cited word for word. Finally, he quotes two short passages from *PM* § 11 concerning the treatment of *androdamas* (ἀνδροδά-μας). Here, Democritus explains how to treat this ingredient with brine, or vinegar and brine, and boil it with sulphur water.[115]

Zosimus appears to neglect *PM* §§ 8 and 10. Instead, he highlights only those passages where Democritus mentions specific liquid substances which, according to Zosimus' intrepretation, should be identified with "water of untouched sulphur."[116] Such a viewpoint explains Zosimus' omission of *PM* § 8, in which Democritus describes the treatment of *klaudianon* using solid ingredients. On the other hand, it is striking that the vegetable juices listed in *PM* § 10, ll. 103–4 did not attract Zosimus' attention, since they could have strongly supported his hermeneutic effort. In any case, despite these omissions, the sequence of quotations confirms the impression that Zosimus was citing the recipes in the same order in which he found them in ps.-Democritus' book. This order is still preserved in *Physika kai mystika* as we have it today, and thus can perhaps be traced back to the version of the book *On Gold* that was available in the third and fourth centuries AD, when Zosimus was composing his own treatises.

Although the other recipes are frequently quoted by later alchemists, none of these quotations helps us understand the recipes' original position within the book on gold-making. Only comparison with the Syriac tradition – particularly the Cambridge codex – seems to confirm that the order in which the recipes have been preserved in the Byzantine manuscripts follows the order they had in the original *Four Books*. Moreover, both traditions confirm that the book on gold-making was divided into two parts by *PM* §§ 15–16 (= *1SyrC* §§ 11–12), after which Democritus claims to switch to a new topic, namely the treatment of dyeing liquids (*PM* § 16, ll. 183–5). We can therefore recognize a clear distinction between *PM* §§ 5–14 (= *1SyrC* §§ 1–10), dealing with dry substances (τὰ ξηρία), and *PM* §§ 17–19 (= *1SyrC* §§ 13–15), dealing with "washes" (ζωμοί, i.e. dyeing liquids). The so-called

---

[113] *Continued*
τῆς κινναβάρεως· τὴν κινναβαριν ποίει λευκὴν δι᾽ ἐλαίου ἢ ὄξους καὶ μέλιτος καὶ τῶν ἑξῆς. Ἐπὶ δὲ τοῦ ἀνδροδάμαντος ὁμοίως πάλιν· ἅλμῃ ἢ ὀξάλμῃ· εἶτα ἐπιφέρει· ἔψει ὕδατι θείου ἀθίκτου, ἵνα γνῷς ὅτι ὕδατα θαλάσσια, καὶ οὖρον, καὶ ὄξος, καὶ τὸ ἐν τῇ κινναβάρει ἔλαιον καὶ μέλι, ὕδωρ θείου ἀθίκτου ἐστιν (scholarly apparatus in Martelli, *Pseudo-Democrito*, 77).

[114] The Syriac translation of *PM* § 7 also specifies which kind of treatment was applied to the pyrites, and adds the verb ܚܠ, "to purify, to wash" (see *1SyrC* § 3).

[115] The same recipe is quoted by Zosimus *CAAG* II 157,20–3.

[116] On the use of similar expressions by Zosimus, see Cristina Viano, "Gli alchimisti greci e l'acqua divina," *Rendiconti dell'Accademia Nazionale delle Scienze detta dei LX. Parte II: Memorie di Scienze Fisiche e Narurali*, 21 (1997): 61–70, and Matteo Martelli, "Divine Water in the Alchemical Writings of Pseudo-Democritus," *Ambix*, 56 (2009): 8–10.

Catalogues of ps.-Democritus probably had a similar structure,[117] as the following passage of Zosimus seems to confirm:

> These are therefore the dyes; the species of the catalogue, both the solid and the liquid ones – <that is> the plants;[118] the solid ones are included from steam [i.e. mercury] to malachite, while the liquid ones are all the species of the catalogue,[119] namely the true water of untouched sulphur.[120]

Although Zosimus does not mention Democritus here, he is almost certainly referring to the ancient alchemist in this passage. As noted above, the first section of the book on gold, concerning dry substances, probably opened with *PM* § 5, which describes how to make mercury (ὑδράργυρος) solid, and concluded with *PM* § 14, dealing with a specific treatment of malachite (χρυσοκόλλη). The same sequence of ingredients occurs in the passage by Zosimus, who specifies that all the solid substances (somehow comparable with ps.-Democritus' τὰ ξηρία) must be included between the citation of "steam" (νεφέλη), usually identified by the alchemists with mercury, and that of malachite. It is not surprising that Zosimus here insists on the centrality of liquid substances, which he had already stressed in the abovementioned passage. Such substances, in particular vegetable juices, played an important role in the second section of the book on gold, now represented by *PM* §§ 17–19 (= *1SyrC* §§ 13–15).

§ 5. The books on silver and precious stones

Unlike the books on gold and purple, the book on silver-making was handed down by the Byzantine tradition as a separate treatise. In fact, the section preserved under the title *On the Making of Silver* (*Peri asēmou poiēseōs* = *AP*) almost certainly represents what remains of the original book *On Silver* cited by Synesius and Syncellus. Here, analysis of the indirect tradition can also help us to understand the original sequence of recipes, although later alchemists did not cite the first recipes of *AP* with any particular reference to their original position. For instance, I have found no explicit reference to the original position of *AP* § 1, and we also lack information on the other early recipes (*AP* §§ 2–5), despite the large number of quotations disseminated through the whole *Corpus alchemicum*.

The situation is different for the later recipes (*AP* §§ 6–9). Although *On the Making of Silver*, at least in its Byzantine form, lists these nine recipes without interruption, some clues preserved by the indirect tradition suggest that *AP* § 6 originally opened a second section of the original book, one more focused on the role of "washes" (ζωμοί) in the transformation of base metals into silver. This division –

---

[117] See *infra*, pp. 26–9.

[118] On the identification of ὑγρά with plants, see Syn. Alch. § 3 n. 5.

[119] Zosimus here probably refers to the catalogue of washes (ζωμοί; Ps.-Dem. Alch. *Cat.* § 2), which included all the liquid substances; we could perhaps add ζωμῶν after the term "catalogue" (καταλόγου).

[120] M, fol. 150ᵛ3–5; B, fol. 134ʳ13–6; V, fol. 118ʳ24–118ᵛ3; A, fol. 123ʳ21–4 (= *CAAG* II 170,6–8): Αἱ οὖν βαφαὶ αὗται· εἴδη τοῦ καταλόγου στερεὰ καὶ ὑγρά, <τουτέστι> βοτάναι· στερεὰ μὲν ἀπὸ νεφέλης ἕως χρυσοκόλλης, ὑγρὰ δὲ πάντα τοῦ καταλόγου, τὸ δὲ ἀληθὲς ὕδωρ θείου ἀθίκτου (scholarly apparatus in Martelli, *Pseudo-Democrito*, 79).

analogous to the one found in the book *On Gold* – is clearly attested by the Syriac tradition, which preserves a theoretical section (*2SyrC* § 5) that functions as a kind of a bridge between the first (*AP* § 1–5) and the last recipes (*AP* §§ 6–9). At the end of this section the author explicitly claims to switch to a new topic, namely "the waters in which both the white and the yellow drugs [lit. herbs] are boiled down." This new topic is also mentioned in *The Chemistry of Moses* (*CAAG* II 310,9) that preserves *AP* § 6 under the title of "Washes [i.e. dyeing liquids] for the silver-making" (᾿Αργυροποιίας ζωμοί), while Zosimus also seems to make a glancing reference to it in the following passage:

> After applying the water of untouched sulphur – the one that is understood in a general sense[121] – with the lead-copper alloy, we roast it for a whole day, according to the instructions of the first recipe of the white washes, but on enveloping fires,[122] according to the instructions of the litharge [recipe].[123]

The reference to the "first recipe of the white washes" (i.e. liquids employed for the making of silver) is puzzling. First, it is not clear exactly which elements of this procedure are taken from the ps.-Democritus recipe; indeed, the treatment of a lead-copper alloy with sulphur water is not the specific topic of *AP* § 6. We may therefore guess that Zosimus invoked this recipe only in relation to the duration (or maybe to the procedure) of the roasting process, since *AP* § 6, ll. 59–60 prescribes dipping a metallic leaf into a dyeing solution, and heating it for a whole day. The similarity between the procedures – both texts referring to the treatment of a metal with a liquid dye – could explain its citation by Zosimus, who perhaps wanted to know exactly how long the roasting process had to last. Either way, we cannot rule out differences between the version of ps.-Democritus that Zosimus quoted and that preserved today in the Byzantine tradition.[124]

Regarding § 7, Zosimus provides a second quotation, which seems to confirm the position in which the recipe is recorded by the Byzantine tradition, when he writes "'... by diluting with the water of white poplar ash': in the second [recipe] of the white washes the term 'ash' has not its ordinary meaning, etc."[125] The sentence quoted by Zosimus is attested in a very similar form in *AP* § 7, which is the

---

[121] See Ps.-Dem. Alch. *Cat.* § 1 n. 2.

[122] See Ps.-Dem. Alch. *AP* §, 2, ll. 12–3.

[123] M, fol. 141ᵛ11–4; B, fol. 119ᵛ4–8; A, fol. 113ʳ22–5 (= *CAAG* II 147,23–6): τὸ ὕδωρ τοῦ θείου ἀθίκτου τὸ ἀπολελυμένον μετὰ τοῦ μολυβδοχάλκου μεταβαλόντες ὀπτοῦμεν ἡμέραν μίαν, καθὼς ἔχει ἐν τῇ πρώτῃ τάξει τῶν λευκῶν ζωμῶν, ἀλλὰ καὶ εἰλικτοῖς, καθὼς ἔχει ἐν τῇ λιθαργύρῳ (scholarly apparatus in Martelli, *Pseudo-Democrito*, 80).

[124] Similar problems arise when we try to identify the second recipe quoted by Zosimus. The term εἰλικτοῖς is attested only in *AP* § 2, where the ingredient λιθάργυρος is not mentioned at all. On the other hand, *AP* § 5, which deals with the treatment of this substance, pays particular attention to the intensity of the fire, even if it does not employ the adjective εἰλικτός. Similar gaps are difficult to justify on the basis of our current knowledge of the *Corpus alchemicum*, since several hypotheses are possible: we could suppose that (1) Zosimus himself made some mistakes when quoting ps.-Democritus (perhaps citing from memory); (2) the manuscript tradition did not preserve the original text by Zosimus; or (3) the recipes quoted by the alchemist differed from the text handed down by the Byzantine epitome.

[125] M, fol. 158ʳ18–20; B, fol. 146ᵛ4–6; V, fol. 107ᵛ8–11; A, fol. 132ʳ20–2 (= *CAAG* II 189,14–5): [...]| ἀναλύσας ἐν ὕδατι σποδοῦ λευκίνων ξύλων· ἐν τῇ δευτέρᾳ τῶν λευκῶν ζωμῶν σποδὸν λευκίνων οὐκ ἔστιν ἁπλῶς κτλ. (scholarly apparatus in Martelli, *Pseudo-Democrito*, 81).

second recipe to use liquid substances in the whitening processes of base metals. The quotation thus confirms that this recipe occupied the same position, at least in the version of ps.-Democritus known to Zosimus.

Zosimus cites the recipe of § 9 several times, often stressing its importance as the only recipe in which ps.-Democritus specifies the exact amount of any single ingredient employed in alchemical procedures. In one place Zosimus writes: "… and the weight of the untouched sulphur is given in the last recipe of the white washes: 'one ounce of orpiment' and so on."[126]

Here Zosimus is quoting the incipit of AP § 9 (l. 79). He always introduces other quotations from the same passage with the expression "in the last recipe of white washes" (ἐν τῇ ὑστέρᾳ τάξει τῶν λευκῶν ζωμῶν),[127] confirming that the current AP § 9 was the final recipe in the second section of the book on silver-making which treats the use of dyeing liquids.

In conclusion, by examining the indirect tradition we can infer that the book on silver – as well as those on gold and on purple – was originally divided into recipes (τάξεις) whose order is retained in (at least in the second part of) the Byzantine epitome (AP §§ 6–9). The sequence of paragraphs (§§ 1–5) in the first part of AP, on the other hand, is corroborated by the Syriac translations – in particular by SyrC. The Syriac tradition further confirms that On Silver was originally structured in two sections. The current §§ 1–5 of AP preserve what survives of the first section, presumably on dry/solid substances, while §§ 6–9 preserve what remains of the second, on liquids. On the basis of the text handed down by SyrC, a theoretical passage (2SyrC § 5) – similar to PM §§ 14–15 – explicitly marked the switch from the first to the second section.

As for the book on the making of precious stones, no excerpts at all are preserved in the Byzantine tradition under the name of Democritus, even though the indirect tradition clearly includes this subject among the main topics discussed by the ancient alchemist. We can nevertheless recognize some quotations from this book in the works of later alchemists. In this respect, the recipe collection edited by Berthelot-Ruelle under the title *Deep Tincture of Stones, Emeralds, Rubies and Jacinths from the Book Taken out from the Sancta Sanctorum of Temples* seems particularly interesting, and merits further examination.[128] Such an inquiry would go beyond the purposes of the present study, which focuses primarily on the ps.-Democritus excerpts in the Byzantine and Syriac transmissions. Consequently, I would like to draw attention especially to SyrC, which preserves four passages on the topic of precious stones (3SyrC §§ 1–4), that are edited here for the first time. As already pointed out, their similarities with some quotations by later

---

[126] M, fol. 145ᵛ 15–7; B, fol. 122ᵛ8–10; A, fol. 115ᵛ3–5 (= CAAG II 152,28–153,2): […] καὶ θείου ἀθίκτου σταθμὸν πεποίηται ἐν τῇ ὑστέρᾳ τάξει |καὶ| τῶν λευκῶν ζωμῶν· ἀρσενικοῦ οὐγγίαν μίαν καὶ τὰ ἑξῆς (scholarly apparatus in Martelli, *Pseudo-Democrito*, 82).

[127] See CAAG II 155,1–3; II 155,17f.; II 161,15f.; II 163,23; II 178,18; II 217,14.

[128] CAAG II 350–64; see Bidez-Cumont, *Mages*, vol. 2, 324 n. 1, and Halleux, *Papyrus de Leyde*, 74–5.

alchemists seem to confirm that these four recipes originally belonged to ps.-Democritus' book *On Stones*.[129]

§ 6. Ps.-Democritus' *Catalogues* in the indirect tradition: Synesius' commentary and the so-called *Chemistry of Moses*

In addition to the recipes that comprised the core of the *Four Books*, later alchemists also attributed several catalogues, or lists of substances, to ps.-Democritus. The distinction between solid and liquid ingredients noted by Zosimus seems to have constituted an important criterion according to which ps.-Democritus divided his recipes into two groups. Zosimus quite often alludes to the same distinction, also in regard to lists of ingredients.[130] For example, "That is why the philosopher [i.e. Democritus] in his catalogue of washes took care to mention steam [i.e. mercury] and sulphur water."[131] Yet among the treatises of the *Corpus alchemicum*, it is the commentary of Synesius that provides the clearest and most complete source of information about the catalogues, which were not included in the Byzantine epitome of the *Four Books*. For instance, in the second paragraph of his commentary, Synesius claims that "it is necessary that we [...] learn what his doctrine is, and in what order his arguments follow one another. It is clear to us that he composed two catalogues, on the white [i.e. on silver] and on the yellow [i.e. on gold]; first he listed the solid substances, afterwards the washes, that is, the liquid substances."[132]

A few paragraphs later, Synesius reasserts that,

> In order to admire the wisdom of this man, look at how he composed two catalogues – for the making of gold (*chrysopoeia*) and for the making of silver (*argyropoeia*) – and how he composed [two catalogues] of washes, the first in the yellow, that is, in [his book on] gold, the second in the white, that is, in [his book on] silver; and he gave to the catalogue of gold the name of "the making of gold" and to the catalogue of silver the name of "the making of silver."[133]

There are some interesting correspondences between this information and the extant portions of the *Four Books*. According to the sources cited above, ps.-Democritus composed two catalogues, dividing his ingredients into solids and liquids. A similar division, as we have seen, separates the recipes in both *On Gold* and *On Silver*. In addition, the catalogue for the book on gold was called

---

[129] See *supra*, p. 9.
[130] See *CAAG* II 169,5–6, where a catalogue is attributed to Hermes and Democritus. In *CAAG* II 241,8–4, Zosimus also refers to different classifications of substances that Democritus employed in order to explain the alchemical procedures.
[131] M, fol. 141ʳ22–4; B, fols. 118ᵛ19–119ʳ2; A, fol. 113ʳ2–4 (= *CAAG* II 147,7–8): Διὰ τοῦτο καὶ ὁ φιλόσοφος ἐν τῷ καταλόγῳ τῶν ζωμῶν μετὰ παρατηρήσεως εἴρηκεν νεφέλην καὶ πάλιν ὕδωρ θείου (scholarly apparatus in Martelli, *Pseudo-Democrito*, 83).
[132] Syn. Alch. § 2, ll. 18–22.
[133] Syn. Alch. § 8, ll. 107–11. See also ll. 122–6.

Chrysopoeia, and that for the book on silver *Argyropoeia*. The same formulae are employed by ps.-Democritus himself with reference to his own works. At the end of the section on gold-making, for instance, he claims (*PM* § 20, ll. 214f.): "The matter for the making of gold (*chrysopoeia*) extends up to these natural substances." A few lines later, he adds (ll. 228f.): "Let us clearly examine also the composition of the species for the making of silver (*argyropoeia*)." In addition, immediately after this sentence, the manuscripts record *On the Making of Silver*, the last paragraph of which opens with the words: "You have received everything useful for gold and silver."

From these similarities, we could infer that the ps.-Democritean works preserved in the Byzantine tradition overlap to a certain extent with the catalogues. However, Synesius' commentary draws a clear distinction between the two sections. After supplying the abovementioned information, the commentator actually devotes extensive discussion to various passages from the catalogues that do not correspond to the text of either *Physika kai mystika* or *On the Making of Silver*. Synesius also clearly distinguishes the incipit of the catalogue on gold from the first recipe of the book on gold-making, these being respectively quoted in two passages:

> Look at what he [i.e. Democritus] said in his introduction to the making of gold [*chrysopoeia*, i.e. the catalogue on gold]: 'Mercury that comes from cinnabar, malachite' [= *Cat.* § 1, ll. 1–2].[134]

> Therefore he said 'the body of *magnēsia*,' that is the mixing of the substances; which is why he said further on in the introduction to [his book on] the making of gold: 'Take mercury and make it solid with the body of *magnēsia*' [= *PM* § 1, l. 67].[135]

Thus the catalogues of substances and the recipes that use those substances were in some way distinct, since Synesius referred both to the beginning of the catalogue and to the incipit of the first recipe. The catalogue merely lists the substances, while the recipes explain how to use them in specific alchemical procedures. It therefore seems likely, as hinted at by Lagercrantz, that ps.-Democritus first listed all the substances useful for making gold and silver (in the section called "Catalogues"), dividing them into solids and liquids, and thereafter explained in the recipes how to process and use those substances.[136]

These lists of substances were not included in the Byzantine epitome of ps.-Democritus. However, the catalogues may be at least partially reconstructed by combining information preserved by Synesius with some sections of the recipe collection called *The Chemistry of Moses*. The earliest witness of *The Chemistry of Moses* is manuscript A (fols. 268ᵛ15–278ᵛ26). Most of its recipes describe metallurgical processes, among which the compiler included excerpts from the *Four Books*. *The Chemistry of Moses*, in fact, includes some ps.-Democritean recipes

[134] Syn. Alch. § 5, ll. 43–4.
[135] Syn. Alch. § 11, ll. 181–4.
[136] Lagercrantz, *Papyrus Graecus Holmiensis*, 109.

along with specific lists of substances considered useful for making gold, silver, and purple (see Appendix, Table 4). By comparing these lists with Synesius' commentary, we can now confidently identify them as the original catalogues of ps.-Democritus, that were not included in either *Natural and Secret Questions* or *On the Making of Silver*.[137] (See Appendix, Table 5.)

Regrettably, while in the abovementioned passages Synesius refers to four catalogues, two for gold-making (on solid and liquid substances respectively) and two for silver-making, *The Chemistry of Moses* preserves only three: (1) *Substances for the Making of Gold* (= Cat. § 1, on solid ingredients), (2) *Substances for the Making of Washes* (= Cat. § 2, on liquids for gold-making), and (3) *Substances for the Making of Silver* (= Cat. § 3, on solid ingredients). The *fourth* catalogue devoted to liquid substances for silver-making seems to be missing. It is worth noting that the same gap is also attested in the commentary of Synesius, who does not mention any wash used for *argyropoeia*, but focuses his hermeneutic effort on the first three catalogues, for which he provides an allegorical and paretymological reading. In this case, however, the gap may be explained by considering the incomplete form in which Synesius' commentary has been handed down.[138] We cannot exclude the possibility that some references to the fourth ps.-Democritean catalogue were originally included in the lost part of his work.[139]

Finally, the original position of the catalogues within the *Four Books* remains uncertain. As previously noted, some clues suggest that the catalogues originally opened the books on gold- and silver-making. This conclusion is partly supported by a passage from Zosimus, who refers to similar lists of ingredients and seems to place these catalogues at the beginning (προοίμιον) of ps.-Democritus' writings.[140] Unfortunately the text of this passage – whose earliest witness is **A**, fols. 252ᵛ-253ʳ – is badly corrupt, and Berthelot-Ruelle proposed a highly conjectural reading of it. At all events, Zosimus clearly refers to *four* catalogues by the ancient alchemist (*CAAG* II 241,16): "but I will not list all the substances of the four catalogues (ἐν τοῖς τέτρασι καταλόγοις); you will find [there] an examination of all the substances for the opportune dyes."[141] This passage is nevertheless open to two interpretations regarding the actual number of catalogues. On the one hand, Zosimus could have in mind the two catalogues for gold-making, plus the two catalogues for the making of silver, in which case, the passage would

[137] Tannery has already stressed the importance of Synesius' commentary for reconstructing some passages of ps.-Democritus excluded from the Byzantine epitome, although he did not take into account *The Chemistry of Moses*: Tanney,"Études sur les alchimistes grecs," 285–6.

[138] See *infra*, p. 49.

[139] A reference to the fourth catalogue seems to be detectable in a later collection of alchemical excerpts in **A** (fols. 136ᵛ–140ᵛ) under the title *On the Stone of Philosophy* (*CAAG* II 198,8–204,7; see Mertens, *Zosime de Panopolis*, LIX-LX). A passage dealing with ps.-Democritus' catalogues (*CAAG* II 199,25–200,6), records that "white waters" (ὕδατα λευκά), "beer" (? ζῦθος) and an unspecified juice (χυλός) were included ἐν τῷ ὑγρῷ τοῦ λευκοῦ, to be understood as "in the section dealing with the liquid substances (ὑγρά) of the catalogue on the white |i.e. on silver|."

[140] *CAAG* II 241,8–24.

[141] On the expression "opportune dyes" (καιρικαὶ βαφαί), which refers to those dyeing tecniques whose success depended on astrological influences, see Mertens, *Zosime de Panopolis*, 62–3.

corroborate the information provided by Synesius. On the other hand, it is possible that *each* of the original four books of ps.-Democritus included a catalogue: a hypothesis partially supported by the list of substances for purple dyeing preserved in both *Physika kai mystika* (§ 2) and *The Chemistry of Moses*. Unfortunately, this last catalogue does not appear in the same position in both witnesses. In *Physika kai mystika* it follows a recipe on purple dyeing, while in *The Chemistry of Moses* it is included among the other ps.-Democritean catalogues. Since the indirect tradition provides no information on the original position of this list, I have preferred to edit it within the *Natural and Secret Questions*, our earliest source for what remains of the original books *On Gold* and *On Purple*.

§ 7. Dating of ps.-Democritus' *Four Books*

Many works were probably circulating under the name of Democritus during the Hellenistic and Roman periods. As we have already seen, in the second century AD Aulus Gellius (X 12,8) complained about the "many fictions" (*multa commenta*) attributed to the atomist, whose production had already become associated with a kind of Eastern – specifically Persian and Egyptian, – wisdom.

Among these pseudonymous works are included the *Four Books*, which are preserved only in an incomplete and epitomised form.[142] According to some elements visible in the *Corpus alchemicum*, these books seem to date back to the second half of the first century AD. The earliest alchemist who clearly refers to them is Zosimus of Panopolis, whose own works date to the end of the third and the beginning of the fourth century AD, thus providing an important *terminus ante quem*. Yet ps.-Democritus himself provides some important clues for dating his work. *PM* § 8 describes an alchemical procedure that requires a specific ingredient called *klaudianon* (κλαυδιανόν).[143] Its presence within the original *Four Books* is confirmed by later commentators, who often mention this substance with reference to ps.-Democritus.[144] In Berthelot's opinion, the term indicates a particular metallic alloy similar to some kinds of copper discussed by Pliny the Elder.[145] The name *klaudianon* is likely to stem from the name of the Emperor Claudius, who reigned between 41 and 54 AD. Similar terminology is also employed with reference to an important mining area in Egypt, which began to be significantly exploited during the reign of Nero (54–68 AD); in fact, the place where an important marble quarry was opened is usually called *Mons Claudianus*.[146] It therefore seems very likely that the *Four Books* date from after the first half of the first century AD.[147]

---

[142] Letrouit, "Chronologie," 85 supposed that the four original books were epitomised around the eighth–ninth century.

[143] The term is also attested in the masculine (Κλαυδιανός) by the *Corpus alchemicum*.

[144] See Zos. Alch. *CAAG* II 159,9 e 187,6; Syn. Alch. § 13, ll. 204f. The same substance is listed by ps.-Democritus himself in *Cat.* § 1, l. 4.

[145] *NH*, XXXIV 3–4: the alloys here are the *aes Marianum* and *aes Livianum*.

[146] See Michel Stéphanidès, "Petites contributions à l'histoire des sciences," *Revue des études grecs*, 31 (1918): 203–4; Ps.-Dem. Alch. *PM* § 8 n. 33.

[147] Festugière, *Révélation d'Hermès*, vol. 1, 225 also focused his attention on the term λακχά, attested in Ps.-Dem. Alch. *PM* § 1. This ingredient, the name of which probably comes from India, could be available only in a period when

Further evidence points to the reign of Nero. On the one hand, Diels has proposed that the alchemist Pammenes, mentioned in *PM* § 20,[148] should be identified with an Egyptian astrologer who lived during Nero's time, and who is mentioned by both Tacitus (*Ann.* XVI 14) and Aelianus (*NA* XVI 42).[149] Tacitus reports that during the consulate of Caius Svetonius and Lucius Telesinus (66 AD), the praetor Antistius Sosianus, exiled for composing satires on Nero, tried to befriend a famous astrologer named Pammenes who was in the same condition, wishing to use Pammenes to recover the emperor's favour.[150] Aware of the relationship between Pammenes and Publius Anteius, Sosianus

> intercepted a letter from Anteius, stole in addition the papers concealed in Pammenes' archives, which contained his horoscope and career, and, lighting at the same time on the astrologer's calculations with regard to the birth and life of Ostorius Scapula, wrote to the emperor that, could he be granted a short respite from his banishment, he would bring him grave news conducive to his safety; for Anteius and Ostorius had designs upon the empire, and were peering into their destinies and that of the prince.[151]

Aelianus provides less rich information, citing Pammenes only with regard to his treatise *On Wild Animals*:

> Pammenes in his book *On Wild Animals* states that there are scorpions in Egypt, winged and with two stings (and he claims not to base his account on hearsay, but in accordance with his personal investigation), and two-headed snakes which have two legs under their tail.[152]

Unfortunately the identification of these three Pammenes – who are presented as experts in three different fields, namely alchemy, astrology and zoology – is not supported by any conclusive evidence. It is possible that a single author, interested in *physika*, worked on all three topics, but we know of no source that can confirm such a hypothesis. In any case, the match between the chronological data provided by Tacitus, and those inferred from the term *klaudianon*, supports dating the *Four Books* to around the time of Nero's reign.[153]

---

[147] *Continued*
there was free trade between Egypt and India: that is, according to Festugière's opinion, between the first and second centuries. On the identification of λακχά, see Ps.-Dem. Alch. *PM* § 1 n. 3.

[148] The presence of Pammenes in the original books by ps.-Democritus is confirmed by Zos. Alch. *CAAG* II 148,15–6. According to Syncellus (297,24–298,1 Mosshammer), Pammenes was a pupil of the Persian magus Ostanes in Egypt, together with Democritus.

[149] See Hermann Diels, *Antike Technik*, second edition (Leipzig: Teubner, 1924), 134 n. 1; Bidez-Cumont, *Mages*, vol. 2, 312 n. 2; Festugière, *Révélation d'Hermès*, vol. 1, 226 n. 1.

[150] Tac. *Ann.* XVI 14, 12–9.

[151] Translation by John Jackson, *Tacitus, Annals 13–16* (Cambridge, MA and London: Loeb, 1937), 357.

[152] *NA* XVI 42, 1–5.

[153] Several scholars seek to date the *Four Books* to the first century AD: see in particular Lippmann, *Entstehung und Ausbreitung der Alchemie*, vol. 1, 27–9, and Wellmann, *Bolos Demokritos*, 68–9. On the other hand, Steele, Bidez-Cumont, and Letrouit argued that the books cannot have been written prior to this period: Robert R. Steele, "The Treatise of Democritus on Things Natural and Mystical," *Chemical News*, 61 (1890): 88 [see also Stanton J. Linden, *The Alchemy Reader. From Hermes Trismegistus to Isaac Newton* (Cambridge: Cambridge University Press, 2003), 38]; Bidez-Cumont, *Mages*, vol. 1, 198 n. 2; Letrouit, "Chronologie," 74. Hammer Jensen, *RE* Suppl. IV (1930), 222, supposed that similar collections of alchemical recipes were attributed to Democritus around the fifth

A final source – although problematic – suggests the same dating. Seneca, in his ninetieth epistle, criticised Posidonius' ideas about the relationship between philosophy and the practical sciences (*artes*/τέχναι). Seneca disagreed with Posidonius, who had stated that wise men first discover the practical arts and their tools, and afterwards present them to craftsmen who make use of such techniques.[154] For Seneca, "wisdom's seat is higher; she trains not the hands, but is mistress of our minds."[155] To support his criticism, the Latin philosopher quotes many fragments from Posidonius that attribute technological discoveries to ancient philosophers, including Democritus:

> Posidonius again remarks: 'Democritus is said to have discovered the arch, whose effect was that the curving line of stones, which gradually lean toward each other, is bound together by the keystone.' I am inclined to pronounce this statement false. For there must have been, before Democritus, bridges and gateways in which the curvature did not begin until about the top. It seems to have quite slipped your memory that this same Democritus discovered how ivory could be softened, how, by boiling, a pebble could be transformed into an emerald – the same process is used even today for colouring stones which are found to be amenable to this treatment![156]

Diels-Kranz include this passage among Democritus' spurious fragments. The first part preserves a quotation from Posidonius (fr. 488 Theiler = 284 Edelstein-Kidd), who attributed the discovery of the arch to the Abderite philosopher. In the second, Seneca himself adds further details – no longer taken from Posidonius – about other strange discoveries said to have be made by Democritus. The reference to the making of artificial emeralds recalls the third alchemical book (mentioned by Synesius and Syncellus as *On Stones*), which probably dealt with similar techniques, well-attested in particular by the Stockholm papyrus.[157] Seneca could have extracted this information from a wider corpus of ps.-Democritean technical writings, which probably started circulating from the late Hellenistic period. The next section will be devoted to the analysis of these technical writings, which have often been linked to the controversial figure of Bolos of Mendes, and their possible relationships with the *Four Books*.

---

[153] *Continued*
century AD: in which case it would be difficult to explain the exact quotations from ps.-Democritus' recipes made by Zosimus.

[154] *Ep.* XC 25: "Omnia, inquit |scil. Posidonius], haec sapiens quidem invenit: sed minora quam ut ipse tractaret, sordidioribus ministris dedit" ("But, says Posidonius, the wise man did indeed discover all these things; they were, however, too petty for him to deal with himself and so he entrusted them to his meaner assistants"). Translation by Richard Mott Gummerre, *Seneca, Epistles 66–92* (Cambridge, MA and London: Loeb, 1917–25), 413–5|.

[155] *Ep.* XC 26: "Sapientia altius sedet nec manus edocet: animorum magistra est"; translation by Mott Gummere, *Seneca Epistles*, 415.

[156] Transl. by Mott Gummere, *Seneca Epistles*, 420.

[157] See Halleux, *Papyrus de Leyde*, 47–52. Since this book has been not preserved in manuscript, it is difficult to guess its original content, or to establish a close parallel with Seneca's account: see Marco Beretta, *The Alchemy of Glass: Counterfeit, Imitation, and Transmutation in Ancient Glassmaking* (Sagamore Beach, MA: Science History Pubblications, 2009), 100–1.

§ 8. Background to the attribution of the alchemical books to Democritus

Although the *Four Books* cannot have been written prior to the first century AD, their attribution to Democritus may be explained by the Abderite philosopher's reputation within other fields of the so-called "natural sciences," including medicine, pharmacology, and agriculture, already established during previous centuries. In particular, two aspects of his reception during the late Hellenistic and first Imperial period should be taken into account in order to better understand and contextualize his association with alchemy: (1) his supposed relations with Eastern traditions, and (2) the technical skills attributed to him.

8.1. The supposed Eastern tradition

As already noted, various alchemists, including Zosimus and Synesius, link the *Four Books* to the Persian magus Ostanes, the supposed alchemical master of Democritus.[158] A relationship between the fifth-century BC atomist and Eastern wisdom had already been suggested by diverse sources as early as the second to first century BC.[159] This close connection with Middle Eastern traditions left its mark on the pseudepigrapha circulating under Democritus' name. Besides the alchemical writings, these include *On the Sacred Writings in Babylon*, listed by Diogenes Laertius among the Democritean works of uncertain authenticity.[160] The title is not included in the tetralogies and, for this reason, it may not have been present in Thrasillus' catalogue of Democritean works (first century BC/first century AD).[161] This pseudepigraphic work is probably the source for Clement of Alexandria's account, according to which Democritus translated the stele of the wise Ahikar (*Akikaros* in Clement's passage), vizir of the Assyrian king Sennacherib (705–681 BC).[162] Moreover, a certain analogy has been noticed between Clement's account

---

[158] The question is discussed in more detail *infra*, pp. 69–73.

[159] See *supra*, n. 9. Following the conclusions reached by Bidez-Cumont (*Mages*, vol. 1, 167–8), who stressed the important role played by Hermippus' Περὶ μάγων in the invention of similar legendary accounts, Sergio Ribichini confirms that these legends "derivano, per la gran parte, da una tradizione maturata in ambienti ellenistici cosmopoliti, che, per reazione alla storiografia greca nazionalistica, sostenevano l'apprendistato da parte di pensatori greci di talune verità essenziali dai sapienti dell'Oriente": Ribichini, "Fascino dall'Oriente e prime lezioni di magia," in *La questione delle influenze vicino-orientali sulla religione greca. Stato degli studi e prospettive della ricerca. Atti del Colloquio internazionale di Roma, 20–22 maggio 1999*, ed. Sergio Ribichini, Maria Rocchi and Paolo Xella (Rome: ed. del CNRS, 2001), 109. See also Arnaldo Momigliano, *Alien Wisdom. The Limits of Hellenization* (Cambridge: Cambridge University Press, 1975), 146. On the other hand, Wellmann (commenting upon 68 |55| B 300,13 D-K) considered Apion's Περὶ μάγου one of the most important sources for the history of magic in Pliny, *NH* XXX 8–11.

[160] Diog. Laert. IX 49,3 = 68 |55| A 33 D-K.

[161] Tharsyllus, personal astrologer of the Roman emperor Tiberius, enumerated Plato's and Democritus' writings by grouping them in tetralogies (groups of four books); see Harold Tarrant, *Thrasyllian Platonism* (Ithaca, NY and London: Cornell University Press, 1993), 85–98. On the contrary, according to Marek Węcowski, "Pseudo-Democritus, or Bolos of Mendes (n. 263)," in *Brill's New Jacoby*, ed. Ian Worthington (Brill Online, 2013), the titles enumerated by Diog. Laert. IX 49,3 (= t. 1 in Węcowski's entry) would represent a list of Democritean treatises (to be ascribed to Bolos of Mendes) whose authenticity was already questioned by Thrasyllus and his contemporaries.

[162] *Strom.* I 15,69 = 68 |55| B 299 D-K. Ahikar is the heros of a famous ethical folktale, the so-called *Book of Ahikar*, which is already attested by a fifth century BC Aramaic papyrus found in Elephantine: see Eduard Sachau, *Aramäische Papyrus und Ostraka aus einer jüdischen militär-Kolonie zu Elephantine*, (Leipzig: J.C. Hinrichs'sche Buchhandlung, 1911), 147–51. "The book has survived in several versions: Syriac, Arabic, Ethiopic, Armenian, Turkish, and Slavonic |...| It may be subdivided into two parts: (1) the life of Ahikar; (2) the sayings uttered for the benefit of

and Pliny's account in his *Natural History* (XXX, 9), where Democritus is said to have entered various tombs in order to find Dardanus' books.[163] Some scholars have traced these stories back to the writings of Bolos of Mendes (second century BC).[164] Although this attribution is not certain, we cannot exclude the possibility that the ps.-Democritean treatise *On the Sacred Writings in Babylon* dates back to the late Hellenistic period.[165]

On the other hand, from this period onward several treatises were attributed to different Persian *magi*, notably Zoroaster and Ostanes. According to the Byzantine lexicon *Suda* (tenth century AD), Zoroaster is supposed to have composed four books entitled *On Nature,* and a second text entitled *On Precious Stones.* The first dealt also with the powers of herbs,[166] constituting one of the principal sources – probably through the lexicon *On Plants* by Pamphilus of Alexandria[167] – for the Persian names attributed to several plants in ps.-Dioscorides, Pliny the Elder, and in ps.-Apuleius' herbal.[168] Conversely, the content of *On Precious Stones* is harder to reconstruct, although it seems to have left a consistent mark on Pliny's *Natural History*, Book 37.[169] Bidez-Cumont likewise attributed similar areas of interest to Ostanes, whom our sources present as a great expert in the "the secret properties of animals, plants, and stones."[170] The same aphorism about nature, "nature delights in nature," presented in the alchemical tradition as Ostanes' teaching to Democritus, was already known during the Hellenistic period.[171] In fact, according to the Roman astrological writer Firmicus Maternus, it was mentioned by the astrologer Nechepso.[172] Such a formulation, riddling on the rules of sympathy and antipathy thought to regulate interactions among all

[162] *Continued*
Nadan, his adopted son" (*EJ* I, *s.v.* Ahikar, book of); see also Frederick C. Conybeare, James Rendel Harris and Anne S. James, *The Story of Ahikar from the Aramaic, Syriac, Arabic, Armenian, Ethiopic, Old Turkish, Greek and Slavonic Versions*, 2nd ed. enlarged and corrected (Cambridge: Cambridge University Press, 1913).

[163] See, for instance, Walter Leszl, "Democritus' Works: From their Titles to their Contents," in *Democritus: Science, the Arts, and the Care of the Soul, Proceedings of the International Colloquium on Democritus, Paris, 18–20 September 2003*, ed. Aldo Brancacci and Pierre Marie Morel (Leiden: Brill, 2007), 58.

[164] See, for instance, Bidez-Cumont, *Mages*, vol. 1, 172; Max Wellmann, "Zu Democrit," *Hermes*, 61 (1926), 474–5; Węcowski, "Pseudo-Democritus, or Bolos of Mendes (n. 263)," comm. on fr. 1.

[165] After the influential studies of Wellmann, scholars have tended to attribute the greater part of ps.-Democritean production to Bolos of Mendes (see *infra*, pp. 37–44). However, despite the importance of Mendesius, ancient sources do not allow us to consider him the only author writing under the name of Democritus: see, more recently, Letrouit in *DPhA* II *b* 53, *s.v.* Bolos de Mendès; M. Laura Gemelli Marciano, "Le Démocrite technicien. Remarques sur la réception de Démocrite dans la littérature technique," in *Democritus: Science, the Arts, and the Care of the Soul*, 224–5; Leszl, "Democritus' Works," 61–2.

[166] See Bidez-Cumont, *Mages*, vol. 1, 105–27.

[167] On the basis of Gal. XI 763,7–8 Kühn, we know that Pamphylus composed a lexicon in which he recorded all the ὀνόματα Αἰγυπτιακὰ καὶ Βαβυλώνια ("Egyptian and Babylonian names") of the plants (see Bidez-Cumont, *Mages*, vol. 1, 116).

[168] Passages collected by Bidez-Cumont, *Mages*, vol. 2, 164–73 (frr. O 16–36). See also Silvano Boscherini, "L''Erbario' di Apuleio e i precetti dei profeti," *Galenos*, 1 (2007): 113–8.

[169] See Bidez-Cumont, *Mages*, vol. 1, 128–30.

[170] See Bidez-Cumont, *Mages*, vol. 1, 189 and vol. 2, 299–301.

[171] See Bidez-Cumont, *Mages*, vol. 1, 204; Festugière, *Révélation d'Hermès*, vol. 1, 232.

[172] *Mat.* IV 22,2. Fr. 28 (2) in Ernst Riess, "Nechepsonis et Petosiridis Fragmenta Magica," *Philologus*, suppl. 6 (1892): 379. Many Greek astrological works circulated in Ptolemaic Egypt under the authorship of the Egyptian pharaoh Nechepso and his priest Petosiris; see, for instance, Pierre Monat, *Firmicus Maternus. Mathesis, livres I-II* (Paris: Les Belles Lettres, 1992) 15–8.

natural elements, was certainly appropriate for experts in a range of fields who shared a common interest in the properties of natural substances.

8.2. Democritus and the technical arts (technai)

The *Four Books'* focus on specific alchemical practices and techniques contrasts with that of the historical Democritus, usually presented in ancient sources as the founder of atomism. Atoms are never mentioned in the Greek alchemical tradition, however, and according to Gemelli Marciano this philosophical theory was not the only ground for Democritus' fortunes during the late Hellenistic and early Roman periods (first century BC – first century AD).[173] Various sources in fact emphasise the ancient atomist's expertise across several *technai*, or technical arts. Thrasyllus' catalogue stresses the *polymathia* of Democritus, who was supposed to be experienced in every art, and introduces a whole section of his works under the heading of *Technika*.[174] This image of Democritus continues to predominate in Synesius' commentary, where the introduction mentions the investigations Democritus was believed to have conducted into all natural questions.[175] It is this reputation – already well attested in Posidonius – that Seneca countered in his polemic: dismissing as groundless the attribution of technical discoveries to Democritus, who in his view would not have lowered himself to the level of a craftsman.[176] However, Petronius, who was not influenced by the theoretical distinction between *sapientia* and *ars* on which Seneca grounded his discussion, states that "Democritus extracted the juice of every plant on earth, and spent his whole life in experiments to discover the virtues of stones and twigs."[177] Although it is not clear from the context whether these "virtues" investigated by Democritus relate to the pharmacological or the dyeing properties of substances, this source still seems to hint at Democritus' technical inclinations, through his interest in *experimenta*.

Thrasyllus' *Technika* lists several works covering technical fields in which Democritus was considered an expert, at least between the first century BC and the first century AD. There we find treatises on agriculture,[178] military strategy,[179] and especially medicine, to which four works were probably devoted

---

[173] Gemelli Marciano,"Le Démocrite technicien," 209–13.

[174] Diog. Laert. IX 48; see also IX 37,7–8: (scil. Δημόκριτος) περὶ τεχνῶν πᾶσαν εἶχεν ἐμπειρίαν.

[175] Syn. Alch. § 1, ll. 5–11.

[176] *Ep.* XC; see *supra* p. 31.

[177] Petr. 88,23 = 68 |55| B 300,6 D-K; translation by Michael Heseltine, *Petronius* (Cambridge, MA and London: Loeb, 1913), 174.

[178] Diog. Laert. IX 48,21 (= 68 |55| A 33 D-K): Περὶ γεωργίης ἤ Γεωμετρικόν (to be corrected to Γεωργικῶν in Wellmann's opinion; *On Agriculture or Agricultural Questions*). The authenticity of this treatise has been extensively questioned after Wellmann, *Die Georgika des Demokritos*, who attributed it to Bolos of Mendes. This conclusion was criticised by Kroll ("Bolos und Democritos," 230), and some scholars do not currently exclude the possibility that the historical Democritus was interested in similar topics (see, for instance, Gemelli Marciano, "Le Démocrite technicien," 224–36).

[179] Diog. Laert. IX 48,23–24 (= 68 |55| A 33 D-K): Τακτικὸν καὶ Ὁπλομαχινόν (*Tactics and Use of Arms*), almost certainly spurious according to Leszl, "Democritus' Works," 41 and 61.

(according to the attested titles).[180] Although there is no room here for a proper investigation of the reception of Democritean theories in the Hellenistic and Roman medical traditions, it is worth mentioning the conclusions reached by Gemelli Marciano based on analysis of two pseudepigrapha attributed to Democritus, dealing with hydrophobia and elephantiasis: the alleged expertise of Democritus in diseases and medical treatment was seriously taken into account by the followers of Asclepiades as well as by their adversaries.[181]

Unfortunately, only a few difficult-to-intepret sources explicitly invest Democritus with expertise in dyeing techniques prior to the *Four Books*. Yet, in Hershbell's opinion, the acknowledged interest of the historical Democritus in metals and colour theory suffices to explain the later attribution of alchemical writings to him.[182] Taking a similar line, Romano considered that some treatises on colours circulating under the name of Democritus might have constituted an important source both for book 7 of Vitruvius' *De Architectura* and for books 35 and 36 of Pliny's *Natural History*.[183] A possible interest of Democritus in pictorial techniques could also be inferred from the title *On Painting* listed by Thrasyllus among the *Technika*.[184] Leszl suggested that Vitruvius (VII praef. 11)[185] might have had such a treatise in mind when he quoted Democritus with reference to the perspective effects achieved by painters of ancient theatrical scenery. However, there is no evidence that such treatises also dealt with the making of pigments and dyes.

Regarding the latter topic, the most important source is a recipe preserved by the Stockholm papyrus (third-fourth century AD), in which a certain Anaxilaus is said to have attributed a technique for whitening copper to Democritus:

Another recipe.[186] Anaxilaus traces back to Democritus also the following recipe. He rubbed common salt together with lamellose alum in vinegar and formed very fine *kollyria* from these and let them dry for three days in the bath chamber.[187] Then he

---

[180] Diog. Laert. IX 48,17–20 (= 68 |55| A 33 D-K): Πρόγνωσις (*Prognosis*), Περὶ διαίτης ἢ Διαιτητικόν (*On Diet or Dietetics*), Ἰητρικὴ γνώμη (*Medical Opinion*), Αἰτίαι περὶ ἀκαιριῶν καὶ ἐπικαιριῶν (*Causes of Unsuitable and Suitable Times*).

[181] Gemelli Marciano "Le Démocrite technicien," 223: "entre le Ier siècle avant et le Ier siècle après J.C., Démocrite était perçu, non seulement chez les disciples d'Asclépiade, mais aussi chez les médecins qui étaient ses adversaires, comme une autorité dans leur domaine et comme quelqu'un qui pouvait fort bien avoir écrit sur les maladies et leur traitement." Ps.-Democritean medical treatises were known to Rufus of Ephesus [in Orib. XLV 28,1 Raeder (CMG VI/2, 1, p. 184)] and Soranus, at least according to Caelius Aurelianus, who states that Democritus dealt with hydrophobia (with reference to opisthotonus: *Cel. Pass.* III 15,119 and 16,132–3) and elephantiasis (*Tard. Pass.* IV 4,816). See Gemelli Marciano "Le Démocrite technicien," 220–3.

[182] Hershbell, "Democritus," 9. On metals, see in particular Theophr. *Sens.* 62 = 68 |55| A 135 D-K. See also Robert Halleux, *Le problème des métaux dans la science antique* (Paris: Les Belles Lettres, 1974), 74–9. On colour theory, see Theophr. *Sens.* 73–82 = 68 |55| A 135 D-K. See also Gal. *De elem. sec. Hipp.* 2,12 De Lacy (CMG V/1, 2, p. 60 = I 417 Kühn) = 68 |55| A 49 D-K.

[183] Elisa Romano, "I colori artificiali e le origini della chimica," in *Sciences exactes et sciences appliquées à Alexandrie*, Actes du Colloque international qui s'est tenu à Saint-Étienne du 6 au 8 juin 1996, ed. Gilbert Argoud and Jean-Yves Guillaumin (Saint-Étienne: Publications de L'Université de Saint-Étienne, 1998), 115–26, on 118–21. Romano tended to attribute such ps.-Democritean production to Bolos of Mendes.

[184] Περὶ ζωγραφίης in Diog. Laert. IX 48,22 = 68 |55| A 33 D-K.

[185] See Leszl, "Democritus' Works," 41. Diels (59 |46| A 39 e 68 |55| B 15b D-K) connected the passage by Vitruvius with Democritus' writing entitled Ἀκτινογραφίη (*Description of Rays*), which is placed by Thrasyllus among the mathematical treatises (Diog. Laert. IX 48,6 = 68 |55| A 33 D-K).

[186] *PHolm* 2. Usually the pronoun ἄλλο in the title of a recipe refers to the topic discussed in the previous recipe. In this case, PHolm. 1 is entitled Ἀργύρου ποίησις ("the making of silver"). In Halleux's opinion (*Papyrus de Leyde*, 184 n. 6) the first

ground them small, cast copper together with them three times and cooled, quenching in sea water. Experience will prove the result.[188]

Wellmann proposed identifying this Anaxilaus with the neo-Pythagorean magician Anaxilaus of Larissa,[189] who according to St Jerome's *Chronicle* was exiled from Rome in 28 AD.[190] Halleux accepts this hypothesis,[191] and also examines the other fragments attributed to the magician.[192] These describe several *mirabilia*, including how to set a fire that makes light-skinned people as dark as Ethiopians,[193] and how to make symposium participants turn pale.[194] Such literature belongs to the genre of "jocular recipes" (Παίγνια) which has been invoked in relation to both Anaxilaus and Democritus.[195] *P.Lond.* 121 (fourth-fifth century AD) preserves similar jocular recipes, also under the name of the Abderite.[196] Halleux tends to attribute this ps.-Democritean *Paignia* to Bolos of Mendes who, because of his interest in a sort of *"magie amusante,"* could have dealt with similar subjects in the treatise that some sources quoted under the title of *Cheirokmēta* (*Artificial Substances*, Χειρόκμητα).[197] The question, as Kingsley rightly notes, is extremely important for reconstructing an area of interest similar or related to the later alchemical production ascribed to Democritus.[198] Unfortunately, the lack of sources concerning Bolos of Mendes hampers the full understanding of his works. In order to propose some answers to these relevant questions, it is necessary to completely reexamine some of the most important passages concerning this Egyptian author.

§ 9. Bolos of Mendes

Following the important studies of Wellmann and Bidez-Cumont,[199] many scholars agree that some of the ps.-Democritean books circulating in Egypt during the late Hellenistic period should be ascribed to the Egyptian author Bolos of Mendes, whose name is linked to Democritus in many secondary sources.[200] There is also

---

[186] *Continued*
recipe was probably also originally attributed to Democritus (*contra*, see Lagercrantz, *Papyrus Graecus Holmiensis*, 107).

[187] See also Halleux, *Papyrus de Leyde*, 34.

[188] Translation by Earle Radcliffe Caley, "The Stockholm Papyrus: An English Translation with Brief Notes," *Journal of Chemical Education*, 4 (1927): 981 (slightly modified). The text has been republished in *CPF* I/1 43a, 3T (t. II, 28–9).

[189] Wellmann, *Bolos Demokritos*, 51–3.

[190] Pp. 163,26–164,2 Helm: *Anaxilaus Larisaeus Pythagoricus et magus ab Augusto Urbe Italiaque pellitur.*

[191] See Halleux, *Papyrus de Leyde*, 67 and 69–70. However, not all scholars agree. For instance, in Letrouit's opinion, we lack sufficient evidence to confirm this identification: "Tout ce que l'on peut dire, c'est qu'au IIIᵉ siècle de notre ère, des recettes pseudo-démocritéennes étaient colportées par un certain Anaxilaos, ce dernier nom pouvant fort bien être un pseudonyme lui aussi." Letrouit, "Chronologie," 18.

[192] Collected by Wellmann, *Bolos Demokritos*, 77–80.

[193] Wellmann, *Bolos Demokritos*, 78 (fr. 2); the most ancient source is Pliny. *NH* XXXII 141.

[194] Wellmann, *Bolos Demokritos*, 78 (fr. 4). See Pliny. *NH* XXXV 175.

[195] Wellmann, *Die Georgika des Demokritos*, 29 n. 3; *Bolos Demokritos*, 53ff.

[196] Col. V 1–19 (= *PGM* VII 167–85); text reedited also in *CPF* I/1 43a, 1T (t. II, 21–4); see, *infra*, p. 44.

[197] Halleux, *Papyrus de Leyde*, 67–70.

[198] Peter Kingsley, "From Pythagoras to the *Turba philosophorum*: Egypt and Pythagorean Tradition," *Journal of the Warburg and Courtauld Institutes*, 57 (1994): 7 n. 41.

[199] Wellmann, *Die Georgika des Demokritos*, and *Bolos Demokritos*; Bidez-Cumont, *Mages*, vol. 1, 117–8 and 169–74.

a general consensus in dating Bolos' work to the third/second century BC; or, according to a recent *mise à jour* of the question, more narrowly to between 260 and 110 BC.[201] However, some scholars prefer an earlier dating, treating Bolos as contemporaneous with Callimachus (310–240 BC),[202] while others suggest a later date around the second half of the second century BC.[203] These chronological coordinates are inferred primarily from two sources dependent on Bolos' treatises: (1) a passage from Stephanus of Byzantium's *Ethnika* (book I, *s.v.* ἄψυνθος), and (2) the beginning of Apollonius the Paradoxographer's *Marvellous Accounts* (*Mir.* I 1).

Stephanus quotes Bolos where he seems to be making a reference to a passage in Book 9 of Theophrastus' *History of Plants* (IX 17,4) that deals with wormwood.[204] Since this section was added to Books 1–8 only in the second half of the third century BC, Bolos' works must date from after that period.[205] On the other hand, the first paragraph of the *Marvellous Accounts* of Apollonius the Paradoxographer, probably written at the beginning of the first century BC, is preceded by the genitive βώλου, that is, "belonging to Bolos," which some scholars have taken to indicate that Bolos was Apollonius' source.[206] If this interpretation is correct, the same period also marks the *terminus ante quem* for Bolos' writings, and in particular for his book *On Wonders* listed by *Suda* β 482 Adler, considered as a source for Apollonius.[207] Unfortunately this reading is not certain, and other scholars have

---

[200] Height attestations are actually known: (1) Apollon. *Mir.* I 1; (2) Colum. *RR* VII, 5,17; (3) Colum. *RR* XI 3,53; (4) St. Byz. 153,8–13 Meineke = I 578 Billerbeck (*s.v.* ῎Αψυνθος); (5) Theophylactus Simocatta, *Quaest. Phys.* 38,3–7 Massa Positano; (6) *Schol.* in Nic. *Ther.* 764a 4–9 Crugnola; (7) *Suda* β 481 Adler; (8) *Suda* β 482 Adler.

[201] Patricia Gaillard-Seux, "Un pseudo-Démocrite énigmatique: Bolos de Mendès," in *Transmettre les savoirs dans les mondes hellénistique et romain*, ed. Frédéric Le Blay (Rennes: Presse Universitaires de Rennes, 2009), 241.

[202] Eugen Oder, "Beiträge zur Geschichte der Landwirtschaft bei den Griechen, Teil I," *RhM*, 45 (1890): 73–4; see also Suzanne Amigues *Théophraste, Recherches sur les plantes*, 5 vols. (Paris: Les Belles Lettres, 1988–2006), vol. 5, L n. 98.

[203] Peter M. Fraser, *Ptolemaic Alexandria*, 3 vols. (Oxford: Oxford University Press, 1972), vol. 1, 640; Kingsley, "From Pythagoras to the *Turba philosophorum*," 5. Bolos' work is dated to ca. 200 BC by Wellmann, *Bolos Demokritos*, 15; Halleux, *Papyrus de Leyde*, 66; Renato Laurenti, "La questione Bolo-Democrito," in *L'atomo fra scienza e letteratura* (Genoa: Istituto di Filologia Classica e Medievale, 1985), 95.

[204] The ᾿Εθνικά passage (153, 8–13 Meineke = I 578 Billerbeck) reads: ῎Αψυνθος, πόλις Θράκης. ᾿Αψυνθὶς δὲ ἡ χώρα. τὸ ἐθνικὸν ᾿Αψύνθιος καὶ ᾿Αψυνθιάς. ἔστι καὶ εἶδος φυτοῦ, περὶ οὗ Βῶλος ὁ Δημοκρίτειος. ὅτι Θεόφραστος ἐν τῷ περὶ φυτῶν ἐνάτῳ, τὰ πρόβατα τὰ ἐν τῷ Πόντῳ τὸ ἀψύνθιον νεμόμενα οὐκ ἔχει χολήν. διχῶς δ᾽ ἡ γραφὴ καὶ διὰ τοῦ ῡ καὶ διὰ τοῦ ῑ. (I adopt the same punctuation as Meineke's and Billerbeck's editions). As already noted by Letrouit in *DPhA* II *b* 53, the epitome of the text explains the lack of some verbs; the majority of scholars refer the ὅτι at l. 2 to the previous mention of Bolos (see Gailard-Seux, "Un pseudo-Démocrite énigmatique," 225). Therefore, we could propose the following translation: "Apsynthos, a Thracian city. Apsynthis is the area. Apsynthios and Apsynthias are the names indicating this nationality. It is also a kind of plant, about which Bolos the Democritean |said| that Theophrastus in the ninth book of his treatise *On Plants* |has stated| that the animals which in Pontus feed on absinth do not have liver |see Theophr. *HP* IX 17,4|. There is a double spelling, both with long *y* and with long *i*."

[205] Amigues (*Théophraste, Recherches sur les plantes*, vol. 5, VI-LVII) explained that the ninth book of *Historia plantarum* was composed by an anonymous *réviseur*, who gathered together and tried to combine two of Theophrastus' treatises – entitled Περὶ φυτῶν ὀπῶν (= IX 1–7) and Περὶ δυνάμεως ῥιζῶν (= IX 8–19) – and some notes (= IX 20). Neleus himself, who inherited the library of Theophrastus († 287 BC; see Diog. Laert. V 51–57), would have added this compilation to the first eight books of *HP* at the beginning of the third century BC (280–275 BC).

[206] Apollon. *Mir.* I 1 Giannini: <.....> Βώλου. ᾿Επιμενίδης ὁ Κρὴς λέγεται ὑπὸ τοῦ πατρὸς καὶ τῶν ἀδελφῶν τοῦ πατρὸς ἀποσταλεὶς εἰς ἀγρὸν πρόβατον ἀγαγεῖν εἰς τὴν πόλιν. Otto Keller, *Rerum naturalium scriptores Graeci minores*. Vol. 1: *Paradoxographi Antigonus, Apollonius, Phlegon, Anonymus Vaticanus* (Leipzig: Teubner, 1877), previously separated the mention of Bolos from the following Epimenides by punctuating Βώλου· ᾿Επιμενίδης κτλ. There is no evidence that allows us to consider Bolos as the father of the wise Epimenides of Crete (see, for instance, Laurenti, "La questione Bolo-Democrito," 92 n. 61).

emphasised that the defective text of Apollonius' passage makes an exact reading of βώλου impossible.[208] Supporters of the 'Bolos' interpretation also disagree over the extent to which Apollonius may have derived information from Bolos of Mendes: some propose that the whole of *Marvellous Accounts* depends on Bolos,[209] others allow only the first six paragraphs.[210]

To complicate matters further, Gaillard-Seux has recently pointed out that the *Ethnika* and *Marvellous Accounts* apparently refer to two different versions of Theophrastus' *History of Plants*.[211] Thus, while Stephanus mentions the ninth book, Apollonius seems to refer to a version in eight books (books VI and VII being grouped together). This discrepancy would contradict the hypothesis that Bolos of Mendes was the intermediate source through which both Stephanus and Apollonius quoted Theophrastus – in which case, we might have expected both authors to quote similar versions of the *History*. However, this divergence might be explained if the *Marvellous Accounts* relies on a plurality of sources. Apollonius' explicit references to Theophrastus are all confined to the last paragraphs; that is to say, those sections in which Bolos' influence has been judged less important, or even absent.[212]

In any case, a similar *terminus ante quem* can be inferred using a third, less problematic source: a fragment of Crateuas, physician to Mithridates VI Eupator (120–63 BC), which reads: "Anagallis: it is used also in Democritus' prescriptions."[213] The passage does not mention Bolos, but provides some information on the plant anagallis, which was probably included in a collection of pharmacological formulae (based on the powers of herbs and natural substances) ascribed to Democritus. The identification of this Democritus with Bolos is possible if we read this passage as being derived from Bolos' treatise *Physika dynamera* (*On Natural Active Substances*), which the lexicon *Suda* places among the works attributed to the Egyptian

[207] See Keller, *Rerum naturalium scriptores*, 43 n. 1; Wellmann, *Die Georgika des Demokritos*, 9.

[208] Gaillard-Seux, "Un pseudo-Démocrite énigmatique," 239.

[209] See Fraser, *Ptolemaic Alexandria*, vol. 1, 440. Laurenti ("La questione Bolo-Democrito," 94) thought that "mentre nei primi sei paragrafi il testo originario di Bolo era seguito con maggiore attenzione, negli altri tale attenzione si allentava e lasciava spazio ad altre fonti."

[210] See Hermann Diels, "Über Epimenides von Kreta, " *Sitzungsberichte der Kgl. Pr. Akademie der Wissenschaften zu Berlin*, (s.n., 1891), 393–4; Laurenti ("La questione Bolo-Democrito," 92–3) noted that the first six paragraphs, more focused on legendary figures such as Epimenides, Pherecydes and so on, differ from the rest of the treatise, which deals mainly with natural phenomena. However, scholars have also proposed the possible influence of Bolos on other paragraphs, namely §§ XXXI (Laurenti, "La questione Bolo-Democrito," 94), XLVI and XLIX (Wellmann, *Die Georgika des Demokritos*, 10–1). Alessandro Giannini, *Paradoxographorum Graecorum Reliquiae* (Milan: Istituto Editoriale Italiano, 1966), 377, in contrast, is very cautious even with regard to Apoll. *Mir.* I 1–6.

[211] Gaillard-Seux, "Un pseudo-Démocrite énigmatique," 239–40.

[212] In particular §§ XVI, XXIX, XXXI, XXXIII, XLI, XLIII, XLVI-L. We note that § XXXI (Θεόφραστος ἐν τῷ περὶ φυτῶν· τὰ πρόβατα, φησίν, τὰ ἐν τῷ Πόντῳ τὸ ἀψίνθιον νεμόμενα οὐκ ἔχει χολήν) is particularly relevant, since it could depend on Bolos, who would have quoted this passage of Theophratus' *History* according to the above mentioned entry in Stephanus' lexicon (see Kingsley, "From Pythagoras to the *Turba philosophorum*," 5 n. 30). In this case, Apollonius does not explicitly mention from which book of Theophrastus he takes his account: it is still possible to posit a direct dependence on Bolos, who, nevertheless, need not be considered the only source for all the quotations from Theophrastus.

[213] Max Wellmann, *Pedanii Dioscuridis Anazarbei de materia medica libri quinque*, 3 vols. (Berlin: Weidmann, 1906–14), vol. 3, 146 fr. 8 (= 68 |55| B 300,4a D-K): ἀναγαλλίδες· χρῶνται δ᾿ αὐτῇ καὶ εἰς τὰς Δημοκρίτου δυνάμεις.

author. Indeed, two entries in this lexicon constitute our principal sources on Bolos' writings:

> Bolos, Democritus, philosopher. *History* and *Medical Art*:[214] it includes also the natural cures by remedies of nature.[215]

> Bolos, from Mendes, Pythagorean: *On What Attracts our Attention from the Reading of Histories, On Wonders, On Natural Active Substances*: it includes *On Sympathies and Antipathies* of <animals, plants>[216] and stones, in alphabetical order; *On the Signs from the Sun and Moon and Ursa Major and Light and Rainbow*.[217]

The first area of interest ascribed to Bolos is clearly medicine, which seems to be covered by several treatises (or sections of treatises) mentioned by the *Suda*. This information accords with Democritus' reputation as an excellent physician, a reputation he acquired during the late Hellenistic period, as noted above. In particular, the works *Medical Art* and *On Natural Active Substances* almost certainly dealt with the medical and pharmacological properties of natural materials, albeit from a perspective that included interest in *mirabilia* and the secret and magical powers of the ingredients. Such an attitude accords well with Neopythagorean tendencies, also alluded to in the lexicon, which describes Bolos as a Pythagorean.[218]

The attribution to Bolos of the treatise *On Sympathies and Antipathies* – which, according to the *Suda*, is part of *On Natural Active Substances* – is confirmed by an anonymous commentator on Nicander's *Theriaca*:

> Bolos the Democritean in his *On Sympathies and Antipathies* tells that the Persians, who had a poisonous tree (the Περσεία) in their country, transplanted it to Egypt with the intention of killing many people; however, the favourable condition [of the new country] turned the tree into its opposite and it gave very sweet fruit.[219]

In addition, both Columella and Tatian quote an analogous text, which they ascribe to Democritus. Columella reports that Democritus in his *On Antipathies*

---

[214] It is not clear whether Ἱστορία καὶ Τέχνη ἰατρική refers to two different works or a single treatise. In particular, the term Ἱστορία could either be identified with the first writing by Bolos quoted in the second entry of *Suda* (β 482 Adler; see also the comment by Diels to 55 B 300,1 in the first edition of *Pre-Socratics*) or interpreted as "medical investigation" (as supposed by Wellmann in his comment to 68 |55| B 300,1). Hershbell, "Democritus," 5, translated it as: "*Scientific observation and Medical art*"; and Halleux, *Papyrus de Leyde*, 65, as "*Histoire et Art de la médicine*."

[215] *Suda* β 481 Adler. The expression Βῶλος Δημόκριτος has been openly questioned, especially after Wellmann's interpretation (Wellmann, *Die Georgika des Demokritos*, 16) according to which the expression indicates a *Doppelname* of the author. However, the majority of scholars do not accept this hypothesis, reading Δημόκριτος as a reference to some kind of relationship between Bolos and the atomist philosopher (see Kroll, "Bolos und Democritos," 230; Bidez-Cumont, *Mages*, vol. 1, 118; Halleux, *Papyrus de Leyde*, 63–4; Laurenti,"La questione Bolo-Democrito," 88–9).

[216] I have accepted the integration proposed by Wellmann, *Bolos Demokritos*, 11. In fact, as we shall see in the following pages, Bolos' works could have covered different topics and not only the study of stones (see lastly Gaillard-Seux, "Un pseudo-Démocrite énigmatique," 230).

[217] *Suda* β 482 Adler; 68 |55| B 300,1 D-K.

[218] On the tendency, already attested by Thrasyllus, to present Democritus as a Pythagorean, see Leszl, "Democritus' Works," 15 and 21–3; Peter Kingsley, *Ancient Philosophy, Mystery and Magic. Empedocles and Pythagorean Tradition* (Oxford: Clarendon Press, 1995), 325–8 and 335–41.

[219] *Schol. in Nic. Ther.* 764a 4–9 Crugnola (= 68 |55| B 300,4 D-K).

explained that certain insects die when approached by a menstruating woman, bare-foot and with long, unbound hair.[220] Tatian, a harsh critic of Greek medicine and magic, cites Democritus' *On Sympathies and Antipathies* as a representative example of such absurdities, by also introducing the Persian magus Ostanes:[221]

> For what must we say about Democritus' *On Sympathies and Antipathies* but this, that this person from Abdera was, as they say, an 'Abderologos' [i.e. foolish]? But, like the namesake of the city, who is said to have been a friend of Hercules and who was devoured by the horses of Diomedes, so he who boasted of the magus Ostanes will be consigned as fuel to the eternal flames on Judgment day.[222]

As already noted by Bidez-Cumont,[223] the expression "he who boasted of the magus Ostanes" probably refers here to Democritus, mentioned a few lines before. In the following paragraphs, Tatian in fact criticises the strange drugs composed of disgusting mixtures of bodily parts (especially bones and sinews), which were also ascribed to Ostanes and Democritus by Pliny.[224] In addition, we can perhaps infer from this passage that Democritus/Bolos mentioned Ostanes in *On Sympathies and Antipathies*, although Tatian does not specify their reciprocal relationship.[225] Yet Bolos' possible interest in Persia is also suggested in the above quoted *scholium* on Nicander, where we find a clear reference to Egypt: the land in which pseudepigrapha related to Democritus and Persian magi first began to circulate. Unfortunately, the lack of further details hampers better contextualisation of the relationship between the works of Bolos and of Ostanes. Although the *scholium* seems to invoke the traditional hatred of Persians towards Egyptians stressed by other ancient sources (such as Herodotus, who portrayed Cambyses as a desecrator of Egyptian temples[226]), it is not clear whether Bolos was referring to paradoxographical accounts concerning the Achaemenid domination of Egypt.[227]

---

[220] Colum. RR XI 3,64 = 68 |55| B 300,3 D-K.

[221] Tatian also quotes Ostanes in other passages (with particular interest in demonology), which have been collected by Bidez-Cumont, *Mages*, vol. 2, 294 n. 1.

[222] *Orat. ad Gr.* 17 (= 68 |55| B 300,10 D-K); see Bidez-Cumont, *Mages*, vol. 2, 294 (fr. 16).

[223] Bidez-Cumont, *Mages*, vol. 2, 296 n. 5.

[224] Pliny. *NH* XXVIII 5–7 = 68 |55| B 300,13a D-K = Bidez-Cumont, *Mages*, vol. 2, 296–7 (fr. 17); see also Patricia Gaillard-Seux, "Sympathie et antipathie dans l' 'Histoire Naturelle' de Pline l'Ancien," in *Rationnel et irrationnel dans la médecine ancienne et médiévale. Aspects historiques, scientifiques et culturels*, ed. Nicoletta Palmieri (Saint-Étienne: Publications de L'Université de Saint-Étienne, 2003), 120–4.

[225] It may be possible to infer a master-pupil relation between Bolos and Ostanes from a passage of the Ἰατρικὰ φυσικὰ καὶ ἀντιπαθικά by the second-century physician Aelius Promotus. See Max Wellmann, "Aelius Promotus: *Iatrika physika kai antipathētika*," *Sitzungsberichte der preußischen Akademie der Wissenschaften*, 37 (1908): 776, ll. 13–8: Πρὸς πυρετόν· ἔστι βοτάνη ἡλίου ἱερὰ ἢ ἀείζων |...|· εἰς κάμινον δὲ χαλκέως ἢ βαλανεῖον ταύτην τὴν βοτάνην ἐὰν θῆς οὐ καυθήσεται. Παρὰ δὲ Ὀστάνει τῷ διδασκάλῳ ἐθεασάμην ὅτι καὶ μολίβδου ῥίνισμα σὺν τῷ ἀειζώῳ εἰς τὴν κάμινον ὑπετίθει ("Against fever: there is the holy plant of sun, the houseleek plant |...|; if you put this plant in a blacksmiths' furnace or in |a furnace employed for warming| baths, it will not burn. In the master Ostanes |writings| I have seen that he put also lead filings in the furnace together with the houseleek plant"). If, according to Étienne Tourtelle, *Histoire philosophique de la médecine, depuis son origine jusqu'au commencement du 18ᵉ siècle*, 2 vols. (Paris: Levrault, 1804), vol. 1, 415, the same author, Aelius Promotus, was presenting himself as a pupil of Ostanes, already Diels (*Antike Technik*, 137; see also Wellmann, *Bolos Demokritos*, 45) supposed that our physician was here quoting a statement of Bolos, as may be confirmed by one of the Παίγνια handed down by PLond. 121 (*PGM* VII 171f. = 68 |55| B 300,19): see *infra*, p. 44.

[226] See, for instance, Hdt. III 29.

The impression of a relationship between Bolos' production and the writings ascribed to Persian magi (*in primis* Ostanes) is confirmed by surviving information on a second work of Bolos, not mentioned in the *Suda*. According to Columella, Bolos composed a book entitled *Cheirokmēta* (*Artificial Substances*) which also seems to have been included in Diogenes Laertius' list (IX 49) of Democritus' works. As it happens, among the *hypomnēmata* ("notes," ὑπομνήματα) ascribed to Democritus there is a text whose title is handed down in a corrupted form. Several editors after Salmasius have corrected the reading of Byzantine manuscripts into *Cheirokmēta problēmata*.[228] A similar title is mentioned by Columella, who reports that,

> the remarkable Egyptian author Bolos of Mendes, whose inventions[229] – which are called *Cheirokmēta* in Greek – have been published under the pseudonym of Democritus, recommends to check often and carefully the backs of sheep in the case of this disease [pusula]: if we happen to discover such a disease in some of them, we must immediately dig a trench straight out of the sheepfold and bury the sheep – the one that was full of blisters – alive and lying on her back, and we must let the whole flock pass over the buried sheep, so that by so doing the disease is driven away.[230]

The same text, but ascribed to Democritus, is also quoted by Pliny the Elder, who stressed its botanical content, derived from the tradition of the magi:

> That Democritus was the author of the book called *Chirocmeta* is a well-attested tradition; yet in it this famous scientist, the keenest student next to Pythagoras of the magi, has told us of far more marvellous phenomena.[231]

A list of thirteen magical/medical plants[232] and a compound called *Hermesias* follows this introduction. Pliny asserts that these are mostly of Eastern origin (especially Persian), and that Democritus – almost certainly Bolos on the basis of Columella's account – also gave their magical names.[233] It seems quite clear that

---

[227] See Kevin van Bladel, *The Arabic Hermes: from Pagan Sage to the Prophet of Science* (Oxford and New York: Oxford University Press, 2009), 51–3.

[228] Manuscripts have χερνικά (or χερνιβά) προβλήματα. See, for instance, Herbert S. Long, *Diogenis Laertii Vitae philosophorum*, 2 vols. (Oxford: Oxford Classical Texts, 1964), vol. 2, 463. Tiziano Dorandi, conversely, has edited the manuscript reading between *cruces*: Dorandi, *Diogenes Laertius, Lives of Eminent Philosophers* (Cambridge: Cambridge University Press, 2013), 692.

[229] It is not easy to translate the term *commenta*. I have kept here its usual meaning of "invention, fiction" (see *OLD* 363, *s.v. commentum*), previously preferred by other scholars (Laurenti, "La questione Bolo-Democrito," 100; Kingsley, "From Pythagoras to the *Turba philosophorum*," 8; Gaillard-Seux, "Un pseudo-Démocrite énigmatique," 233). However, according to *ThLL* III 1865f., Columella's passage is the most ancient text to use *commentum* with the meaning of *scriptum, liber, oratio* (perhaps corresponding to the Greek ὑπόμνημα, usually translated into Latin as *commentarium*).

[230] *RR* VII 5,17 (= 68 |55| B 300,3 D-K).

[231] Pliny. *NH* XXIV 160 (= 68 |55| B 300,2 D-K); translation by William H.S. Jones, *Pliny, Natural History, Books 24–27* (Cambridge, MA and London: Loeb, 1956), 113.

[232] (1) *Aglaophotis* (from the Persian side of Arabia; used by the magi), (2) *Achaemenis* (from India), (3) *Theombrotion* (from Choaspes, drunk by Persian kings), (4) *Adamantis* (from Armenia and Cappadocia), (5) *Arianis* (from Ariana), (6) *Therionarca* (from Cappadocia and Mysia), (7) *Aethiopis* or *Merois* (from Meroe), (8) *Ophiusa* (from Elephantine and Ethiopia), (9) *Thalassaegle* (from river Indus), (10) *Theangelis* (from Syria, Crete and Persia; drunk by magi), (11) *Gelotophyllis* (from Bactria), (12) *Hestiateris* (from Persia), (13) *Helianthes* (from Cilicia; used as an ointment by the magi and Persian kings).

[233] Pliny. *NH* XXIV 166: "atque harum omnium magica quoque vacabula ponit."

Bolos' *Cheirokmēta* dealt with the medical and magical properties of specific herbs, grounding this knowledge in the Persian-Chaldean tradition presented as characteristic of the magi.[234] Wellmann and Bidez-Cumont assumed that Bolos knew Ps.-Zoroaster's *On Nature*,[235] at least, the part concerning plants and herbs, and perhaps an analogous text ascribed to Ostanes.[236]

A final witness of Bolos' work slightly distances itself from the foregoing passages, instead relating the *Cheirokmēta* to another treatise by the Egyptian author. A passage by Vitruvius reports that,

> I admire also Democritus' *On the Nature of Things,* and his commentary that is entitled *Cheirokmēta,* in which he used a ring to seal on wax everything he had experienced.[237]

Vitruvius, like Pliny, attributes the work to Democritus rather than to Bolos, confirming that during the first century BC some of Bolos' writings were already considered Democritean. The *commentaria* mentioned by Vitruvius were probably a collection of explanatory notes on questions that Bolos wanted to test personally. These problems seem to have been more extensively discussed in a work entitled *De rerum natura,* perhaps to be identified with the *On Natural Active Substances* quoted by the *Suda*.[238] The term *commentarium* corresponds to the Greek *hypomnēma* (ὑπόμνημα),[239] which often describes works that were not intended to be widely published, comprising collections of personal notes for circulation only among a select group of readers.[240] Galen, for instance, often referred to his exegetical works on Hippocrates' treatises by using the term *hypomnēmata* when they were not designed for the publication.[241] The *Cheirokmēta* may also have comprised personal notes, based on the reading of *On Natural Active Substances,*

---

[234] Bidez-Cumont, *Mages,* vol. 1, 30–8, noted that, after Cyrus conquered Babylonia, the magi, namely the priests of the Zoroastrian religion, started to become confused with Chaldeans, so that the Greeks considered them the inheritors of both of the Persian religious tradition and Babylonian scientific knowledge, particularly astrology and astronomy. Momigliano recognised in a fragment of Aristoxenus (fr. 13 Wehrli) the earliest testimony of such an overlap between magi and Chaldeans: *Alien Wisdom,* 143.

[235] See Wellmann, *Bolos Demokritos,* 14–5; Bidez-Cumont, *Mages,* vol. 1, 117–9.

[236] Wellmann (*Bolos Demokritos,* 48ff.) supposed that any passage of the *Natural History* in which Pliny quotes the magi with reference to magical herbs was taken from Bolos' lost treatises. However, it is likely that these scattered quotations do not stem only from a single source (see, for instance, Bidez-Cumont, *Mages,* vol. 1, 120 n. 1).

[237] Vitr. IX praef. 14,6–9 (= 68 |55| B 300,2 D-K).

[238] See, lastly, Gaillard-Seux, "Un pseudo-Démocrite énigmatique," 234. Kingsley considered Vitruvius' passage as a clear example of literary fiction: "From Pythagoras to the *Turba philosophorum*," 8. Halleux, on the other hand, distinguished the χειρόκμητα as "effets merveilleux produits artificiellement, par opposition aux φυσικά, effets merveilleux observés dans la nature," *Papyrus de Leyde,* 68.

[239] As already mentioned, the work Χειρόκμητα can possibly be recognised among the dubious ὑπομνήματα listed by Diog. Laert. IX 48–49 |= 68 (55) A 1 D-K|: see *supra,* p. 41.

[240] With regard to the distinction between ὑπομνήματα and συγγράμματα, see Tiziano Dorandi, *Nell'officina dei classici* (Rome: Carocci, 2007), 65–81. Following this interpretation, it would be possible to consider the Χειρόκμητα as a work that Bolos did not intend for publication (ἔκδοσις). Kroll ("Bolos und Democritos," 230; see also Bidez-Cumont, *Mages,* vol. 1, 118) previously supposed that the circulation of this writing under the name of Democritus could have been not intentional (Bolos is in fact described only as "democritean" by several sources), but arose at a later date. In this respect, Columella says that these books *"sub nomine Democriti falso produntur,"* i.e. "were published under the pseudonym of Democritus," perhaps without Bolos' knowledge. Of the opposite opinion is Kingsley, "From Pythagoras to the *Turba philosophorum*," 8–9, who, nevertheless, thinks that Bolos edited the Περὶ Θαυμασίων (*On Wonders*) under his own name (see also Gaillard-Seux, "Un pseudo-Démocrite énigmatique," 242).

[241] See Dorandi, *Officina dei classici,* 65–7.

concerned with more specific and probably more technical questions which the author sought to solve by personal "experiments." Halleux suggests that Bolos collected a variety of personal experiences related to medicine, magic and craftsmanship in general;[242] in fact, the same term *cheirokmēta*, compounded of *cheir* ("hand," χείρ) and *kamnō* ("to labour," κάμνω), is usually used by ancient authors to refer to handmade products, as opposed to natural ones.[243]

Unfortunately the ancient sources do not help us understand more precisely which kinds of *technai* were covered by Bolos' work, or to evaluate its possible relationship with those subjects that later became characteristic of alchemical production. Wellmann's hypothesis, however, that Bolos composed a treatise entitled *Natural Dyes* or *Dyeing Substances* (*Physikai baphai* or *baphika*)[244] and was almost certainly the author of the four alchemical books of ps.-Democritus epitomised in the *Corpus alchemicum*, cannot stand.[245] First, the alchemical author ps.-Democritus and Bolos lived in different centuries: the former in the first century AD, the latter between the third and second centuries BC. Second, no ancient source explicitly ascribes any recipe on the dyeing of stones, metals, or wool to Bolos, yet these are the main topics covered by ps.-Democritus' alchemical books. Furthermore, the three passages outside the *Corpus alchemicum* that link such interests with the name of Democritus – particularly in the Stockholm Papyrus, Seneca's ninetieth epistle, and the *Paignia* of London Papyrus 121 – never cite either the name of Bolos or any of his works.[246] Likewise, Aulus Gellius, while criticizing Pliny for wrongly attributing many pseudepigrapha to Democritus, never mentions Bolos, even though he stresses the many authors who unduly exploited the great philosopher's name.[247]

While it is not impossible that the three sources just mentioned might rely in some way on Bolos' earlier writings,[248] such a conjecture remains entirely hypothetical. The second recipe of the Stockholm Papyrus is certainly an important piece of

---

[242] Halleux, *Papyrus de Leyde*, 66.

[243] The term is already attested in Aristotle, who, for instance, in *Mete* II 1, 353b 25–28 used the term χειρόκμητα (ὕδατα) to indicate artificial reservoirs, where water had been collected by means of human technology (τέχνης προσ-δεῖται τῆς ἐργασομένης). The role played by the τέχνη is confirmed by Diog. Laert. V 33, who distinguished between σώματα χειρόκμητα, ὡς τὰ ὑπὸ τεχνιτῶν γινόμενα ("the *cheirokmeta* bodies, as those produced by experts/craftsmen") and natural bodies (τὰ δὲ ὑπὸ φύσεως). In addition, Str. III 5,6 used the term with reference to specific landmarks (such as altars, towers, or pillars; ὅροις ... χειροκμήτοις τισὶ βωμοῖς, ἢ πύργοις ἢ στυλίσιν) which had been set up by ancient and mythical travellers (including Heracles and Dionysus) to mark the countries they had been able to reach. Lastly, the "artificial" component indicated by the term χειρόκμητα is well attested by ancient lexicographers: Hsch. χ 297 H-C χειρόκμητα· χειροποίητα, ἤγουν ὑπὸ χειρῶν γεγενημένα; *Suda* χ 253 Adler χειρόκμητα· ὑπὸ χειρῶν γεγλυμμένα ἢ περιεξεσμένα (cf. anche *Suda* λ 692 A. ... χειρόκμητα καὶ τετεχνασμένα).

[244] Wellmann, *Bolos Demokritos*, 68: "wie dieser Titel dem Werke des Alchemisten und Mystikers Demokrit, der in die nachchristliche Zeit gehört, verbleiben muss, do darf man, wie mir scheint, bei Bolus nach Analogie des chemischen Schrift des Hermes Trismegistos und den Titel φυσικαὶ βαφαί oder βίβλοι φυσικῶν βαφῶν oder endlich βαφικά denken."

[245] Kroll, "Bolos und Democritos," 230, was the first to criticize Wellmann's hypotheses, noting that no ancient sources mention a work entitled Βαφικά. Any alchemical element in Bolos' work is denied by Hershbell, "Democritus," 8, and Letrouit in *DPhA* II b 53, s.v. Bolos de Mendès.

[246] *PHolm.* 2 (see *supra*, pp. 35–6), Sen. *Ep.* XC 32–33 (see *supra*, p. 31), and the Παίγνια of *PLond.* 121.

[247] Gellius X 12,8 = 68 |55| B 300,7 D-K; Pliny. *NH* XXIV 160.

[248] Such dependence was proposed, for instance, by Halleux, *Papyrus de Leyde*, 67–72. See also Wellmann, *Die Georgika des Demokritos*, 29–33.

evidence, for it preserves a metallurgical recipe that, it claims, was attributed to Democritus by Anaxilaus (identifiable with the first century BC magus Anaxilaus of Larissa). This source therefore demonstrates that some metallurgical recipes were already circulating under the name of Democritus during the first century BC, a period when the philosopher was considered an expert in various *technai*. However, ancient sources do not allow us to trace this text back to Bolos. Seneca's account, a century later, highlights Democritus' supposed interest in techniques used for counterfeiting precious stones, a topic that is also covered by the *Four Books* that date from the same period. Once again, however, any possible relationship with Bolos is purely hypothetical, and we cannot rule out the possibility that Seneca was drawing upon those very ps.-Democritean alchemical writings that were later epitomised into the *Corpus alchemicum*. Finally, *PLond.* 121, dating to the fourth/fifth century AD, is a collection of excerpts whose origins are difficult to identify with any degree of certainty:

> Democritus' table tricks: To make bronzeware look as though it is made of gold: mix unburnt sulphur with chalky soil and wipe it off. To make an egg become like an apple: boil the egg and smear it with a mixture of egg-yolk and [red] wine. To make the chef unable to light the burner: set a houseleek plant on his stove, etc.[249]

The first two "tricks" are about dyeing, even if reduced to a brief record of some extravagant effects (or *mirabilia*). The following trick seems to depend on a passage from Aelius Promotus that is probably a quotation from Bolos' work: however, the version given by *PLond.* 121 is simplified, since the houseleek plant is here placed on a simple domestic stove rather than the furnaces more specifically quoted by Aelius Promotus.[250] But this is not striking, since the so-called "tricks" of Democritus seem to be a collection of recipes gathered from different sources, among which the exact role played by Bolos cannot be discerned.[251]

9.1 *The problematic meaning of cheirokmēta*

Kingsley has recently tried to reaffirm the possible alchemical aspects of Bolos' work by emphasizing how the term *cheirokmēta* is well attested in the *Corpus alchemicum* (especially in Zosimus) and also appears in reference to the titles of alchemical writings.[252] However, the use of this term, particularly by Zosimus, is not always clear, and is sometimes based on corrections proposed by various editors. For instance,

---

[249] Col. V 1–19 (= *PGM* VII 167–85); translation by Hans Dieter Betz, *The Greek Magical Papyri in Translation, Including the Demotic Spells* (Chicago and London: University of Chicago Press, 1986), 119 (slightly modified). The following items include: "To be able to eat garlic and not stink [...]; To keep an old woman from either chattering or drinking too much [...]; to make the gladiators painted [on the cups] 'fight' [...]; to make cold food burn the banqueter [...]; to let those who have difficulty intermingling perform well [...]; to be able to drink a lot and not get drunk [...]; to be able to travel [a long way] home and not get thirsty [...]; to be able to copulate a lot."

[250] Ἰατρικὰ καὶ ἀντιπαθικά in Wellmann, "Aelius Promotus," 776,13–8, see *supra*, n. 225.

[251] See Kroll, "Bolos und Democritos," 231.

[252] Kingsley, "From Pythagoras to the *Turba philosophorum*," 8. See also Halleux, *Papyrus de Leyde*, 69, who stressed the importance of *Suda* ζ 168 Adler, which mentions the title of χειρόκμητα with reference to Zosimus' alchemical writings (Χημευτικά; see Mertens, *Zosime de Panopolis*, XCVIIf.).

Zosimus mentions one of his own writings where he refers to "the filtered lye I have spoken of in [the section on] carding in my *Cheirokmēta*."[253] Manuscript **M** – the only witness of this passage – gives the reading *cherotmētōn* (χεροτμήτων) which was corrected by Mertens to *cheirokmētōn* (χειροκμήτων) on the basis of *Suda* ζ 168 Adler, in which the alchemical treatises by Zosimus are said to have been called *Cheirokmēta*.[254]

More problematic still is a second example where Zosimos seems to introduce the term, for the *Corpus alchemicum* preserves different versions of it. The first version is handed down by **A** (fols. 251^v–256^r; *CAAG* II 239–46), while the second is transmitted through the indirect tradition via quotations by Olympiodorus and by the anonymous compiler of the excerpt *The Airy Water*.[255] In the first instance, the passage, which constitutes the incipit of Zosimus' treatise *First Book of the Final Quittance*, reads:

> For the so-called divine art, the art that is grounded for the most part on general principles and sophistic arguments, has been entrusted to the guardians [of the temples? i.e. Egyptian priests?] for their sustenance; not only that art, but at the same time (?) also the so-called four arts and the *cheirotmēmata*.[256]

The second version of this text, as quoted by Olympiodorus, reads:[257]

> For the so-called divine art, that is, the art based on general principles which occupies all those who seek out all the *cheirotmēmata*[258] and the worthy arts, I mean the four arts, those that seem to produce some effect, have been entrusted only to the priests.

Both versions give the reading *cheirotmēmata* which, as noted by Letrouit,[259] constitutes a *hapax* composed of *cheir* ("hand") and *tmēma* ("cut"). The fact that almost all the manuscripts preserve this reading casts doubt on the correction to *cheirokmēta* proposed by Festugière. However, we should note that the two terms do not seem too different in their meaning. In each, the prefix *cheir* refers to a kind of technical and "manual" procedure made more specific by the verbs that follow: on the one hand, the more generic *kamnō*, "to work, to labour," and on the other, *temnō*, "to cut." It is clear from the Syriac tradition that Zosimus used

---

[253] IV 55–6 Mertens.

[254] On the frequent mix-up between χειρόκμητα and χειρότμητα in Byzantine manuscripts, see Kingsley, "From Pythagoras to the *Turba philosophorum*," 6 n. 36, and Mertens, *Zosime de Panopolis*, LXXXVIIf.

[255] *CAAG* II 90,14–6 and II 209,10–20; the latter Berthelot and Ruelle falsely attributed to Zosimus. See Letrouit, "Chronologie," 36 n. 93.

[256] Text edited by Festugière, *Révélation d'Hermès*, vol. 1, 363 on the basis of A, fols. 251^{v}25–252^{r}4 (= *CAAG* II 239,5–9), where Festugière corrected the final χειροτμήματα into χειρόκμητα. A provides a very corrupt text, difficult to understand in some passages; for scholarly apparatus, see Martelli, *Pseudo-Democrito*, 110.

[257] Olympiodorus' text is preserved by M, fol. 171^{v}12–6; V, fol. 21^{r}1–6; A, fols. 206^{v}25–207^{r}3 (see Festugière, *Révélation d'Hermès*, vol. 1, 363 and Letrouit, "Chronologie," 19) and it is quite similar to the version in *The Airy Water* (Τὸ ἀέριον ὕδωρ, A, fol. 111^{r}11–5 = *CAAG* II 209,11–4). All the manuscripts have χειροτμήματα (only V has χειρότμητα), corrected by Festugière into χειρόκμητα. For scholarly apparatus, see Martelli, *Pseudo-Democrito*, 110.

[258] Kingsley "From Pythagoras to the *Turba philosophorum*," 6 n. 36 (following Saffrey's indications) explains that in manuscript M we find "a cross above and between the two letters οτ, indicating that someone suspected the reading was wrong." Actually, the manuscript presents a simple correction of the accentuation: the accent that had been put initially on the letter ο (χειρότμηματα) has been erased by a cross and a second (correct) accent has been put on the η.

[259] Letrouit, "Chronologie," 19 n. 38.

both terms, which seem to be in some way related to alchemical practices. For instance, after mentioning the first book of *chēmeia*, revealed to mankind by fallen angels, he emphasises the variety of topics included in this book. Having mentioned techniques for dyeing any base metal,[260] he specifies (**SyrC**, fol. 50ʳ 18–21):[261]

ه‌أستنبكا أومحنا اليحناللا مجحتالا مأمحنيا أومحنيا مجحتاللا. حما وهه أامحرين

وبحس كحم حوكم ومحماحدنم حابرا أه محافحممم. ومحامنم حتهكومحما.

> But there are many other arts [in the book of *chēmeia*]; these arts are as many as we are able to find out by means of what is hand-made [i.e. by means of the *cheirokmēta*] and by means of what is hand-cut, that is called *cheirotmēta* [or *cheirotmēmata*].[262]

In the passage, next to the term *cheirokmēta* that has been translated by the expression حابرا ومحماحدنم وكحه (lit. "what is made by hand"),[263] Zosimus also mentions *cheirotmēta* (equivalent to *cheirotmēmata*), which has been translated into Syriac with the expression (وكحه و محافحممم (حابرا)): "[what] is cut [by hand]." This attestation of both nouns, together with the use of *cheirotmēmata* in the abovementioned Greek passages, suggests that the term must be retained rather being amended to *cheirokmēta*.

The presence of such terminology in passages where Zosimus describes the contents of the first book of *chēmeia* or the four dyeing arts – almost certainly to be identified with the making of gold, silver, precious stones, and purple[264] – deserves particular attention. The terms *cheirotmēmata* and *cheirokmēta* apparently refer to specific ways of processing the substances used to dye base metals, stones, and textiles, based on procedures that can be identified by verbs such as "to cut" (grinding, mincing, chopping and laminating processes) or "to make." However, similar procedures were also characteristic of other *technai* concerned with processing and transforming natural ingredients (plants, herbs, stones, animal products) to different ends. The same substances could be employed across a range of arts – including pharmacology, cooking, and dyeing – that shared a large pool of technical knowledge applied to particular crafts.[265] According to the first passage of Zosimus cited above on p. 45,[266] the alchemist dealt with the making of soap (an artificial

---

[260] With regard to this passage, see *infra*, pp. 60–1.

[261] Berthelot-Duval's translation is rather misleading and does not make the presence of the two terms explicit in the passage (*CMA* II 239): "Là sont exposés des arts nombreux, de telle sorte que nous pouvons y trouver ces (opérations) faites à la main, ces expériences que l'on appelle χειρότμητα."

[262] The Syriac term *kyrwṭwmyt'* may be the transcription of both Greek terms. The Greek form χειρότμητα is more common and is also attested in non-alchemical works (LSJ⁹ 1986, *s.v.* χειρότμητος): the two terms seem to be synonymous.

[263] I have read all the expressions as a translation of the Greek plural form τὰ χειρόκμητα, where the pronoun وكحه renders the Greek article τά; otherwise it is possible to consider وكحه as referring to the implied term أامحنبا, "arts."

[264] See Festugière, *Révélation d'Hermès*, vol. 1, 275 n. 7, and Letrouit, "Chronologie," 20.

[265] Syn. Alch. § 10, l. 153 also employs the verb τέμνω in order to make a comparison between alchemists' practices and the ways in which other craftsmen, such as the λιθοξόος ("stonecutter") and τέκτων ("carpenter"), worked different materials (ὕλη).

[266] Zos. Alch. IV 55–6 Mertens.

substance whose use was of course not confined to alchemical practices) in a specific section of his work possibly entitled *Cheirokmēta*.[267]

In conclusion, the terms *cheirokmēta* and *cheirotmēmata*, although part of Zosimus' vocabulary, do not seem to have had a specifically 'alchemical' meaning. The above-mentioned passages do not therefore allow us to fix a specific connotation to the title (*Cheirokmēta*) of Bolos' work. In addition to the chronological gap between Bolos and Zosimus,[268] we must also consider the broad semantic spectrum covered by the two expressions. It is likely that Bolos' *Cheirokmēta* dealt with making artificial substances; however, the few fragments of Bolos preserved by the indirect tradition – which refer particularly to medical/magical questions – do not allow us to identify these substances with specific arts or technical fields.[269]

Having examined a range of problems that still resist clear and sure solutions, we may infer two points from the passages under discussion:

(1) The chronological data do not allow us to identify Bolos of Mendes, whose activity dates to the third/second century BC, with the ps.-Democritus who wrote the four alchemical books on dyeing that date to the first century AD.[270] The putative proto-alchemical content of Bolos' *Cheirokmēta* is conjectural and problematic, and the technique for whitening copper recorded in the second recipe of the Stockholm Papyrus cannot be ascribed with certainty to him. This text merely shows that by the first century BC some metallurgical practices were already connected to the name of Democritus, whom several contemporaneous sources present as an able *technitēs*. Only in later works, the *Four Books* epitomised into the *Corpus alchemicum*, is Democritus clearly presented as interested in specific procedures related to making gold, silver, and precious stones, and dyeing wool purple. These four technical fields are recognizable as the contents of an organic and coherent treatise thereafter considered one of the fundamental texts of early alchemy.[271]

---

[267] In the light of the abovementioned Syriac text, we cannot rule out the possibility that in this passage of Zosimus (IV 55–6 Mertens the reading of M (χεροτμήτων) should be standardized as χειροτμημάτων or χειροτμήτων.

[268] See Hershbell, "Democritus," 16 n. 13: "But Zosimos is several centuries after Bolus, and the little evidence there is in Columella and Pliny about the χειρόκμητα of Bolus suggests it dealt with herbal remedies, and not gilding and silvering."

[269] See Gaillard-Seux, "Un pseudo-Démocrite énigmatique," 233–4.

[270] This identification, proposed by Wellmann, *Bolos Demokritos*, 69 (see also Berthelot, *Origines de l'alchimie*, 99) has been accepted by several scholars: see, for instance, Holmyard, *Alchemy*, 25–6; Multhauf, *Origins of Chemistry*, 94–101; Jack Lindsay, *The Origins of Alchemy in Graeco-Roman Egypt* (New York: Barnes & Noble, 1970), 90–110; Georgia L. Irby-Massie and Paul T. Keyser, *Greek Science of the Hellenistic Era: A Sourcebook* (London and New York: Taylor & Francis Routledge, 2002), 235. Conversely, it has been refused by Kroll, "Bolos und Democritos," 231; Fraser, *Ptolemaic Alexandria*, vol. 1, 440ff.; Hershbell, "Democritus," 8 and 15; Letrouit, "Chronologie," 17 and DPhA II b 53, *s.v.* Bolos de Mendès. Festugière (*Révélation d'Hermès*, vol. 1, 223–33), however, suggested that an older section of the *Four Books* could date to the Hellenistic period and have been composed by Bolos of Mendes; the presence of elements that cannot be dated prior to the first century AD being explained in light of later revisions and integrations. However, according to this interpretation we would have to consider three different steps in the compilation of a work that ancient sources present as a coherent treatise (at least in its original form): (1) the Hellenistic version by Bolos; (2) the first-century version by an anonymous redactor; (3) the Byzantine version based on the epitome of four original books.

[271] See *infra*, pp. 57–63.

(2) The figure of Bolos of Mendes is, however, very helpful in explaining some mechanisms operating in the production of the pseudepigrapha ascribed to Democritus. First, Bolos reused some literature circulating under the names of the Persian magi Zoroaster and Ostanes, and played an important role in associating the name of Democritus with a certain kind of Eastern wisdom. Similar elements are also emphasised in the *Corpus alchemicum*, which presents Democritus as a pupil of Ostanes.[272] Bolos' interest in the secret properties of substances (including plants, herbs, stones, and even body parts) and their medical applications also fits the image of Democritus presented by later sources, namely as an expert in such fields.[273] This kind of medical background is also apparent in some sections of the *Four Books*, which refer to the cure of wounds and certain diseases (especially ophthalmic problems),[274] and note similarities between the preparation of dyeing and healing *pharmaka*.[275] As we have already noted, the same substance is often common to several different *technai*. According to what we know of Bolos' works, it is quite probable that these already presented Democritus as a "natural philosopher" (φιλόσοφος φυσικός) interested in identifying and applying the powers of nature regulated by secret rules of sympathy and antipathy. These rules were considered essential for explaining the relationships between vegetable, animal, and mineral substances, and could be applied to diverse fields including magic, pharmacology, medicine, and dyeing. These arts provide a common thread that connets several later treatises attributed to the philosopher Democritus, and which contributed to the development of his reputation as an expert in various *technai*.

§ 10. A commentary on ps.-Democritus' alchemical work: the dialogue between Synesius and Dioscorus

The *Four Books* became a fundamental authority in late antique and Byzantine alchemy, frequently referred to and quoted by later alchemists to support their own theories and to explain both abstract principles and practical matters. The *Corpus alchemicum* refers to commentaries on these books by ancient alchemists, including a work attributed to the alchemist Petasius, mentioned in the following passage:

> Petasius in his *Democritean Commentaries* shows what [it] is that they [i.e. the alchemists] call herbs; he writes word for word: 'he [i.e. Democritus] calls herbs the egg yolks.'[276]

According to this short note, Petasius seems to have written a work entitled *Democritean Commentaries* in which he tried to interpret and clarify a complicated alchemical vocabulary.

---

[272] See *infra*, pp. 69–73.
[273] See *supra*, pp. 34–6.
[274] Ps.-Dem. Alch. *PM* § 20.
[275] Ps.-Dem. Alch. *PM* §§ 15–6.
[276] *CAAG* II 356,1–3. The alchemist Petasius is often quoted in the *Corpus alchemicum* (*CAAG* II 95, 15–6; 97,15–7, etc.): he is presented as a commentator on ps.-Democritus' passages in *CAAG* II 282,7–9 (≈ 278,17; see Bidez-Cumont, *Mages*, vol. 1, 208–10) and associated with Synesius by the Byzantine alchemist Christianus in *CAAG* II 416,15; see Letrouit, "Chronologie," 47–8.

Similarly, a commentary in dialogue form appears under the name of Synesius. It has been handed down in the manuscript tradition under the title *The Philosopher Synesius to Dioscorus: Notes on Democritus' Book*.[277] Immediately after these words, the work opens with the subtitle: "With God's approval, the philosopher Synesius greets Dioscorus, priest of the great Serapis in Alexandria." The earliest and most important witnesses of the treatise are manuscripts M, B and A; manuscript V presents a fragmentary text.[278] All these manuscripts hand down only an incomplete form of the commentary, as may be inferred from its final sentence (Syn. Alch. § 9, ll. 320f.): "With God's aid we shall start our commentary." Therefore, the extant part of the dialogue seems to be a kind of introductory section to the commentary proper, which was not included in the Byzantine anthologies. Perhaps it is possible to recognize a section of this lost part (at least in an epitomized form) in a short excerpt entitled *Peri leukōseōs* (*On Whitening*),[279] which appears in manuscript immediately after a collection of Zosimus' *excerpta* that has been recently edited by Mertens under the general title of *Mémoires authentiques*.[280]

Unfortunately, the *Corpus alchemicum* does not provide detailed information on Synesius: his name is merely cited in the list of alchemists in M,[281] and omitted altogether from the passage by the philosopher Anonymous which traces the history of alchemy from its origins to the Byzantine period.[282] We might have expected to find Synesius mentioned alongside the important alchemical commentators Olympiodorus and Stephanus. However, Anonymous, who certainly knew Synesius' commentary,[283] does not mention him in this section, perhaps because Synesius was not considered to have attained the same stature as the other authorities, namely Olympiodorus and Stephanus, whom Anonymous explicitly identifies with the renowned Neoplatonic commentators of Plato and Aristotle.[284] This lacuna seems to have been partially filled in the list of alchemists handed down by manuscript A, even if this list is probably derived from the abovementioned passage by the philosopher Anonymous.[285] The manuscript in fact mentions both Synesius and Dioscorus: "Synesius, Dioscorus, priest of the great Serapion in Alexandria."

I shall follow chronological order in presenting the scant information about Synesius and Dioscorus, the two interlocutors of the dialogue, that is preserved in the *Corpus alchemicum*:

---

[277] Συνεσίου φιλοσόφου πρὸς Διόσκορον εἰς τὴν βίβλον Δημοκρίτου ὡς ἐν σχολίοις (mss. M and V give σχολείοις).

[278] M, fols. 72ᵛ9–78ʳ4; B, fols. 20ʳ19–31ᵛ13; A, fols. 31ʳ23–37ᵛ15; V, fols. 79ʳ4–91ʳ5. The fragmentary text in V has been partially completed by a later copyist (fols. 82ʳ–83ᵛ, now unreadable due to humidity, and fol. 87).

[279] M, fol. 118ʳ2–14; B, fols. 90ᵛ18–91ʳ9; A, fol. 92ʳ16–26. Ms. A preserves the same section also at fols. 14ᵛ20–30 and 250ᵛ13–21.

[280] See Letrouit, "Chronologie," 37 and Mertens, *Zosime de Panopolis*, LIV.

[281] M, fol. 7. His name appears third in the second column: see *infra*, n. 338.

[282] *CAAG* II 424,6–425,9.

[283] See *CAAG* II 432,12.

[284] *CAAG* II 425,4–5: οἰκουμενικοὶ πανεύφημοι φιλόσοφοι καὶ ἐξηγηταὶ τοῦ Πλάτωνος καὶ Ἀριστοτέλους.

[285] The list appears twice in the manuscript: once at fol. 195ᵛ and again at 294ʳ.

1. Zosimus – at least in the Greek chapters (κεφάλαια) that have been edited – mentions neither Synesius nor Dioscorus. Berthelot-Ruelle's edition is here misleading, since an excerpt edited in the Zosimus section seems to mention the name of Synesius; however, this excerpt, entitled *On the Stone of Philosophy*, is not ascribable to Zosimus, since it includes several quotations from alchemists dating *after* the alchemist of Panopolis.[286]

2. The earliest alchemist who clearly refers to Synesius is Olympiodorus, who probably lived during the sixth century AD, and who may be identified with the Neoplatonic commentator of the same name.[287] He quotes two extended passages from the dialogue between Synesius and Dioscorus, introducing these with the formula "Synesius writing to Dioscorus."[288]

3. The Byzantine alchemist Christianus (*CAAG* II 416,15) mentions Synesius in regard to the interpretation of an ingredient called Pontus' rhubarb.

4. The latest alchemical writings that preserve information on Synesius are the treatises ascribed to the philosopher Anonymous. If we accept the conclusions reached by Letrouit,[289] who recognized two different authors (both dating to the eighth-ninth century AD) under this 'pseudonym,' we have the following division:

   A. The so-called 'philosopher Anonymous 1' cites Synesius just once, introducing him with the formula "as the great Synesius clearly stated."[290] The adjective "great" (μέγας) seems to suggest that the Byzantine alchemist holds Synesius in high regard.[291]

   B. The so-called 'philosopher Anonymous 2' (who would be the author of the short history of alchemy mentioned above) quotes Synesius once, simply mentioning the dialogue between the commentator and Dioscorus (*CAAG* II 432,12).

While these passages suggest that Synesius' commentary was reasonably well known, they unfortunately preserve very few details useful for reconstructing the identity and chronology of its two interlocutors. Since Dioscorus is presented as a priest of the Serapeum, Synesius' work probably dates to before 391 AD, when the temple was destroyed. We may certainly have reservations about his priestly status, which recalls the common *topos* of an ancient 'esoteric' literature replete with references to Egypt and Egyptian temples.[292] However, even if we

---

[286] *CAAG* II 199,19.

[287] See Cristina Viano, *La matière des choses. Le livre IV des Météorologiques d'Aristote et son interprétation par Olympiodore* (Paris: Vrin, 2006), 199–206 (with exhaustive bibliography; in particular, 199 n. 1).

[288] *CAAG* II 90,20: Συνέσιος πρὸς Διόσκορον γράφων; *CAAG* II 102,10: Συνεσίου πρὸς Διόσκορον γράφοντος.

[289] Letrouit, "Chronologie," 63–5; see *supra*, n. 5.

[290] *CAAG* II 440,9: καθὼς ὁ μέγας Συνέσιος διεσάφησεν.

[291] This consideration seems to support the distinction proposed by Letrouit: indeed, it would be strange for a single author first to call Synesius ὁ μέγας, and afterwards to not even mention him in his history of alchemy. This short passage on the history of the alchemical art probably belongs to a different author, whom Letrouit calls "the philosopher Anonymous 2."

[292] We should however bear in mind that the involvement of Egyptian priests in alchemical practices is often mentioned by ancient alchemists. Besides the passages by ps.-Democritus, taken into account in the following part (see *infra*, pp.

doubt whether an Egyptian priest was indeed a pupil of sorts to a commentator on alchemical treatises, the reference to an old temple with a very important annexed library – also mentioned by Zosimus[293] – inserts an important chronological element.

A comparison of Zosimus' and Synesius' writings does not reveal any explicit cross-reference of one alchemist to the other. Although this silence may result from the epitomized and incomplete form in which both treatises have been preserved,[294] at our present state of knowledge it seems significant, particularly when we consider that authors of alchemical treatises are usually keen to quote ancient authorities in support of their own assertions. On the other hand, some sections of Synesius' commentary are quite similar to passages from Zosimus preserved in both the indirect tradition and Syriac manuscripts. Among the thirteen books handed down by SyrC under the name of Zosimus,[295] the ninth – whose incipit reads "Ninth treatise on the letter Ṭet^h; the letter Ṭet^h includes a full account of the making of mercury; it is the key of everything" – highlights several properties of mercury which are emphasized in a similar way by Synesius, §§ 9–11.[296] Synesius' explanation of how mercury penetrates and binds with the metallic bodies it was expected to dye[297] also presents a strong lexical analogy with a passage from Zosimus on the so-called *tetrasōmia*, or "four [metallic] bodies," quoted by the alchemist Olympiodorus.[298] Unfortunately, the lack of cross-references keeps us from knowing whether these correspondences are due to the influence of one author on the other, or simply to the similar interests of the two alchemists who independently focused their research on analogous topics. Indeed, in the indirect tradition, the earliest references to both Zosimus and Synesius come from the later alchemist Olympiodorus, who juxtaposes passages from their works, placing one after the other. We can speculate that the works of the two authors date to approximately the same period and were influenced by a similar cultural *milieu* – perhaps that of Alexandria[299] – in which both studied

---

[292] *Continued*

63–5), we should at least recall the Egyptian priest Nilus, often mentioned and criticized by Zosimus in both the Greek (*CAAG* II 191,3–18) and Syriac traditions (*CMA* II 228). Zosimus also attacks traditional Egyptian medicine, which was usually practised by priests, by contrasting it with secular medicine, based on the study of specific illustrated handbooks (I 175–85 Mertens).

[293] I 84 Mertens.

[294] In this case, too, the Syriac tradition is unhelpful: according to Berthelot-Duval's edition, there is no mention of Synesius in the *Corpus Syriacum*, although some sections handed down by SyrL under the name of Democritus (*CMA* II 83–4) suggest analogies with Synesius' commentary (see *CMA* II 84 n. 2). However, the same sections are preserved in SyrC under the name of Zosimus (*CMA* II 243–4), to whom they probably should be ascribed.

[295] See Matteo Martelli, "Medicina e alchimia. 'Estratti galenici' nel Corpus degli scritti alchemici siriaci di Zosimo," *Galenos*, 4 (2010): 208–11.

[296] Mm. 6.29, fols. 94^r–96^v; partial French translation in *CMA* II 270–2. Part of this book is handed down by SyrL under the name of Democritus. I have discussed some sections of this book in notes of commentary on Synesius' dialogue.

[297] Syn. Alch. § 10, ll. 160–3.

[298] *CAAG* II 96,6–4; see Syn. Alch. § 10 n. 54.

[299] With regard to Zosimus, see Mertens, *Zosime de Panopolis*, XIV.

and practised alchemy. If so, Synesius' dialogue would then date to the first half of the fourth century AD.

10.1 *Synesius the alchemist and Synesius of Cyrene*

Despite a lack of reliable historical information, some scholars have suggested identifying the Synesius of the alchemical dialogue with the renowned rhetorician and philosopher Synesius of Cyrene, Bishop of Ptolemais. Berthelot wrote that "there is nothing surprising in the notion that Synesius [of Cyrene] really wrote on alchemy, save to take away perhaps certain interpolations due to later copyists in the works attributed to him."[300] However, the assumptions upon which this hypothesis rests are problematic, and scholars are now more cautious about accepting this identification.

Berthelot first drew attention to the obvious homonymy between the names of the two authors and between Dioscorus, priest of the Alexandria's Serapeum, and the omonymous brother of the bishop Synesius, to whom Synesius addressed several letters. Berthelot even suggested that Dioscorus and his brother Synesius converted at some point to Christianity after their initial adherence to paganism.[301] However, this hypothesis is unsupported by any historical source; on the contrary, we note that no authors in the *Corpus alchemicum*, despite their tendency to attribute treatises to well known authorities, ever attempted to identify the alchemist Synesius with the scholar Synesius of Cyrene.[302]

Berthelot also emphasized the philosophical and scientific training that Synesius of Cyrene received in Alexandria by attending the classes of Hypatia, daughter of the mathematician Theon Alexandrinus (ca. 335–405). It is well known that Synesius of Cyrene addressed several letters to his teacher. Berthelot focused his attention especially on *Epistle* XV, in which Synesius, after falling ill, asked Hypatia to send him a hydrometer (or aerometer), providing her with a detailed description of the instrument. However, although this letter is presented by Lacombrade as evidence for the breadth of scientific investigation conducted in Hypatia's school,[303] Raïos has more recently noted that Synesius' detailed description suggests that Hypatia

---

[300] Berthelot, *Origines de l'alchimie*, 190. My translation from the original French.

[301] Berthelot, *Origines de l'alchimie*, 191.

[302] It is difficult to share the conclusion reached by Garth Fowden, who argues that the simple references to Dioscorus and Alexandria which appear in the title of the alchemical dialogue reveal that ancient alchemists considered Synesius to be the Neoplatonic philosopher: see, Fowden, *The Egyptian Hermes. A Historical Approach to the Late Pagan Mind* (Princeton, NJ: Princeton University Press, 1993), 179 n. 108. On the contrary, such details must be understood in light of alchemists' tendency to connect their treatises to Egypt and Egyptian temples. The mention of the Serapeum was in all probability not meant to recall Synesius of Cyrene, especially when we consider that the philosopher went to Alexandria only after the destruction of the temple. Indeed, the Neoplatonic philosopher never mentions the temple in his own writings. According to Christian Lacombrade, *Synésios de Cyrène. Hymnes* (Paris: Les Belles Lettres, 1978), XV n. 3, we can perhaps detect a reference to the desecration of this holy place in *Insomn.* 12; however, no recent scholar has highlighted such an allusion in commentaries on the same passage: see, for instance, Antonio Garzya, *Opere di Sinesio di Cirene* (Turin: UTET, 1989), 585; Jacques Lamoureux and Noël Aujoulat, *Synésios de Cyrène. Opuscules I* (Paris: Les Belles Lettres, 2004), 293.

[303] Christian Lacombrade, *Synésios de Cyrène. Hellène et Chrétien* (Paris: Les Belles Lettres, 1951), 43 n. 29 wrote about the *Ep.* XV: "On a deviné sans peine, à ce minutieux dessin, la première ébauche de notre aréomètre, ce qui laisse à penser que, plus accessible à notre curiosité, l'école d'Hypatie nous eût encore réservé bien d'autres surprises."

did not herself have deep knowledge of the instrument.[304] While Hypatia is indeed famous for her mathematical and astronomical investigations,[305] ancient sources do not allude to equal skill in other disciplines such as medicine or pharmacology. According to extant sources, the hydrometer was used to measure the density of liquids.[306] Although this instrument can have several practical applications, ancient texts usually mention it in medical contexts, and it is not clear whether Graeco-Alexandrian alchemists would have made use of it.[307]

Finally, scholars have often recognized the presence of some hermetic and esoteric elements in the works of Synesius of Cyrene: an important feature of various Neoplatonic schools of the time. For instance, the philosopher cites the *Chaldean Oracles*,[308] refers to Orphic writings,[309] explicitly mentions Hermes and

[304] Dimitris R. Raïos, *Archimède, Ménélaos d'Alexandrie et le «Carmen de ponderibus et mensuris»: contributions à l'histoire des sciences* (Ioannina: Panepistēmio Iōanninōn, 1989), 131.

[305] See *Suda* υ 166 Adler, *s.v.* Ὑπατία, which mentions three works by Hypatia all related to similar fields: Ἔγραψε ὑπόμνημα εἰς Διόφαντον, τὸν ἀστρονομικὸν κανόνα, εἰς τὰ Κωνικὰ Ἀπολλωνίου ὑπόμνημα κτλ., ("She wrote a commentary on Diophantus [third-century mathematician], the *Astronomical Canon* and a commentary on Apollonius' *Conics* [i.e. Apollonius of Perga, Greek geometer and astronomer of the second century BC]"). Scholars do not agree on the identification of the second writing. Lacombrade (*Synésios de Cyrène*, 42) corrected the text by adding <εἰς> before τὸν ἀστρ. καν.; allowing it to be considered as a commentary on a work by Ptolemy. On this problem and for a general interpretation of this entry, see Gemma Beretta, *Ipazia d'Alessandria* (Roma: Editori Riuniti, 1993), 48–50. Hypatia also contributed to the edition of the commentary on the *Almagest* written by her father Theon (see Beretta, *Ipazia d'Alessandria*, 41–5).

[306] All references to hydrometer date from after the fourth century: see Dimitris R. Raïos, "L'invention de l'hydroscope et la tradition arabe," *Graeco-Arabica*, 5 (1993): 275–86. Besides the abovementioned letter by Synesius, we mention the *Carmen de ponderibus et mensuris* (vv. 103–6) by Remmius Favinus, which probably dates back to the fourth century: see D.R. Raïos, *Recherches sur le «Carmen de ponderibus et mensuribus»* (Ioannina: Panepistēmio Iōanninōn, 1983), 27–45; and Ch. 7 of the first lecture of the *Kitāb Mīzān al-ḥikma (Book of the Balance of Wisdom)* by al-Ḫāzinī, in which the invention of this device is attributed to the philosopher الفوفس الرومى (al-fūfus al-rūmī), usually identified with the Greek mathematician Pappus of Alexandria: see, for instance, Nicolas Khanikoff, "Analysis and Extracts of the Book of the Balance of Wisdom, An Arabic Work on the Water-Balance Written by 'Al-Khâzinî in the Twelfth Century," *Journal of the American Oriental Society*, 6 (1858): 40; more recently Raïos, "L'invention de l'hydroscope," 281–2, has proposed the Greek physician Rufus. Finally, we can recognize another reference to the hydrometer in the prose version of the *Carmen de ponderibus* (preserved in mss. *Parisini lat.* 7530 and 11478), dating to the seventh/eighth century: see D.R. Raïos, "Autour de la paraphrase du «Carmen de ponderibus et mensuris»," in *Science antique, science médiévale (Autour du manuscrit d'Avranches 235). Actes du Colloque international, Mont-Saint-Michel, 4–7 septembre 1998*, ed. Louis Callebat and Olivier Desbordes (Hildesheim, Zürich, New York: Olms-Weidmann, 2000), 297–318.

[307] Hydrostatics was well developed in Antiquity, and dealt not only with density of fluids, but also with the specific weight of solids. For instance, this science was applied to solving the problem of the golden crown of King Hiero II. According to Vitruvius (*De Arch.* IX praef. 9–12), Plutarch (*Non posse suav. vivi* 1094 B-C) and the *Carmen de ponderibus and mensuris* (vv. 125–34), Hiero asked Archimedes to analyse the composition of his crown, to determine whether the gold had been adulterated with the addition of silver. The *Carmen* also preserves an interesting description of the hydrometer (vv. 136–44) stemming from a treatise on metallic alloys written by the Alexandrian mathematician Menelaus (first century AD). This writing, lost in its original Greek form, has been preserved in Arabic translation (Escurial, Arabic ms. 955 [960]; see Joseph Würschmidt, "Die Schrift des Menelaos über die Bestimmung der Zusammensetzung von Legierungen," *Philologus*, 80 (1925): 377–409; Anton M. Heinen, "The Treatise on Alloys by Menelaos of Alexandria. An Example of an Ancient Greek Text Lost in the Original but Preserved in an Arabic Translation," in *L'eredità classica nelle lingue orientali*, ed. Massimiliano Pavan and Umberto Cozzoli (Florence: Istit. Encicl. Treccani, 1986), 179–80. The possible relationship between the scientific studies conducted in Alexandria since Archimedes' time and Graeco-Egyptian alchemy has yet to be systematically investigated: we must note, however, that neither the Leiden and Stockholm papyri (see Halleux, *Papyrus de Leyde*, 52) nor the *Corpus alchemicum* mention docimastic tests based on the use of similar instruments. Only in the medieval tradition does the *Mappae clavicula* refer to hydrostatics for checking the composition of metallic alloys.

[308] *Ep.* XLIII, L, CXCVI; *Insomn.* 4, 7, 9.

[309] *Dion.* 5 e 7.

Zoroaster,[310] and devotes part of his treatise *On Dreams* to an explanation of the theory of sympathy. This last text deserves particular attention, since it develops the idea of a 'universal kinship' relating all the elements of the world, considered as different limbs of the same living body:

> Let the foregoing be proof that divinations are amongst the best vocations of man; if all things are signs appearing through all things, inasmuch as they are brothers in a single living creature, the cosmos, so also they are written in characters of every kind, just as those in a book of which some are Phoenician, some Egyptian, and others Assyrian. The scholar reads these, and he is a scholar who learns by his natural bent ... It must needs be, I think, the parts of this great whole, since all shares one feeling and one breath, belonging to each other. They are, in fact, limbs of one entire body, and may not the spells of the magicians be even such as these? Obviously, for charms[311] are cast from one part of it to another as signals are given, and he is a sage who understands the relationship [συγγένεια] of the parts of the universe.[312]

Synesius continues by giving examples of this universal sympathy. Regarding medicine, he claims that "as when the bowel is in pain, another part suffers also with it, so a pain in the finger settles in the groin, although there are many organs between these parts which feel nothing" (*Insomn.* 2,25–7). He also notes that, in the natural world, herbs and stones have an affinity with the divine since they are connected to particular gods or heavenly powers (*Insomn.* 2,29–31). Scholars commenting upon this passage have often highlighted its connections with Neoplatonism, drawing attention particularly to the notion of the cosmos conceived as a living body.[313] A similar idea is also often presupposed in alchemical treatises, where, for instance, the formula *hen to pan* ("one the all," ἓν τὸ πᾶν) almost certainly evokes a kind of unity into which the different constituents of the universe resolve.[314] Alchemists also referred to sympathy in order to elucidate interactions between ingredients. Often they explained metallurgical practices by drawing clear analogies with organic phenomena (both animal and vegetable), apparently

[310] *Dion.* 10. Synesius harshly criticized those who, without having received philosophical training, sought to distinguish themselves by proposing stupid and meaningless doctrines, writing: "Let us say to them, for it would be well worth it: O boldest of all men, if we had known that you had been so fortunate as to hold the estimate of the soul which was that of Amus, or Zoroaster or Hermes Trismegistus, or Antonius, we should not have presumed to teach you or to conduct you through a course of study, endowed as you would be with a greatness of mind to which even conclusions are but premises," *Dion.* 10,26–31, translation by Augustine Fitzgerald, *The Essays and Hymns of Synesius of Cyrene* (Oxford: Oxford University Press, 1930), 168, slightly modified. Synesius here lists four singular figures. The identification of Amus (Ἀμοῦς) is uncertain (see Garzya, *Opere di Sinesio*, 684 n. 47; Aujoulat in Lamoureux-Aujoulat, *Synésios de Cyrène*, 162 n. 75): this could refer either to an anchorite who lived in Nitria during the fourth century AD or to the mythic Egyptian pharaoh Thamus, mentioned by Plato (*Phaedr.* 274c-e). Antonius must be identified with the homonymous saint of the Thebaid. As for Zoroaster and Hermes, who are already associated in Zos. Alch. I 41 Mertens (see Mertens, *Zosime de Panopolis*, 3 n. 27), these 'mythical' figures are related to the Persian and Egyptian traditions respectively; see Lacombrade Christian, "Le *Dion* de Synésios de Cyrène et ses quatre sages barbares," ΚΟΙΝΩΝΙΑ, 12 (1988): 17–26, for a deeper analysis of the passage.
[311] On the difficult interpretation of the term ἴυγξ, see Davide Susanetti, *Sinesio di Cirene. I sogni, introduzione, traduzione e commento* (Bari: Adriatica editrice, 1992), 96 n. 12.
[312] *Insomn.* 2,1–23; translation by Fitzgerald, *Essays and Hymns*, 237–8.
[313] See Lacombrade, *Synésios de Cyrène*, 151–2; Susanetti, *Sinesio di Cirene. I sogni*, 96 n. 12; Aujoulat in Lamoureux-Aujoulat, *Synésios de Cyrène*, 200–1. See also Syn. *Aeg.* II 7.
[314] See Paul Plass, "A Greek Alchemical Formula," *Ambix*, 26 (1982): 69–73.

applying a biological model to the study of the physical world even with regard to phenomena that would now be viewed as inorganic.[315]

However, such affinities between Synesius of Cyrene and alchemical literature are insufficient to support the philosopher's supposed interest in alchemy. Interest in the sympathy theory was shared by many philosophers, astrologers, physicians, alchemists and magicians of the period, and does not require expertise in the alchemical art, something that Synesius of Cyrene never mentions in his writings, even when focusing on concrete practices related to the sympathetic theory. On the contrary, Synesius condemns the most practical aspects of the divinatory art, which, in his eyes, provide important spiritual training for approaching the divine.[316] His book *On Dreams* focuses in particular on human 'psychology,' explaining which parts and functions of the human soul are supposed to be most involved in divination.[317]

All these aspects of Synesius of Cyrene's thought undoubtedly deserve deeper analysis. Possible relationships between the development of Alexandrian alchemy (for which the dialogue between Synesius and Dioscorus represents an important source) and various aspects of Neoplatonic philosophy and related theurgic practices should also be more deeply considered.[318] Such an inquiry should not be restricted to the basic problem of identifying the authors of alchemical and Neoplatonic treatises, but should also investigate their possible mutual relations by considering whether a kind of practical knowledge – related, for instance, to the working of metals which were used in making statues and idols – carries over into more philosophical writings.[319]

Such an inquiry would of course exceed the scope of this section, which is primarily concerned with the identification of Synesius, author of the alchemical dialogue, with Synesius of Cyrene. In this regard, chronological data offers the most important evidence against this identification.[320] Synesius of Cyrene was born around 370 AD and received his first education in his hometown. According to his letters, he

[315] In the above quoted passage by Synesius, it is interesting to note his use of the term συγγένεια, which alchemists already employed in the earliest writings with reference to specific affinities between substances: see Ps.-Dem. Alch. *PM* § 17, l. 196, *AP* § 1, l. 8 and § 2, l. 39; Pebichius in Syn. Alch. § 10, ll. 163–4 and in Olymp. Alch. *CAAG* II 91,3; Ostanes in Zos. Alch. *CAAG* II 197,5–18; *PHolm.* 63,2.

[316] *Insomn.* 12. According to Aujoulat's interpretation (in Lamoureux-Aujoulat, *Synésios de Cyrène*, 227–30), Synesius' criticism is addressed especially against the most practical aspects of theurgy. On the rationalistic tendencies of the philosopher of Cyrene, see also Beretta, *Ipazia di Alessandria*, 68–82.

[317] Synesius focused his attention especially on the φαντασία and on the πνεῦμα, which he considered as a semi-corporal vehicle (ὄχημα) of the soul: see Robert Ch. Kissling, "The OXHMA – ΠΝΕΥΜΑ of the Neo-Platonists and the *De Insomniis* of Synesius of Cyrene," *American Journal of Philology*, 43 (1922): 318–30; Garzya, *Opere di Sinesio*, 28–30.

[318] See Fowden, *Egyptian Hermes*, 116–53.

[319] A significant example is represented by Proclus, who in an interesting passage on the hieratic art (*CMAG* VI 158–61, commented on by Festugière, *Révélation d'Hermès*, vol. 1, 133–6) claimed that those who have been initiated are able to recognize the sympathy that connects what is visible with what is not visible, since they are aware of the "chains" which bind everything together, from the lowest extremity (the visible) to the highest (the invisible). This theoretical explanation is followed by a list of substances (each with its celestial correspondences) that were very probably used for theurgic practices. However, despite his complete adherence to the theory of universal sympathy and its related theurgic applications, Proclus explicitly condemned alchemy in *Remp.* II 234,14–5 Kroll, where he stated that it is impossible to replicate by the art of mixing different species (ἐκ μίξεώς τινων εἰδῶν) that which nature produces as one single species (τὸ εἶδος ἕν).

[320] See also Lacombrade, *Synésios de Cyrène*, 70–1.

returned to Cyrene in 395 after his first visit to Alexandria.[321] The young scholar had been attracted to Alexandria particularly by the reputation of the teacher Hypatia.[322] We may therefore date this first visit to around 392–395, since Hypatia did not start teaching until after the destruction of the temple of Serapis, following measures taken by the Emperor Theodosius against paganism in 390–391 in order to strengthen his alliance with Ambrose, Bishop of Milan. These measures gave authority to Theophilus (patriarch of Alexandria from 385 to 412), who first converted the Temple of Dionysus in Alexandria into a church.[323] The destruction of the Serapeum followed, despite the efforts of the philosopher-priest Olympus who bravely defended the temple. Sources usually place Hypatia's teaching after these tragic events. Socrates Scholasticus interrupts his account of the destruction with a section on Hypatia's work,[324] and according to Damascius' *Life of Isidore* (frr. 92–102 Zintzen), Hypatia succeeded Olympus, the priest-philosopher who came from Cilicia to Alexandria to administer the Serapeum and who was one of the protagonists of the Hellenic resistance during the 390–391 conflicts.[325] Accordingly, if Synesius first visited Alexandria after the destruction of the Serapeum, it would be very difficult to identify him as the author of the alchemical dialogue, which was almost certainly composed while the temple was still standing. Synesius the alchemist, in conclusion, was not Synesius of Cyrene.

10.2 *Synesius the alchemist and Synesius of Philadelphia*

Lacombrade proposes a different identification – with Synesius of Philadelphia – on the basis of the following entry in the lexicon *Suda*:

> Androclides, son of Synesius of Philadelphia in Lidia: he taught in the time of the philosopher Porphyry, since he mentioned him in his work *On the Writers of the Art of Rhetoric of that Time*.[326]

The entry is problematic in many respects. The subject of the second clause is Androclides, who is claimed to have been a contemporary of Porphyry in the precedent part. The pronoun "him" probably refers to Porphyry; indeed, the chronological data given in the text seems to be based on Androclides' reference to Porphyry.[327] However, the title of the work is puzzling, especially in regard to the

---

[321] See in particular *Ep.* CXLV, where Synesius mentioned Heraclian (l. 12) who, according to the *Codex Theodosianus* XI 24,3, must be identified with a senior official of Egypt in 395 AD (see Lacombrade, *Synésios de Cyrène*, 24 n. 2).
[322] See *Ep.* CXXXVII and Lacombrade, *Synésios de Cyrène*, 38.
[323] See Sozomenus, *Hist. Eccl.* VII 15 and Beretta, *Ipazia di Alessandria*, 20.
[324] *Hist. Eccl.* V 6. According to the lexicon *Suda* (υ 166,3 Adler, s.v. Ὑπατία: [...] ἤκμασεν ἐπὶ τῆς βασιλείας Ἀρκαδίου), Hypatia flourished during the reign of the emperor Arcadius (395–408).
[325] See Beretta, *Ipazia di Alessandria*, 25.
[326] α 2180 Adler (= Porph. fr. 423T Smith). See Lacombrade, *Synésios de Cyrène*, 71.
[327] I have preferred this reading, which is supported by Francesco Romano, "Porfirio technologos?," *Siculorum Gymnasium. Rassegna semestrale della Facoltà di Lettere e Filosofia dell'Università di Catania*, 31 (1978): 518 – to Bidez's interpretation, which takes Porphyry as the subject of the verb "he mentioned" (μέμνηται) and consequently reads the pronoun "him" (αὐτοῦ) as referring to Androclides: Joseph Bidez, *Vie de Porphyre, le philosophe néo-platonicienne* (Ghent: van Goethem, 1913), 73.

term *technologoi* ("writers of the art of rhetoric," τεχνολόγων), that is conjectural.[328]

Despite these doubts, it is still difficult to agree with the interpretation proposed by Lacombrade, who reads the word *technologos* (translating it as "amateur d'arguties") as a possible reference to alchemy.[329] Although we do find a mention of Porphyry in the *Corpus alchemicum*, it is clear that the *Suda* entry is to be understood as strictly related to rhetoric, since it contains no element suggestive of alchemy.[330] Synesius of Philadelphia, father of Androclides, is thus apparently just another example of homonymy with the alchemist. Were we to identify him with the alchemist Synesius, we would also have to date the alchemical dialogue to the beginning of the third century AD: several years before Zosimus composed his own works. Such a chronology would be risky, given that the alchemist of Panopolis never mentions Synesius in his writings.

§ 11. The *Four Books* and the definition of alchemy

The exact definition of the content of the Leiden and Stockholm papyri is still debated by scholars.[331] Owing to their specific focus on technical procedures, these recipe books are not considered by some interpreters as the expression of a 'mature alchemy,'[332] which, according to Festugière's definition, should include a doctrinal component besides the simple description of metallurgical techniques.[333] In contrast, the *Four Books* of ps.-Democritus, while covering the same range of technical questions, must be considered a significant and very early example of alchemical literature. The treatises included in the *Corpus alchemicum* frequently cite Democritus, who thus became the philosopher *par excellence*, his authority invoked to support of a variety of alchemical doctrines. This tendency is apparent, for instance, in the writings of Zosimus, who used the ps.-Democritean treatises as the starting point for discussing several issues in his *Chapters to Eusebeia*.[334]

---

[328] The manuscript tradition has, in fact, the participle τεχνολογῶν, from the verb τεχνολογέω, "to dissertate, to cavil". For a deep textual analysis of this entry, see Romano, "Porfirio technologos."

[329] Lacombrade, *Synésios de Cyrène*, 71 n. 35. Lacombrade also thought that the father-son relationship mentioned in the lexicon *Suda* could hint at a "tradition familiale," characteristic of the esoteric sciences. However, any element seems to support a similar interpretation, and the same degree of kinship is stated again in *Suda* φ 296 Adler: Φιλαδελφέως υἱὸς Ἀνδροκλείδης.

[330] *CAAG* II 25,12; 205,14. See Romano, "Porfirio technologos," 519.

[331] See Halleux, *Papyrus de Leyde*, 24–30.

[332] See, for instance, Ingeborg Hammer Jensen, "Deux papyrus à contenu d'ordre chimique," in *Oversigt over det Kgl. Danske Videnskabernes Selskabs Forhandlinger* (Copenhagen, 1916), 279–302, and *Die älteste Alchymie*, 40–1; Festugière, *Révélation d'Hermès*, vol. 1, 221–2.

[333] The definition of alchemy given by Festugière, *Révélation d'Hermès*, vol. 1, 218 reads: "L'alchimie gréco-égyptienne, d'où ont dérivé toutes les autres, est née de la rencontre d'un fait et d'une doctrine. Le fait est la pratique, traditionnelle en Égypte, des arts de l'orfèvrerie. La doctrine est un mélange de philosophie grecque, empruntée surtout à Platon et à Aristote, et de rêveries mystiques."

[334] On this work, see Mertens, *Zosime de Panoplis*, LIV-LX. It is worth mentioning that the Arabic tradition ascribes to Zosimus a commentary on ps.-Democritus' work, entitled *Kitāb Mafātiḥ aṣ-ṣanʿa, Book on the Keys of the Art* (see Sezgin, *Geschichte des arabischen Schrifttums*, vol. 4, 75 n. 6 and Ullmann, *Die Natur- und Geheimwissenschaften im Islam*, 163). Moreover, in the Arabic *Book of Pictures*, ascribed to Zosimus, Zosimus claims: "I did not desire anything from the sages except to become one of their students. However, I particularly selected Democritus for this, and I became a student of his, although there are 660 years between us" (Abt-Fuad, *Book of Pictures*, 346).

Several commentaries on ps.-Democritus' works were writtten, of which Synesius' dialogue is the most important example extant in the Byzantine tradition.[335] This preoccupation with Democritus continued in the work of later alchemists; the extant remains of alchemical works by Olympiodorus, Stephanus, and Christianus are packed with references to ps.-Democritus.[336] The short history of alchemy written by the Byzantine alchemist Anonymous[337] sets Democritus among the founders of the art, and he also appears in the lists of the most important alchemists preserved in **M** and **A**.[338]

Alongside this ancient and long lasting tradition that emphasizes ps.-Democritus' contribution in laying the foundations of alchemy, we must attribute the same relevance to the four topics covered by his *Four Books*, namely gold- and silver-making, dyeing of stones, and purple dyeing of wool. However, this fourfold division of alchemy – also deployed in the Leiden and Stockholm papyri – contrasts with a narrower definition of the art, which is often identified with the mere making of precious metals. This contrast, which seems to imply a progressive narrowing of those topics considered alchemical, is not easy to explain. If we extend our analysis to the works of other ancient alchemical authorities, we encounter the same range of techniques covered by ps.-Democritus. According to the Syriac tradition, the alleged teacher of Democritus, the Persian magus Ostanes, wrote on similar topics. In letters between the Egyptian alchemist Pebichius and the Persian wise man Osron,[339] the former reports that he translated all of Ostanes' writings into both Greek and Egyptian, and describes them as follows:[340]

ܩܕܡܝܐ ܡܟܬܒܐ ܘܡܟܬܒܐ ܬܪܝܢܐ ܘܬܠܝܬܐ ܘܪܒܝܥܐ . ܘܚܡܝܫܝܐ

ܘܫܬܝܬܐ ܘܫܒܝܥܐ . ܘܬܡܝܢܝܐ ܘܬܫܝܥܝܐ ܘܥܣܝܪܝܐ . ܘܚܕܥܣܪ

ܘܬܪܥܣܪ . ܘܬܠܬܥܣܪ . ܘܐܪܒܥܣܪ . [...] .

[334] *Continued*
     On this work's uncertain dependency on Zosimus' original writings, see Bink Hallum, "The Tome of Images: an Arabic Compilation of Texts by Zosimos of Panopolis and a Source of the Turba Philosophorum," *Ambix*, 56 (2009), 76–88; Theodor Abt and Salwa Fuad, *The Book of Pictures by Zosimus of Panopolis. Corpus Alchemicum Arabicum* II.2 (Zurich: Living Human Heritage, 2011), 73–138.

[335] See *supra*, pp. 48–9.

[336] I list all passages in which Zosimus or later alchemists refer to the *Four Books* (at least in the form preserved by the Byzantine tradition) in the apparatus under the translation of the relevant ps.-Democritean excerpts.

[337] See also *supra*, p. 49.

[338] **M**, fol. 7ᵛ (*CAAG* I 110). Ὀνόματα τῶν φιλοσόφων τῆς θείας ἐπιστήμης καὶ τέχνης: |1ˢᵗ column| Μωσῆς, Δημόκριτος, Συνέσιος, Παύσηρις, Πηβίχιος, Ξενοκράτης, Ἀφρικανός (Ἀφρί- **M**), Λουκᾶς (-κάς **M**), Διογένης, Ἵππασος, Στέφανος, Χίμης, Χριστιανός |2ⁿᵈ column| Μαρία, Πετάσιος, Ἑρμῆς, Θεοσέβεια, Ἀγαθοδαίμων, Ἰσίδωρος (Ἠσι- **M**), Θαλῆς (-λῆς **M**), Ἡράκλειτος, Ζώσιμος, Φιλάρετος, Ἰουλιανή, Σέργιος. The list (*CAAG* II 25,6–26,4) is preserved in two different versions at **A**, fols. 195ᵛ4–19 and 294ʳ1–13.

[339] Bladel, *Arabic Hermes*, 48 n. 117, supposed that the name Osron (ܐܘܣܪܘܢ), referring to a Persian philosopher and magician, could stem for the middle Persian term *āsrōn*, "priest."

[340] French translation in *CMA* II 310 (see also Bidez-Cumont, *Mages*, vol. 2, 337–8): the manuscript in this part is difficult to read owing to its poor state of conservation.

I opened his [Ostanes'] book and I found every art: astrology and astronomy and philosophy and philology[341] and magic and the art of mysteries and sacrifices and that art which is terrible for many peoples, but which is absolutely necessary, i.e. the making of gold ... And the whole book [included also?] the stones and the purples and the divine dyeing of glasses.[342]

This passage, which stresses the polymathy of the Persian magus, includes gold-making, the working of precious stones, the colouring of glass, and the purple dyeing of fabric: the very same topics covered by the four ps.-Democritean books.[343]

An excerpt preserved in A and B under the title of *Which Species Concern the Deep Dyeing of [Precious] Stones and How to Process Them*,[344] also mentions the dyes employed for colouring stones by Ostanes and another ancient alchemist, Maria the Jewess, who according to Syncellus' account was, like Democritus, a pupil of Ostanes.[345] In the chapter entitled *Another Excerpt on Stones*, ps.-Democritus and Maria are quoted together, explaining how to use liquid ingredients in mordant processes.[346] The passage specifies that similar procedures were employed for the cold dyeing of purple.[347]

In conclusion, despite the fragmentary state of our sources, we can partly reconstruct a wider definition of alchemical production that was concerned with a variety of dyeing techniques applied to metals, precious stones, and fabrics, and attributed in the *Corpus alchemicum* to the earliest authors (Ostanes, Democritus, Maria, Pibechius), namely, to the very founders of the alchemical art. Sadly, the greater part of this production is now lost since it was not included in the Byzantine anthologies. Of the *Four Books* themselves, only the sections focused on gold and silver have been preserved. This selection suggests the development over time of a narrower conception of alchemy, concerned particularly with metallic transmutation, a conception well attested, for instance, in the definition given by the Byzantine lexicon *Suda*: "*Chēmeia*: the preparation of gold and silver; Diocletian looked for the books on this subject and burned them."[348]

---

[341] The Syriac expression is a word for word translation of the Greek term φιλολογία (see *CMA* II 310): ܠܡܐ corresponds to φίλος in composition (see *ThSyr* II 3883–4) and ܡܠܬܐ corresponds to λόγος (or λόγιον; see *ThSyr* I 2114).

[342] SyrC, fol. 131ʳ15–131ᵛ3. I have omitted a few lines of the passage (quite fragmentary at this point), which do not seem to add any important information about the contents of Ostanes' treatises. Berthelot-Duval (*CMA* II 310) translates the last line as "et des teintures divines des pierres précieuses"; however, the term ܙܓܘܓܝܬܐ usually refers to "glass" (corresponding to the Greek ὕαλος; see *ThSyr* I 1081).

[343] Pebichius is also presented as one of Ostanes' pupils in a Greek alchemical fragment in Bodleian MS *Arch. Seld*. B 18 (f. 192ᵛ), edited in *CMAG* VI 44. Zosimus also mentions the alchemist several times, placing his work in relation to ps.-Democritus' writings (see *CAAG* II 155,16–7; 184,18–20): see also Letrouit, "Chronologie," 21–2.

[344] *Τίνα τὰ εἴδη τυγχάνουσι τῆς τῶν λίθων καταβαφῆς καὶ πῶς οἰκονομεῖται*, edited in *CAAG* II 351–3 within the recipe-book entitled *Deep Tincture of Stones, Emeralds, Rubies and Jacinths from the Book Taken out from the Sancta Sanctorum of Temples*; new edition of the passage in Bidez-Cumont, *Mages*, vol. 2, 323–4.

[345] Syncell., 297, ll. 275. Mosshammer (see *infra*, p. 69). Letrouit, "Chronologie," 21 suggested that Maria's work postdates Democritus, since she quotes the aphorism on natures.

[346] *Ἄλλο κεφάλαιον περὶ λίθων*, *CAAG* II 354–8.

[347] See *supra*, pp. 9–10. Similar cold procedures are also mentioned in the Leiden and Stockholm papyri: *PLeid.X*. 95,5 (καὶ ἔσται ψυχροβαφής); *PHolm*. 106 (πορφύρας ψυχροβαφὴ ἀληθῶς γινομένη), 121 (φαιῶν ψυχροβαφαί) and 126 (χρυσανθῆ ποιῆσαι ψυχροβαφῇ).

[348] χ 280 Adler.

The reference here to Diocletian is significant. The emperor's role is described more fully in the next part of the lexicon where he is said to have burnt all the books in Egypt on the *chēmeia* of gold and silver (τὰ περὶ χημείας χρυσοῦ καὶ ἀργύρου … βιβλία) to prevent the striking of false coins, debasing the coinage, and revolts against Roman authority. The source for this information is probably the chronographer John of Antioch (seventh century AD),[349] who also provides an alchemical interpretation of the myth of the Argonauts. In this reading, the Golden Fleece represents a parchment that explains how to produce gold by means of *chēmeia*.[350] The definition of this last term given in the *Suda* is evidently related to John's earlier account of the books burnt by Diocletian, which were said to deal with the "*chēmeia* of gold and silver." The lexicon replaces the more difficult and exotic term *chēmeia* (χημεία) with the easier *kataskeuē* (κατασκευή), or "preparation." Unfortunately, the exact meaning of the word *chēmeia* is not clear and scholars disagree over its etymology.[351] Its first known appearance is in an alchemical context, within the writings by Zosimus. According to Syncellus, Zosimus' account was taken from the "Holy Scriptures" (αἱ ἱεραὶ γραφαὶ ἤτοι βίβλοι) and a treatise of Hermes entitled *Physika*. These sources described how chemical secrets were revealed to mankind by fallen angels, who, becoming infatuated with human women, gave them a book entitled *Chēmeu* (Χημεῦ). The account clearly belongs to the pseudepigraphic tradition of Enoch,[352] which was reworked by Zosimus as follows:

> The same scriptures say that from them [i.e. the fallen angels] the giants were born. So theirs is the first teaching concerning these arts handed down by Chēmeu. He called this book Chēmeu, whence also the art is called Alchemy (*chēmeia*).[353]

The authenticity of the passage is confirmed by the Syriac translation:[354]

ܡܢܗܘܢ ܗܟܝܠ ܐܡܪܝܢ ܟ̈ܬܒܐ. ܗܠܝܢ ܕܐܬܝܠܕܘ ܥ̈ܒܕܐ. ܘܐܝܬܝܗ̇ ܡܫܡܠܝܘܬܐ ܩܕܡܝܐ ܕܝܠܗܘܢ. ܥܠ ܗܠܝܢ
ܐܘܡܢܘܬ̈ܐ. ܩܪܝܢ ܗܘܘ ܠܟܬܒܐ ܕܝܠܗܘܢ. ܡܢ ܗܪܟܐ ܘܐܦ ܟܘܡܝܐ ܡܬܩܪܝܐ.

Therefore, these books claim that from them [i.e. the fallen angels] the works [of nature?] arose[355] and theirs is the first exposition concerning these arts. They called

[349] Fr. 248 Roberto = *FHG* IV, fr. 165 Müller. The same information is found in the *Acts of Saint Procopius* (sixth century AD; in *Acta Sanctorum Julii* 1721, t. II, 557, § 4).

[350] Fr. 26,3 Roberto = *FHG* IV, fr. 165 Müller. See also *Suda* δ 250 Adler, *s.v.* δέρας· βιβλίον ἐν δέρμασιν γεγραμμένον, περιέχον ὅπως δεῖ γίνεσθαι διὰ χημείας χρυσόν.

[351] See Robert Halleux, *Les textes alchimiques* (Turnhout: Brepols, 1979), 45–7 for a lucid analysis of the different interpretations.

[352] See James R. Partington, *A History of Chemistry*, vol. I/1: *Theoretical Background* (London: Macmillan & Co, 1970), 173–7. On Zosimus' Gnostic reading of this myth, see Kyle A. Fraser, "Zosimos of Panopolis and the Book of Enoch: Alchemy as Forbidden Knowledge," *Aries*, 4 (2004): 125–45.

[353] Syncell. 14,12–4 Mosshammer, translated by William Adler and Paul Tuffin, *The Chronography of George Synkellos, A Byzantine Chronicle of Universal History from the Creation* (Oxford: Oxford University Press, 2002), 19 (slightly modified). The expression "handed down" is not present in the Greek text and has probably been added by the translator in order to make explicit the function of *Chēmeu* within the sentence. The Syriac translation of this passage does not include the name in this sentence.

[354] SyrC fol. 49v3–5.

[355] The Syriac text differs at this point from the version quoted by Syncellus: where Syncellus has τοὺς γίγαντας γεγενῆσθαι the Cambridge manuscript records ܐܬܝܠܕܘ ܥ̈ܒܕܐ; if the verb ܝܠܕ translates the Greek γίγνομαι, then the term ܥ̈ܒܕܐ [where the point above the *bet*[h] indicates that it is the plural of ܥܒ݂ܕܐ (ʿ*bodo*) "work, occupation" (*ThSyr* II 2773)]

these books *Kumu* (≈ Χημεῦ), whence also alchemy (*kumya* ≈ χημεία)[356] takes its name.

The Syriac text continues by giving more details about the structure of the treatise *Kumu*, which is supposed to have been divided into twenty-four books, as well as of the various commentators who only made matters worse by trying to clarify the contents. While discussing the importance of having a book of *kumya*, Zosimus complains that commentators have focused their attention only on the transformation of silver into gold, even though the original books dealt with the properties of all metals and their mutual relationships.[357] Even if Zosimus did not restrict himself to gold-making, his account highlights the centrality of metallurgical practices within the art of *chēmeia* (χημεία/ܟܘܡܝܐ). In addition, we should note that the alchemical work entitled *Isis to his son Horus*, which pre-dates Zosimus' writings,[358] already connects angelic revelation with the knowledge of "the preparation of gold and silver," an expression using the same words as the lexicon *Suda* to define *chēmeia*.[359] It is difficult, however, to gauge the extent to which the alchemical reworking of the Enochian myth led to a more restricted definition of alchemy, focused on metallurgical practices or the making of gold and silver, while in the earlier versions of the Enochian myth the fallen angels taught several arts to humankind including the working of precious stones and dyeing procedures.[360]

[355] *Continued*
   does not correspond to the Greek γίγαντας. This different reading is hard to account for, especially if we consider that the presence of the giants is confirmed by comparison with the Enochian literature. We might suppose a misreading by the Syriac translator, or a corruption in the transmission of the text. A simple correction would be ܥܒܕܐ ('*abdo*) "servants," but this term would be, in any case, quite different to γίγας (see Rubens Duval, *Lexicon Syriacum auctore Hassano bar Bahlule*, 3 vols. (Paris: Imprimerie Nationale, 1888–1901), vol. 1, 482, *s.v.* ܓܢܒܪܐ).
[356] French translation in *CMA* II 238–9.
[357] SyrC, fol. 50[r-v]; French translation in *CMA* II 239.
[358] Michèle Mertens dates the work to the second/third century AD: Mertens, "Une scène d'initiation alchimique: la *Lettre d'Isis à Horus*," *Revue de l'histoire des religions*, 205 (1988): 4. Some parts of this text are quoted by Zosimus (according to Syncell. 14 Mosshammer and Olymp. Alch. CAAG II 89), who ascribed them to Hermes (see Festugière, *Révélation d'Hermès*, vol. 1, 254–5).
[359] See *CAAG* II 29,8 and 33,17–22 (see also Festugière, *Révélation d'Hermès*, vol. 1, 256–7). In Isis' work two angels are attracted by the beauty of the goddess; the name of the first is not mentioned, while the second one is called Amnaël ('Αμναήλ). This name is not attested in the extant remains of the treatises ascribed to Enoch (see the following note). According to Michèle Mertens, "Pourquoi Isis est-elle appelée 'prophetis'?," *Chronique d'Égypte*, 64 (1989): 391, Amnaël could correspond with the angel Anaël ('Αναήλ) associated in astrological manuscripts with the planet Venus; see also M. Mertens, *Un traité gréco-égyptien d'alchimie: la 'Lettre d'Isis à Horus'* (Liège: Mémoire de licence inédit, Université de Liège, 1984), 78–82.
[360] The edition and analysis of the manuscripts discovered in the Qumran caves has confirmed that part of the literature handed down under the name of the Biblical patriarch Enoch dates back to the third century BC. The legend of the fallen angels who revealed the arts to mankind is attested in several versions of the so called *Book of Enoch* (or *1Enoch*; preserved in Aramaic, Greek, and Ethiopic, especially in the first part usually referred as the *Book of Watchers*. Within this account, the angel Azaël ('Αζαήλ) is particularly interesting for our inquiry. The Greek version reads: *Apocalypsis Henochi* VIII 1 Black [in Matthew Black and Albert-Marie Denis, *Pseudepigrapha Veteris Testamenti Graece* (Leiden: Brill, 1970), 22] ἐδίδαξεν τοὺς ἀνθρώπους 'Αζαὴλ μαχαίρας ποιεῖν καὶ ὅπλα καὶ ἀσπίδας καὶ θώρακας, διδάγματα ἀγγέλων, καὶ ὑπέδειξεν αὐτοῖς τὰ μέταλλα καὶ τὴν ἐργασίαν αὐτῶν, καὶ ψέλια καὶ κόσμους καὶ στίβεις καὶ καλλιβλέφαρον καὶ παντοίους λίθους ἐκλεκτοὺς καὶ τὰ βαφικά (≈ Syncell. 12,13–8 Mosshammer), "Azaël taught men to make swords, weapons, shields and armour – an angelic teaching – and showed them minerals (perhaps metals) and how to work them, and (he showed them) armlets, *parures*, the *fard*, the make up for eyes, every kind of choice stone and colouring tincture." For the Aramaic version, see Jazef T. Milik, *The Books of Enoch. Aramaic Fragments of Qumrân Cave 4* (Oxford: Clarendon Press, 1976), 168–9.

These sources seem to witness the coexistence of two different views of alchemy. The first is a broader conception that includes four different areas of expertise related to dyeing, while the second, narrower view focuses on the making of precious metals. However, this duality is not restricted to alchemy's early stages. The adjective *chymeutikos* (χυμευτικός) is also employed in the *Corpus alchemicum* to describe dyeing processes for precious stones.[361] A later fragment preserved in **A** states that:

> The present book is called [the] book on metals <and> alchemy, on the making of gold, on the making of silver, on the solidification of mercury – it includes vapours, tinctures and impressions[362] on lead[363] – as well as emeralds, rubies and all the other colours, and pearls, and the red dyeing of royal leathers. All these [techniques?] are based on sea water, on eggs, according to the metallurgical art.[364]

In **A** this section immediately follows a treatise by the philosopher Anonymous, corresponding to the work handed down by **M** (fols. 181ʳ–184ʳ) under the simple title of "by the philosopher Anonymous."[365] Manuscript **A** preserves the treatise in an incomplete and interpolated form, and Berthelot-Ruelle, who followed **A** without taking **M** into account, incorrectly published it in the section devoted to Zosimus' writings. But, as Berthelot himself noted, the above quoted passage post-dates Zosimus by many centuries, and actually represents the title or the preface of "a Byzantine manual of chemistry" dating to between the eighth and the tenth century.[366] The passage in the manuscript **A** is, indeed, clearly separated from the previous work by a *dikolon* (:), usually employed by the copyist to mark a break between different sections.

The original content of such a Byzantine handbook is difficult to reconstruct. On the basis of the topics listed in the surviving preface, Berthelot suggested that it included several recipes on the working of precious metals and stones, the making of glass and artificial pearls, and purple dyeing. Such a collection might provide a possible source for the technical sections preserved in **A** and **B**[367] and, to a lesser extent, in **M**. This interpretation is certainly tempting, particularly since it posits the existence of a recipe book covering the same areas of expertise encountered in medieval collections, such as the *Mappae clavicula*, but unfortunately we know of no source to support it. At all events, the title allows us to infer that, even during

---

[361] See *CAAG* II 353,19 and 26.

[362] The term φούρμα, stemming from the Latin *forma*, is always attested in the plural by some later alchemical recipe-books. According to *CAAG* II 326,12–26, it would refer to the impression left on a melted metal (see also *CAAG* II 375,11–376,24): see *GMIG* II 1697, *s.v.* φούρμα, equivalent of the Greek τυπάριον, "small figure, image." The term already appears in different forms (φόρμη, φούρμη, φώρμη) in the third-fourth century AD (LSJ⁹ 1951, *s.v.* φόρμη).

[363] The term βροντήσιον indicates a specific kind of bronze, whose making is described in *CAAG* II 376,25–377,6. See also Stéphanidès, "Petites contributions," 202–3.

[364] Translation based on the revised Greek text published in Martelli, *Pseudo-Democrito*, 71 (= A 240ʳ24–240ᵛ5; *CAAG* II 220,11–6).

[365] M, fol. 181ʳ1 Ἀνεπιγράφου φιλοσόφου. This treatise is titled *On the Making of Gold* (Περὶ χρυσοποιίας) in the index of the manuscript (M, fol. 2ᵛ5). See *supra*, n. 5 and Letrouit, "Chronologie," 63.

[366] See *CAAG* III 360. We must note that this preface uses a late vocabulary that recalls other Byzantine texts in the *Corpus alchemicum*, including recipes edited in *CAAG* II 326, 375 and 377–9.

[367] See in particular the section on making precious stones, coloured glasses and artificial pearls, preserved in B, fols. 152ᵛ–173ʳ and A, fols. 141ʳ–159ʳ.

the Byzantine period, a broad range of technical competence could be treated within a book titled *The Book on Metals and Alchemy* (βίβλος μεταλλικὴ <καὶ> χυμευτική), including them, the same crafts covered by the *Four Books* of ps.-Democritus, and confirming the continuing vitality of a more diverse conception of alchemy that included a wider range of dyeing techniques.

§ 12. The Egyptian background

Several elements in the extant parts of the *Four Books* position the work in relation to Egypt. First, ps.-Democritus himself distinctly states, probably at the very start of his book on gold-making (*PM* § 4), that "I too have come to Egypt to deal with natural substances, so that you may disregard many captious questions and the confused matter." In Egypt he could indeed have made his teaching available, aiming to organize and understand craft traditions that he perceived as incoherent and preoccupied with marginal questions.[368] At the end of the same book, the author again mentions Egypt by introducing the alchemist Pammenes. Unfortunately, this passage is somewhat uncertain, especially when compared with the Syriac tradition that preserves quite a different text. In the two versions the excerpt reads:

This is ⟨the method⟩ of Pammenes, which he taught the Egyptian priests. The matter for the making of gold extends up to these natural substances.[369]	Since it overcomes nature, this [nature?] is greatest that I taught the Egyptians, after their priests came and made me swear to teach them about the power of this book. In fact many are the substances for the yellow [i.e. for making gold] and for the white [i.e. for making silver].[370]

We know from Zosimus' account (*CAAG* II 148,15–6) that the method ascribed to Pammenes by ps.-Democritus comprised a specific treatment of lead that probably coincides with the technique described in *PM* § 19. According to *PM* § 20, Pammenes himself explained this treatment to Egyptian priests. However, the abovementioned § 4 and the Syriac tradition (which gives ܐܠܦܬܐܠܘܢ, "that I taught") may lead us to correct the Greek ἐπεδείξατο ("he taught") into ἐπεδειξάμην ("I taught"). According to this reading, ps.-Democritus himself would have taught this specific technical procedure, taken from the alchemist Pammenes, to the Egyptian priests – although the coherence of the Byzantine manuscripts in this point requires us to be circumspect regarding this possibility.[371] Either way, both versions point to Egypt and its temples as the site where the two alchemists expounded their alchemical methods to the priests.

Finally, we may infer a third element linking ps.-Democritus to the land of the Pharoahs. In *PM* § 15 (l. 155), Democritus addresses his interlocutors as ō *symprophētai* (ὦ συμπροφῆται). The title *prophētēs* is often bestowed on ancient

---

[368] See also Martelli, *Pseudo-Democrito*, 135–48.
[369] *PM* § 20, ll. 213–5.
[370] *1SyrC* § 16, ll. 1–3. I have noted the different readings of the London Syriac manuscripts in the footnotes to the edition of the Syriac text.
[371] For a deeper discussion of this passage, see *1SyrC* § 16 nn. 29–30.

alchemists, including Chymes, Isis, and Moses, so it is not surprising that ps.-Democritus adopts this term for the group of scholars he regarded as peers.[372] Indeed, during the Imperial age, *prophētēs* meant "no longer just 'interpreter of oracles' alongside *mantis* (μάντις) as was the case in Classical Greece ... nor 'member of the highest priestly class,' as in Hellenistic Egypt,[373] but rather 'revealer of every truth in immediate contact with its god.'"[374]

Democritus himself is called *prophētēs* by two later sources that emphasize his connection to Egypt. Zosimus, in the excerpt entitled *First Book of the Final Quittance*,[375] claims that in Egypt the mining and working of precious metals (especially for minting coins) is strictly controlled by the kings, and jealously guarded by the priests. Democritus and other alchemists were forced to keep their knowledge secret, "since they were friends of Egyptian kings and held a high position among *prophētai*."[376] In a passage by the philosopher Anonymous, ps.-Democritus himself addresses his teaching to the (presumably Egyptian) kings, *prophētai*, and priests:

> Democritus addressing himself to us and to the kings said: 'You must know this point, o king, sovereigns, priests and prophets, that if you do not gain a deep understanding of substances ... you will work in vain and labour without any profit.'[377]

This passage of ps.-Democritus, even if quoted by a later author, places a strong emphasis on the knowledge of natural substances, recalling some passages from the ps.-Democritean excerpts preserved in Byzantine manuscripts. The influential figures invoked in the first line also belong to the same historical and cultural *milieu* as that described by Zosimus.[378]

Such details have aroused the suspicion of some scholars who point to the lack of clear evidence to support the supposed Egyptian origin of alchemy.[379] For instance, Fowden has drawn attention to the standardized form of such narratives, yet insists

---

[372] *CAAG* II 183,22; II 28,20–1 and 33,7; II 353,19.

[373] See also Mertens, "Pourquoi Isis," 260–1.

[374] Andrè-Jean Festugière, "L'arétalogie isiaque de la Korè Kosmou," in *Mélanges d'archéologie et d'histoire offerts à Charles Picard*, 2 vols. (Paris: Presse universitaire de France, 1949), vol. 1, 380; my translation from the original French. In addition, also some magicians mentioned in papyri are called προφῆται and the same title was probably attributed to the anonymous authors of botanical writings (perhaps herbals) which started circulating during the Hellenistic period and which are mentioned by Dioscorides, Pliny the Elder and ps.-Apuleius: Mertens, "Pourquoi Isis," 264–5; Boscherini, "L'Erbario di Apuleio," 113–8.

[375] *CAAG* II 239–46. New edition by Festugière, *Révélation d'Hermès*, vol. 1, 363–8.

[376] *CAAG* II 240,12 = Festugière, *Révélation d'Hermès*, vol. 1, 364, ll. 24–5: φίλοι ὄντες τῶν βασιλέων Αἰγυπτίου καὶ τὰ πρωτεῖα ἐν προφητικῇ αὐχοῦντες.

[377] Tranlastion based on the revised Greek text published in Martelli, *Pseudo-Democrito*, 161–2 (= *CAAG* II 427,2–6).

[378] See also the passage attributed by Berthelot-Ruelle to Zosimus (on the problems of its authenticity, see Letrouit, "Chronologie," 36), where Democritus is again presented as adressing his teaching to Egyptian priests (*CAAG* II 158,3): καὶ ταῦτα μὲν οὕτως πρὸς τοὺς Αἰγυπτίους προφήτας γράφει ὁ Δημόκριτος κτλ.

[379] See, for instance, Bidez-Cumont, *Mages*, vol. 1, 205: "Tandis que s'accrédite et se répand, malgré les hésitations des spécialistes, l'idée que l'alchimie est née en Égypte, les restes de notre apocryphe, dans leur ensemble, serviront à montrer que, si courante que soit cette opinion, née à l'ombre des pyramides, elle doit son prestige à un préjugé." Both scholars admit, however, that the *Four Books* of ps.-Democritus were composed in Egypt (Bidez-Cumont, *Mages*, vol. 1, 199).

on an important distinction between the so-called philosophical Hermetic texts and the technical ones:

> The various references made by the philosophical *Hermetica* to priests, conversations in temples and so forth strike one, it is true, as more decorative than essential; and the genre itself is unrelated to anything we know about the Thoth literature. But the technical Hermetica are related to Thoth literature ... and their allusions to the priestly milieu compel more attention.[380]

Moreover, we should bear in mind that the abovementioned alchemical passages allude to several elements that seem to have been characteristic of Egyptian craftsmanship and metallurgy since Pharaonic times. As stressed by Aufrère, the mining of metallic ores was strictly controlled by the Egyptian kings, since digging into the mountains was perceived as the violation of a sacred space permitted only to a select *équipe* of experts, probably belonging to a group of leading dignitaries.[381] Particular craftsmen, who previously worked in Egyptian temples alongside the priests, were also responsible for making ritual statues crafted from precious metals and stones—sacred objects that were afterwards consecrated and animated by means of the 'opening of the mouth' ritual. Traunecker identified an example of the sacred space where this ritual was performed (the *Château de l'or*) in a room built by Pharaoh Thutmosis II during the renewal of the Temple of Amun in Karnak.[382] On the basis of his reading of the hieroglyphics engraved on the walls of this room, Traunecker suggested that both craftsmen and priests were initially involved in the ritual.[383]

Although ps.-Democritus and Zosimus lived many centuries after these early sources, we should note the persistence of similar traditions into later times. For instance, the hieroglyphic inscriptions that decorate a room in the Dendera temple complex, probably built in the first century BC or first century AD, preserve excerpts from an earlier technical text dealing with the making of statues and some procedures for gilding them. According to Derchain's translation, the first part of the inscription lists the following craftsmen and experts engaged in making idols for the *Château de l'or*:

> Experts in preparing moulds (for casting statues), 2 people; engravers, 2 people; inlayers, 2 people; stonecutters, 2 people; sculptors, 2 people; jewellers, 2 people. In all, 12 people working for a month; or 48 people in total. They have not been initiated in front of the god. They cause the statues to come into existence ... They will cast all the gold and silver

[380] Fowden, *Egyptian Hermes*, 166.

[381] Sidney H. Aufrère, *L'univers minéral dans la pensée égyptienne*, 2 vols. (Cairo: IFAO, 1991), vol. 1, 59–82. Egyptian military control over mining and the working of precious metals – explicitly mentioned by Zosimus in the abovementioned passage – is confirmed by some passages of Agatharchides, preserved by Photius (*Bibl. cod.* 250, 447b6–449a10 and 457b35–458b1); the same passages are also preserved in the alchemical manuscripts *Marcianus gr.* 299 and *Parisinus gr.* 2327 (see Letrouit, "Chronologie," 66–8) and by Diodorus Siculus III 12,1–6; see Robert Halleux Halleux, "L'affinage de l'or, des origines aux premiers alchimistes," *Janus*, 62 (1975): 79–102.

[382] Claude Traunecker, "Le Château de l'Or de Thoutmosis III et les magasins nord du temple d'Amon," *CRIPEL* 11 (1989): 89–111; it is the second room of the north part of the complex.

[383] Traunecker, "Le Château de l'Or," 108.

jewellery with precious stones, which have to touch the divine bodies. But, as far as the
very secret work is concerned, it will involve those officiants who have been initiated in
front of the god, those who are part of the clergy.[384]

The passage draws an interesting distinction between those who were responsible
for making ritual objects and the priests who alone had access to the *Château de
l'or*.[385] Despite their different duties, craftsmen were aware of the sacred nature of
their work, as suggested by the handbooks engraved on temple walls.[386] In the
second/third century AD, Clement of Alexandria still recalled the ancient procedures
for making Egyptian statues.[387] A century later, Zosimus mentioned the Egyptian
priest Nilus who knew and practised metallurgical techniques and, following the
Pharaonic religion, still believed, together with his pupils, that the statues were
animated.[388]

This tradition may also inform the legend that relates the Greek philosopher Democritus to the kings and priests of Egypt. Such beliefs seem to underlie the development
of the notion of a "sacred and holy art" (ἱερὰ καὶ θεία τέχνη), which should be learned
and preserved but also kept secret from the uninitiated.[389] The close connection of the
Egyptian *technai* with the sacred and religious *milieu* of the temples and sanctuaries
may have helped ennoble traditional craft techniques, driving authors such as
ps.-Democritus to investigate a wide range of dyeing techniques, and to recognise in
them the action of a nature (φύσις) that was considered in some way divine.

We should also recall that the kinds of technical expertise covered by the *Four
Books* were already well developed in Egypt's long-established metallurgical tradition.[390] During the Hellenistic and Imperial periods, Egypt developed as a
major cultural crossroads and trading centre for precious raw materials and luxuries

---

[384] English text based on Derchain's French translation of the inscription; see Philippe Derchain, "L'Atelier des Orfèvres
à Dendara et les origines de l'Alchimie," *Chronique d'Égypte*, 65 (1990): 233–4.

[385] François Daumas, "Quelques textes de l'Atelier des Orfèvres dans le temple de Dendara," in *Livre du centenaire:
1880–1980* (Cairo: IFAO, 1980), 115–7.

[386] The inscription provides also a kind of explanatory commentary on an older handbook describing how to make
ritual objects and statues. According to Derchain's translation (Derchain, "L'Atelier des Orfèvres," 235–6), the
first part reads: "S'il dit |i.e. the older handbook| d'un dieu que la matière en est le bois et l'or, sans préciser le
nom du bois, il veut dire que c'est du jujubier, plaqué d'or fin ... S'il dit d'un dieu que la matière en est le cuivre,
il veut dire que c'est du bronze noir |see *infra*, p. 67|. S'il dit d'un dieu que la matière en est l'électrum, il veut
dire que c'est du bois – ce bois c'est le jujubier – plaqué d'or fin. S'il dit d'un dieu que la matière en est l'or fin, il
veut dire que l'intérieur en est d'argent et, pareillement (à la notice précédente), le placage d'or fin. S'il dit d'un
dieu que le placage en est d'or fin, c'est que ce placage est d'or de l'épaisseur d'une (coquille) d'œuf d'ibis." In
addition, Aufrère, *L'univers minéral*, vol. 1, 330–5, mentions a hiéroglyphic recipe for the making of a black ointment, engraved on a wall of the Edfu temple. The ointment – the composition of which was kept secret – was supposed to animate the statues.

[387] *Prot.* IV 48,4–6. See Beretta, *Alchemy of Glass*, 45.

[388] See, in particular, *CAAG* II 191,3–18, where Nilus is said to have performed a procedure for dyeing a Cu-Pb alloy by
means of the minerals of κωβάθια (perhaps arsenic sulphides). Zosimus' writings preserved in the *Corpus Syriacum*
include more information on this priest (see, for instance, *CMA* II 228). In these works, Zosimus attributed several
metallurgical techniques to Egyptian priests: see *CMA* II 226 and 228. In VII 8–10 Mertens, he also claimed to have
seen a goldsmith's oven in the temple of Memphis.

[389] The *Corpus alchemicum* usually refers to alchemy in this way.

[390] See Berthelot, *Origines de l'alchimie*, 29; Alfred Lucas and John Richard Harris, *Ancient Egyptian Materials and
Industries*, 4th ed. (London: E. Arnold, 1962), 195–269; Paul T. Nicholson and Ian Shaw, *Ancient Egyptian
Materials and Technology* (Cambridge: Cambridge University Press, 2000), 148–76.

from the Middle and Far East.[391] Along with these goods, experts in diverse arts also travelled to Egypt.[392] In this regard the Leiden and the Stockholm papyri, which scholars have often viewed as handbooks addressed to Egyptian craftsmen, are important sources that offer an overview of the various crafts known and practised in Egypt during the first centuries AD.[393]

With regard to metallurgy, various methods are attested for producing coloured metals or alloys, especially specific surface treatments that produced various colours through artificial patination processes. Such techniques were probably used to make the "red or purple gold" characteristic of some goods found in Tutankhamun's tomb,[394] and were almost certainly employed in producing the so-called "black bronze,"[395] a particular black, patinated copper-gold-silver alloy already known during the Middle Kingdom as ḥmty km (lit. "black copper")[396] and later evoked by Zosimus in his sixth book.[397] Another black alloy, commonly called niello, is explicitly attributed to Egyptian craftsmen by Pliny the Elder: "The people of Egypt stain their silver so as to see portraits of their god Anubis in their vessels, and they do not engrave but paint their silver."[398]

The procedure consists of roasting a silver-copper alloy with sulphur: the alloy is darkened by the production of silver and copper sulphides on its surface. Other

---

[391] An important source of information about the trade between India and some areas around the Red Sea is the first-century treatise *Periplus Maris Erythraei*. Ch. 6 mentions, among other products imported to the harbour of *Adylis*, the λάκκος χρωμάτινος, possibly to be identified with the ingredient λακχά quoted by ps.-Democritus (*PM* § 1 n. 7). Halleux (*Papyrus de Leyde*, 49) also notes that *PHolm*. 47 (= 83) describes a technique for changing quartz into beryl by means of Indian indigo (ινδικόν), which was also attributed to *Indi* by Pliny, *NH* XXXVII 79.

[392] See Halleux, *Papyrus de Leyde*, 30; Carlo Zaccagnini, "Patterns of Mobility among Ancient Near Eastern Craftsmen," *Journal of Near Eastern Studies*, 42 (1983): 245–64, showed that several craftsmen in the Middle East were already moving between regions in the second millennium BC.

[393] See, for instance, *CAAG* I 3–6; Holmyard, *Makers of Chemistry*, 30 ("They appear to be the recipe-books of an Egyptian chemist"); Partington, *Short History of Chemistry*, 17 ("It is possible that the work represents the note-book of a fraudulent goldsmith"); Leslie B. Hunt, "The Oldest Metallurgical Handbook. Recipes of a Fourth Century Goldsmith," *Gold Bulletin*, 9 (1976): 24–7. For a more cautious position, see Halleux, *Papyrus de Leyde*, 53–6.

[394] The golden goods are covered by a thin purple layer made by processing them with ochre, orpiment and ferrous pyrites. Similar results were also attained by making specific metallic alloys of gold with a small amount of copper: see Robert W. Wood, "The Purple Gold of Tut'ankhamūn," *Journal of Egyptian Archaeology*, 20 (1934): 62–5; Deborah Schorsch, "Precious-Metal Polychromy in Egypt in the Time of Tutankhamun," *Journal of Egyptian Archeology*, 87 (2001): 67–9. We should also note that ps.-Democritus several times mentions specific treatments for giving such colours to gold (called χρυσοκόραλλος or χρυσοκογχύλιον in the *PM*).

[395] See Paul T. Craddock, "Gold in Antique Copper Alloys," *Gold Bulletin*, 15 (1982): 69–72; P.T. Craddock and Alessandra Giumlia-Mair, "Ḥsmn-Km, Corinthian Bronze, Shakudo: Black-Patinated Bronze in the Ancient Word," in *Metal Plating and Patination: Cultural, Technical and Historical Developments*, ed. Paul T. Craddock and Susan La Niece (Oxford: Butterworth-Heinemann Ltd., 1993), 102–27. The identification of "black copper" with the famous "Corinthian bronze" proposed by the two scholars has been questioned: see in particular David M. Jacobson and Michael P. Weitzman, "Black Bronze and 'Corinthian Alloy,'" *The Classical Quarterly*, 45 (1995): 580–3.

[396] See Alessandra Giumlia-Mair and Stephen Quirke, "Black Copper in Bronze Age Egypt," *Revue d'égyptologie*, 48 (1997): 95–108 with bibliography.

[397] As preserved in the Syriac manuscript Mm. 6.29 (fols. 33x^{r}5–33^{v}2). Edition of the Syriac recipes by Erica C.D. Hunter, "Beautiful Black Bronzes: Zosimos' Treatises in Cam. Mm.6.29," in *I bronzi antichi: produzione e tecnologia. Atti de XV Congresso internazionale sui bronzi antichi organizzato dall'Università di Udine, sede di Gorizia, Grado-Aquileia, 22–26 maggio 2001*, ed. Alessandra Giumlia-Mair (Montagnac: Monique Mergoil, 2002), 655–9.

[398] Pliny. *NH* XXXIII 131, translation by Harris Rackham, *Pliny, Natural History, Books 33–35* (Cambridge, MA and London: Loeb, 1971), 99.

processes for gilding metals (as well as other substances, such as stones and wood) by applying thin golden leaves had been well known since the Pharaonic period.[399] Recent studies of the Egyptian collection of the Cleveland Museum of Art have stressed the role of gypsum in glueing the leaves to their support, while Aufrère mentions egg white.[400]

Procedures for the purple dyeing of wool are also well covered by the Leiden and Stockholm papyri, and have been connected to the fashion for dyed clothes that became popular in Egypt between the third and eighth century.[401] The *Four Books* list several dyes that are quite similar to the ingredients cited by the two papyri, enabling us to trace this trend back to at least the first century AD. Pliny the Elder mentions the expertise of Egyptian dyers, who developed specific mordant procedures.[402] By treating white cloth with different mordants, they could obtain different shades even if the cloth was plunged into a cauldron of the same dye.[403] Processes for dyeing textiles were already known in Pharaonic Egypt, where linen cloth was the first to be dyed, although woollen textiles also seem to have been used.[404] Ps.-Democritus and the Leiden and Stockholm papyri give particular emphasis to wool as the basis for the dyeing techniques they describe.

Finally, the various procedures for making precious stones, again well attested by the two papyri, would have been similar, according to Halleux, to "the old Egyptian method of glazing quartz"; these techniques operated below the melting point of the stones and simply coated them with a thin, coloured layer.[405] Such techniques may be related to the ancient tradition of working with vitreous paste that was long established in Egypt. The expertise of Egyptian glassmakers has been recently

---

[399] The dyeing of metallic πέταλα is also well attested in *PM* §§ 17–18 and *AP* §§ 6–9. Among the craftsmen mentioned by the Dendera inscription (see *supra*, pp. 65–6), we find also "inlayers, 2 people." Based on analysis of several documentary papyri, Burkhalter noted that the making of gold leaves (πετάλωσις), used to decorate statues and parts of the temple, was an occupation of Egyptian goldsmiths (called χρυσοχόοι by the papyri) during the Graeco-Roman period; see Fabienne Burkhalter, "La production des objets en métal (or, argent, bronze) en Égypte hellénistique et romaine à travers les sources papyrologiques," in *B.C.H. suppl.* 33: *Commerce et artisanat dans l'Alexandrie hellénistique et romaine. Actes du colloque d'Athènes, 11–12 décembre 1988*, ed. Jean-Yves Empereur (Athens: École française d'Athènes, 1998), 125–33.

[400] Patricia S. Griffin, "The Selective Use of Gilding on Egyptian Polychromed Bronzes," in *Gilded Metals. History, Technology and Conservation*, ed. Terry Drayman-Weisser (London: Archetype Publ., 2000), 49–72, and Aufrère, *L'univers minéral*, vol. 1, 377, respectively.

[401] See Halleux, *Papyrus de Leyde*, 43.

[402] Pliny. *NH* XXXV 150.

[403] Enrico Renna, "Ricette per i succedanei della porpora in due papiri greci," in *La porpora. Realtà ed immaginario di un colore simbolico*, ed. Oddone Longo (Venice: Istituto veneto di scienze, lettere ed arti, 1998) 136 related this passage to the different kinds of mordant mentioned and listed in the Leiden and Stockholm papyri (see, for instance, the catalogue of *PLeid.*X. 92).

[404] See Lucas-Harris, *Ancient Egyptian Materials*, 150–4; see Grace Mary Crowfoot and Norman de Gary Davies, "The Tunic of Tut'ankhamūn," *Journal of Egyptian Archaeology*, 37 (1941): 113–30, for a close analysis of textiles founded in Tutankhamun's tomb, and their comparison with other archaeological evidence. The notion that ancient Egyptians could not use wool because of their religious beliefs (developed by scholars on the basis of few Greek sources, *in primis* Hdt. II 82) is now seriously questioned: see Nicholson-Shaw, *Ancient Egyptian Materials*, 269. Joachim F. Quack, "Les Mages Égyptianisés? Remarks on Some Surprising Points in Supposedly Magusean Texts," *Journal of Near Eastern Studies*, 64 (2006): 280, comments, "I have even found text fragments from a Demotic technical treatise on the dyeing of textiles ascribed to the Egyptian god Ptah."

[405] Halleux, *Papyrus de Leyde*, 49.

stressed by Beretta, who recognizes in this tradition one of the important technical components that contributed to the early development of alchemy.[406]

§ 13. Persian elements: ps.-Democritus and Ostanes

An Egyptian setting also frames the account of Democritus' initiation by the Persian magus Ostanes, at least in the paragraph that opens Synesius' commentary, which represents the first clear expression of the legend within the *Corpus alchemicum*:

> Therefore we set forth right now to say who was that famous man, the philosopher Democritus: he came from Abdera and as a natural philosopher he investigated all natural questions and composed writings about all natural phenomena. Abdera is a Thracian city, but he became a very wise man when he went to Egypt, and was initiated in the temple of Memphis along with all the Egyptian priests by the great Ostanes. He took his basic principles from him and composed four books on dyeing, on gold, silver, [precious] stones and purple. I stress this point: he wrote by taking his basic principles from the great Ostanes. For he was the first to write that nature delights in nature, and nature masters nature, and nature conquers nature, and so on.[407]

The same information is also given by the Byzantine chronographer Syncellus, who probably knew a *Corpus* of alchemical writings quite similar to the anthology preserved for us in the Byzantine manuscripts:[408]

> Democritus of Abdera, the natural philosopher, was flourishing. In Egypt, Democritus was initiated into the mysteries by Ostanes the Mede, who had been dispatched to Egypt by the Persian kings of that time to take charge of the temples in Egypt. He was initiated in the temple of Memphis along with other priests and philosophers, among them a Hebrew woman of learning named Mariam, and Pammenes. Democritus wrote about gold and silver, and stones and purple, but in an oblique way.[409]

The relationship of such accounts to the *Four Books* is not entirely clear. Various scholars have supposed that this alchemical work probably opened with such a story, and they consider Synesius' and Syncellus' versions to be in some way related to *PM* § 3.[410] However, as already noted, the absence of Ostanes' name and the presence of various elements that are not mentioned either by Synesius or Syncellus leaves some room for doubt concerning the authenticity of the *PM* paragraph.[411] On the basis of such questions, some scholars have proposed that Synesius was himself the first author to introduce the legend; in particular, Hammer-Jensen

---

[406] Beretta, *Alchemy of Glass*, 8–22 (Ancient Egypt) and 40–7 (Hellenistic Egypt).

[407] Syn. Alch. § 1, ll. 5–17.

[408] Riess, *RE*, s. v. "Alchemie," 1341, 33ff.

[409] Syncell. 297, 24–8 Mosshammer (= 68 |55| 300,16 DK); transl. by Adler-Tuffin, *Chronography of George Synkellos*, 361.The passage has been edited also by Bidez-Cumont, *Mages*, vol. 2, 311 fr. A3, on the basis of the manuscripts *Parisini Graeci* 1711, fol. 147 and 1764, fol. 93. The two scholars stressed the strong similarities with Synesius' account, so that they supposed a possible relation between the two sources (p. 311 n. 1).

[410] For instance, Tannery, "Études sur les alchimistes grecs," 283.

[411] See *supra*, p. 20.

suggested that the commentator could have taken the story from *PM* § 20.[412] Hammer-Jensen actually proposed correcting the name of the alchemist Pammenēs, mentioned in *PM* § 20, l. 123, to *pammegethēs* ("greatest, most important," παμμεγέθης), treating this as a kind of title that ps.-Democritus bestowed upon himself at the end of his book on gold-making.[413] Synesius, however, would have read the adjective as an allusion to Ostanes, as can be inferred from the frequency of the adjective "great" (μέγας) used in Synesius' dialogue to describe the Persian magus.[414] Such an exegesis would have been suggested to Synesius by the various legends about the supposed relationships between the atomist philosopher Democritus and the Persian magus Ostanes which had already been circulating in the Hellenistic period, and which Synesius would have read as referring to alchemy.

Although the available sources do not allow us to answer the question definitively, particularly given the lack of any reference to Ostanes in the *Four Books*, two important elements seem to have played a key role in the origin of the legend of Democritus as a pupil of Ostanes. Alchemical sources clearly stated that Democritus took the aphorism about the power of natures from Ostanes, who is said by Synesius to have first discovered and formulated it. However, Democritus' dependence upon the Persian magus had also been previously underscored by Zosimus, who gave different details of their supposed teacher–pupil relationship in a passage from the excerpt *On the Body of Magnesia and its Treatment*:

What does Ostanes want to say? In fact he speaks about the mixture of the substances that flee [i.e. that volatilize] and that do not flee: 'Once more pyrite stone has affinity with copper.' Ostanes indeed did not speak of mercury, but of complete dissolution, so that when dissolved it [the pyrite stone?] did not have any residue, but all its parts became completely water. You must understand the species of waters and how to dissolve into them; the philosopher [i.e. Democritus] in his [section on] washing and dissolving processes clearly explained how to dissolve substances by saying: 'In order that it becomes water.' And the philosopher says again: '*Magnēsia* and magnetite have an affinity with iron.'[415] The teacher [i.e. Ostanes] says again: 'Once more mercury has an affinity with tin.' And the pupil says: 'Mercury makes an amalgam with tin'; and he says: 'This whitens any [metallic] body';[416] 'Once more lead has an affinity with pyrite; the annual stone (?)[417] with lead.' Imitating these statements, the philosopher said about our art: 'Nature delights in nature.'[418]

---

[412] Hammer Jensen, *Die älteste Alchymie*, 88.

[413] See *1SyrC* § 16 n. 29.

[414] Vereno, *Studien zum ältesten alchemistischen Schrifttum*, 58 and 95, agreed with Hammer-Jensen's interpretation, although he did not accept the correction of Παμμένης; in his opinion, Synesius misread the name by interpreting it as the adjective πάμμεγας (or παμμεγέθης), which he applied to the magus Ostanes.

[415] Ps.-Dem. Alch. *AP* § 1, l. 7–8.

[416] Ps.-Dem. Alch. *AP* § 2, ll. 13–5.

[417] The substance called ὁ λίθος ὁ ἐτήσιος is often mentioned within the *Corpus alchemicum*: it was supposed to be a special stone that has been related by Bidez-Cumont (*Mages*, vol. 2, 323 n. 3) to the so-called philosophers' stone. On the interpretation of the adjective ἐτήσιος, see Michel Stéphanidès, "Notes sur les textes chymeutiques," *Revue des études grecs*, 35 (1922): 311.

[418] Translation based on the revised Greek text published in Martelli, *Pseudo-Democrito*, 168–9 (= *CAAG* II 197,5–18). See also Bidez-Cumont, *Mages*, vol. 2, 322 fr. A9.

Zosimus' passage confirms the teacher (διδάσκαλος) and pupil (φοητής) relationship between the Persian magus and the Greek philosopher by emphasising Democritus' intellectual debt to Ostanes, from whom he derived the theoretical background for his four books. Zosimus actually compares several sentences from the works of the two authors, highlighting the close similarities in the vocabulary and formulae used by each. These connections underscore the role of an affinity (συγγένεια) linking the nature (φύσις) of different ingredients, which provides the basis of the aphorism associated with both alchemists. Similar references to this affinity recur within the books of ps.-Democritus,[419] although Zosimus' quotations are not attested in any works of Ostanes preserved in Byzantine manuscripts. Despite this lacuna, the abovementioned passages indicate a close relationship between ps.-Democritus' inquiries, concerned with the identification of a single nature underlying the plurality of substances used by alchemists, and Ostanes' writings that emphasised the affinities between different ingredients.[420] This relationship cannot be reduced to Democritus' straightforward quotation of the aphorism on nature, usually ascribed to Ostanes. In addition to the passages from Synesius and Syncellus, we must also consider the explanation given by Zosimus that the aphorism was the formalization of a broader inquiry into the properties and interactions of natural substances that underpinned the alchemical work of both Democritus and Ostanes.

Further evidence suggests that ps.-Democritus inserted quotations from Ostanes into his own works. Zosimus mentions some words that Democritus took from the Persian magus.[421] Synesius' commentary refers to two longer sentences of Ostanes quoted by the Greek philosopher: the first concerning the secrecy of alchemical teaching, and the second dealing with the dissolution of solid substances useful for making the dyeing "drug" (φάρμακον).[422] The focus on grinding and dissolving processes is also apparent in the Zosimus passage, which seems to support the central role of such topics in defining relationships between Ostanes' and ps.-Democritus' alchemical writings.[423] Although the fragmentary status of the sources keeps us from understanding the precise extent of this dependency, the abovementioned elements do suggest that Democritus took both theoretical and practical principles from the Persian magus, from whose writings he also quoted some passages in parts of the *Four Books* that are now lost.

Their relationship received more complete treatment from Synesius and Syncellus, who situated Democritus' initiation in the Egyptian temple of Memphis. The origin

---

[419] Particularly in the books on the making of silver, as already seen in the translation of Zosimus' passage; see also Ps.-Dem. Alch. *PM* § 17, ll. 195–6 and *AP* § 4, ll. 38–9.

[420] This feature of the *Four Books* is more extensively analysed in Martelli, *Pseudo-Democrito*, 135–48.

[421] Zos. Alch. *CAAG* II 148,12–3: διὰ τοῦτο καὶ τὸν διδάσκαλον (*i.e.* Ὀστάνην) φάσκει (*scil.* Δημόκριτος) λέγοντα· πάσας τὰς οὐσίας βάπτοντα, "that is why he (i.e. Democritus) says that also his master claimed: |the ingredients?| which dye all the substances."

[422] Syn. Alch. § 4, ll. 40–2 and § 1, ll. 26–9.

[423] See Syn. Alch. § 2 n. 3.

of this legend, at least in the form attested by the *Corpus alchemicum*, is difficult to reconstruct. On the one hand, ps.-Democritus himself clearly refers to Egypt when mentioning his teaching and the people (Egyptian kings and priests) to whom this teaching was addressed.[424] On the other, several sources claim that the philosopher travelled widely to Persia, Egypt, India, and Ethiopia in order to be educated by Eastern wisemen.[425] According to Bidez-Cumont, from the Hellenistic period on, historians stressed the debt of ancient Greek philosophers to foreign wisdom, reacting against an overly Hellenocentric view of classical historiography.[426] Among other philosophers, Democritus was presented as a pupil of Persian Chaldeans and as one who travelled over a large part of Asia. However, such accounts make no mention of Ostanes. The only occasion for a meeting between the two philosophers is inferable by comparing a passage of Diogenes Laertius with the one that opens the thirtieth book of Pliny's *Natural History*, on the history of magic:

> He [Democritus] was a pupil of certain Magi and Chaldaeans. For when King Xerxes was entertained by the father of Democritus he left men in charge, as, in fact, is stated by Herodotus.[427]

> The first man, so far as I can discover, to write a work on magic was Osthanes, who accompanied the Persian king Xerxes in his invasion of Greece, and sowed what I may call the seeds of this monstrous craft, infecting the whole world by the way at every stage of their travels.[428]

Despite the similarity of the two passages, Pliny does not claim that Democritus was educated by Ostanes or by Xerxes' entourage to which Ostanes would have belonged. Indeed, the Latin author claims in the following paragraph that Democritus travelled widely to learn magic, thereby discovering the hidden books of Apollobex the Copt and Dardanus the Phoenician. Therefore, no extant sources explicitly state that Ostanes was among the Persian wisemen left by Xerxes in Abdera. Even if we were to accept such a hypothesis, the result would be a different version of the legend in which Ostanes educated Democritus during his childhood in his native city.

Identifying the sources of the two abovementioned passages is also problematic. Diogenes Laertius quotes Herodotus (perhaps Hdt. VII 109 and VIII 120), but we cannot rule out that the author also had in mind other sources that are not explicitly mentioned.[429] Pliny's account is considered by Wellmann and Bidez-Cumont to depend upon Apion's *On the Magus*.[430] These scholars regard Apion as an

---

[424] See *supra*, pp. 63–5.

[425] See Berthelot, "Des origines de l'alchimie et des œuvres attribuées à Démocrite," 519 and 525; see also *supra*, pp. 32–3.

[426] Bidez-Cumont, *Mages*, vol. 1, 167–9.

[427] Diog. Laert. IX 34 (68 |58| A 1 D-K), translation by Robert Drew Hicks, *Diogenes Laertius, Lives of Eminent Philosophers*, 2 vols. (London: Heinemann, 1925), vol. 2, 443–5.

[428] Pliny. *NH* XXX 8 (68 |58| B 300, 13 D-K) Translation by William H.S. Jones, *Pliny, Natural History, books 28–32* (Cambridge, MA and London: Loeb, 1963), 283 (slightly modified).

[429] See Bidez-Cumont, *Mages*, vol. 1, 167 n. 1.

[430] Wellmann, *Bolos Demokritos*, 67–9; Bidez-Cumont, *Mages*, vol. 1, 171 n. 3; vol. 2, 11 n. 2 and 267 n. 1. See *Suda* π 752,7f. Adler; *FGrH* 616 F 23.

important source from whom Pliny obtained crucial information about a certain kind of ps.-Democritean literature that had, since the Hellenistic period, stressed the importance of medical and magical knowledge supposedly acquired by Democritus from Eastern 'scholars.' Whatever the actual role played by Apion,[431] it is evident that part of this literature should be ascribed to Bolos of Mendes, whose writings probably emphasised such a debt to Ostanes and Persian wisdom.[432]

In particular, we should recall Bolos' Egyptian origins, which show how already during the late Hellenistic period some pseudepigrapha describing a relationship between Democritus and Ostanes were circulating in Egypt. According to Syncellus, Ostanes was "dispatched to Egypt by the Persian kings of that time to take charge of the temples in Egypt";[433] while the list of alchemists preserved in manuscript **A** mentions "Ostanes from Egypt" ('Οστάνης ἀπ' Αἰγύπτου), anticipating the presentation of the magus in the later Arabic tradition which makes Ostanes an *Alexandrian* alchemist.[434] The progressive transformation of the Persian magus into an Egyptian scholar has been recently related by Quack to a wider penetration of Egyptian cultural elements into the literature ascribed to Persian magi during the Hellenistic period.[435] Egypt was indeed a crucial crossroads for those cultures so often evoked in syncretic legends, including the account of Democritus and Ostanes. Although we cannot identify the earliest formulation of this story, it is likely that the central nucleus was already known during the third/second century BC, probably connected to Bolos' production. This legend was later inherited and reworked by alchemical authors, perhaps within the *Four Books* of ps.-Democritus. The attribution of these books to Democritus itself points to this tradition. We therefore cannot rule out the possibility that the author of the *Four Books* sought in this way to explain his debt to doctrines ascribed to Ostanes, and thus introduced into his books a narrative section similar to the third paragraph of the *Physika kai mystika* as it has come down to us.

---

[431] To my knowledge, the only evidence that could confirm that Apion knew some pseudepigrapha ascribed to Democritus is in Pliny, *NH* XXIV 167, if we accept the identification of the *celeber arte grammatica*, quoted by Pliny, with the rhetorician Apion (Bidez-Cumont, *Mages*, vol. 2, 169 n. 8).

[432] See *supra*, pp. 40–3.

[433] See *supra*, p. 69. This information, questioned by Bidez-Cumont, *Mages*, vol. 1, 168–9, is considered reliable by Bladel, *Arabic Hermes*, 52–3: "It is not impossible that a figure like Ostanes could have been sent to Egypt and have participated in the Egyptian priesthood under the Achaemenids etc."

[434] *CAAG* I 25. See in particular the entry on Ostanes of al-Nadīm's *Fihrist*, which has been translated and commented on by Johann W. Fück, "The Arabic Literature on Alchemy According to An-Nadīm (A.D. 987): A Translation of the Tenth Discourse of the Book of the Catalogue (Al-Fihrist) with Introduction and Commentary," *Ambix*, 4 (1951): 91 § 5.

[435] Quack, "Les Mages Égyptianisés?," 267–82 (in particular 276–81, devoted to Ostanes).

CRITERIA OF EDITION

Greek text

I have taken into account four manuscripts (**MBAV**) that have been judged useful for the *constitutio textus*. These four codices seem to belong to two different branches of the tradition—**MV** and **BA,** respectively—and present several examples of contamination (see Martelli, *Pseudo-Democrito*, 3–54). In choosing between alternative readings provided by the four manuscripts (or correcting them), I have taken into account both the indirect and the Oriental traditions. Where this analysis has failed to provide decisive arguments, I have opted for the reading I judged to fit best into the context. In compiling the critical apparatus I have adopted the following criteria:

(a) I have recorded the most significant cases in which the Syriac tradition (both **SyrC** and **SyrL**) preserves interesting readings, which have been translated into Latin.

(b) The different spellings of the names of substances are always recorded, even when they may be explained by Byzantine pronunciation. In the edition, I have tried to impose a certain consistency on the variant spellings attested by the manuscripts, which often seem to reflect a kind of Byzantine *mise à jour* of the nomenclature. I normally adopt the forms that were used in the period in question, particularly in the case of excerpts from the *Four Books*, where the original form dates to the first century AD.

(c) In order to clarify the solutions proposed in the text, I have reproduced all the alchemical signs used in the manuscripts to indicate the names of the substances, their quantity, and specific operations. The edited texts are based on the interpretation of signs that are actually difficult to solve in regard both to their exact meaning and the grammatical case in which the corresponding terms should be inflected.

When the alchemical signs indicate the case in which to inflect the corresponding terms, I have reproduced the sign followed by its interpretation in square brackets. In addition, I have recorded the name of the first interpreter: e.g. $\textit{δ}^υ$ [i.e. χρυσοῦ BeRu] **MB**. Otherwise, I have proposed my own interpretation: e.g. $\textit{δ}^υ$ [i.e. χρυσοῦ] **MB**. When there is no reference to the case, I have first proposed the interpretation of the sign (followed by the name of the first interpreter) and afterwards reproduced the sign itself: e.g. χρυσὸς BeRu : $\textit{δ}$ **MB** (see Appendix, table 6 for the list of alchemical signs).

Syriac text

The Syriac translation of ps.-Democritus' *Four Books* preserved by **SyrC** is here edited for the first time. Since the work is transmitted by a *codex unicus*, damaged by humidity in many passages, I have sought to carefully reproduce its text, both by following its punctuation and by giving account of all my corrections and integrations in textual notes. In addition, the text has been systematically compared with the translation handed down by **SyrL** (edited by Berthelot-Duval in *CMA* II 10-4) and all the differences have been recorded in the critical notes. Where the two translations differ significantly—namely in *1SyrC* §§ 1 and 2—**SyrL** text has been reprinted (on the basis of Berthelot-Duval's edition) after the corresponding versions of **SyrC**.

Latin text

After the Greek and the Syriac texts I have edited a Latin translation of the Greek excerpts completed by Matthaeus Zuber in 1606. Since this translation was based on a copy of **M**, it does not include the *Catalogues*, which have been preserved only by **A**. In my edition I reproduce three specific features of this handwritten translation preserved by Vindobonensis Lat. 11427:

(a) First, Zuber left several blank spaces in the manuscript, particularly for passages that he could not understand or interpret.
(b) Second, he sometimes used an asterisk to refer to marginal notes containing his personal comments and suggested corrections. These marginal notes have been recorded in the footnotes.
(c) Third, he did not "translate" the alchemical signs used in the Greek manuscripts, but simply reproduced them. I have therefore reproduced these signs in my edition and explained their meaning in square brackets.

CONSPECTUS SIGLORUM

MANUSCRIPTS

M	Marcianus Graecus 299 (tenth/eleventh century)
B	Parisinus Graecus 2325 (thirteenth century)
V	Vaticanus Graecus 1174 (fourteenth/fifteenth century)
A	Parisinus Graecus 2327 (fifteenth century)
Z	Vindobonensis Latinus 11427 (seventeenth century; trans. Matthaeus Zuber)
SyrL	Syriac translations in London, British Library MS Egerton 709 (sixteenth century) and *Oriental* 1593 (fifteenth/sixteenth century); ed. Marcelin Berthelot and Rubens Duval, *La chimie au Moyen Âge, 2. L'alchimie syriaque* (Paris, 1893), 12–4.
SyrC	Syriac translation in Cambridge University Library, MS Mm. 6. 29 (fifteenth century), fols. 90ᵛ–96ᵛ.

EDITORS AND INTERPRETERS

BeRu	Marcellin Berthelot, and Charles-Émile Ruelle, *Collection des anciens alchimistes grecs* (Paris, 1888), vol. 2, 41–53.
BiCu	Joseph Bidez, and Fernand Cumont, *Les mages hellénisés* (Paris, 1938), vol. 2, 317–8.
D-K	Herman Diels, and Walther Kranz, *Die Fragmente der Vorsokratiker* (Berlin, 1951), 68 [55] B 300, 18.
Fabr	Johann Albert Fabricius, *Bibliotheca Graeca* (Hamburg, 1717), vol. 8, 233–48.
Fal	Vittorio De Falco, "Proposte di correzioni a testi alchimistici," *Athenaeum*, 26 (1948): 97–101.
Garzya	Antonio Garzya, *Opere di Sinesio di Cirene* (Torino, 1989), 801–21.
Lag	Otto Lagercrantz, *Papyrus Graecus Holmiensis (P. Holm.). Recepte für Silber, Steine und Purpure* (Uppsala, 1913), 112–4.
Pizzim	*Democritus Abderita De arte magna sive de rebus naturalibus. Nec non Synesii, et Pelagii, et Stephani Alexandrini, et Michaelis Pselli in eundem commentaria.* Domenico Pizzimentio Vibonensi interprete (Padua, 1573), fols. 5–11.
Ruska	Julius Ruska, *Turba philosophorum. Ein Beitrag zur Geschichte der Alchemie* (Berlin, 1931).
Zur[1]	Carlo Oreste Zuretti, "*Proposte di lettura a luoghi della Collection des anciens alchimistes grecs publiée par M. Berthelot,*" *BZ*, 30 (1929/30): 678–9.
Zur[2]	Carlo Oreste Zuretti, "*Proposte di lettura a luoghi della 'Collection des anciens alchimistes grecs publiée par M. Berthelot,*" *RRIL*, 54 (1931): 198–9.

COMPENDIA

a.c.	ante correctionem	inser.	inseruit/inseruerunt
add.	addidit/addiderunt	iter.	iteravit
alt.	alterum	leg.	legendum
cod	codex	legit.	legitur
codd	codices	litt.	littera/litterae
coll.	collato/a	om.	omisit/omiserunt
coni.	coniecit/coniecerunt	p.c.	post correctionem
del.	delevit/deleverunt	pr.	prior/prius
des.	desinit/desinunt	praec.	praeceptum
dubit.	dubitanter	praecc.	praecepta
ed.	edidit/ediderunt	prop.	proposuit/proposuerunt
fort.	fortasse	secl.	seclusit/secluserunt
i.e.	id est	seclude.	secludendum
imag.	imago/imaginem	tert.	tertium
inc.	incipit/inciperunt	trad.	tradidit/tradiderunt
in mg.	in margine	trib.	tribuit/tribuerunt
in parenth.	in parenthesi	ut vid.	ut videtur
in ras.	in rasura	vol.	voluit/voluerunt

Δημοκρίτου περὶ πορφύρας καὶ χρυσοῦ ποιήσεως· φυσικὰ καὶ μυστικά

1   1. Βαλὼν εἰς λίτραν μίαν πορφύρας †διοβολοῦ λίτραν† σκωρίας σιδήρου εἰς οὔρου δραχμὰς ἑπτά, ἐπίθες ἐπὶ πυρᾶς ὥστε λαβεῖν βράσματα. Εἶτα λαβὼν ἀπὸ τοῦ πυρὸς τὸ ζέμα, βάλε εἰς λεκάνην, προβαλὼν τὴν πορφύραν καὶ ἐπιχέας τὸ ζέμα
5 τῇ πορφύρᾳ· ἔα βρέχεσθαι νυχθήμερον ἕν. Εἶτα λαβὼν βρύων θαλασσίων λίτρας δύο, βάλε ὕδωρ ὡς εἶναι ἐπάνω τῶν βρύων τετραδάκτυλον καὶ ἔχε ἕως ἂν παχυνθῇ· καὶ διυλίσας τὸ διύλισμα, θέρμανον· καὶ συνθεὶς τὴν ἐρέαν, κατάχεε. Χαυνο- τέρα δὲ συντεθήτω, ὥστε φθάσαι τὸν ζωμὸν ἕως τοῦ πυθμένος,
10 καὶ ἔασον νυχθήμερα δύο. Εἶτα λαβὼν μετὰ ταῦτα, ξήρανον ἐν σκιᾷ· τὸν δὲ ζωμὸν μὴ ἐκχέῃς. Εἶτα βαλὼν εἰς τὸν αὐτὸν ζωμὸν βρύων λίτρας δύο, βάλε ἐν τῷ ζωμῷ ὕδωρ ὡς γενέσθαι τὴν ἀναλογίαν τὴν πρώτην· καὶ ἔχε ὡσαύτως ἕως ἂν παχυνθῇ. Εἶτα ὑλίσας, βάλε τὴν ἐρέαν ὡς τὸ πρῶτον καὶ ποιησάτω νυχθήμερον
15 ἕν. Εἶτα λαβών, ἀπόπλυνον εἰς οὖρον καὶ ξήρανον ἐν σκιᾷ. Εἶτα λαβὼν λακχάν, τρῖψον· καὶ λαβὼν λαπάθου λίτρας τέσσαρας,

M, fol. 2ʳ (ubi titulus legitur) + fols. 66ᵛ27-71ʳ6

B, fols. 8ᵛ10-17ʳ16

V, fols. 33ᵛ13-35ᵛ16 (= ll. 1-63) + fols. 1ʳ1-7ʳ16 (= ll. 61-229) et Vᵃ, fol. 149ᵛ1-12 (= ll. 184-96)

A, fols. 24ᵛ5-29ᵛ3

CAAG II 41-9

Tit. Δημοκρίτου — μυστικά **M** 2ʳ : Δημοκρίτου φυσικὰ καὶ μυστικά **M** 66ᵛ **BA** : ἐκ τῶν δημοκρίτου περὶ πορφύρας φυσικῆς **V** ‖ **1** hic compendium libri de purpura incipit ‖ εἰς λίτραν μίαν BeRu : εἰς λιᵗ α′ **MV** : εἰ λίτραν μίαν **B** : εἰς λίτραν μιᾶν **A** ‖ διοβολοῦ **MV** : διοβοˡ **B** : διοβόλου **A** : διαβολοῦ Zuber (**Z** 62ᵛ1) : δύο ὀβολοὺς Zur¹ : fort. διωβόλου ‖ alt. λίτραν del. Zur¹ ‖ **2** σκωρίας **BA** : σκόρεαν **MV** ‖ σιδήρου BeRu : ♂ **MV** : ♃ **BA** ‖ εἰς **MBVA** : καὶ Zur¹ ‖ δραχμὰς BeRu : ♈ **MBVA** : ζ′ **MV** : β′ **BA** ‖ **6** ὡς **MV** : ὥστε **BA** ‖ **7** ἔχε ἕως **M** : ἔ. ὡς **A** et **B** ut vid. (ἔ[...]ς) : ἔ. ἄχρις **V** ‖ **8-9** χαυνοτέρα **BAV** : χαυνοτερα (sic) **M** : χαυνότερα BeRu ‖ **11** μὴ ἐκχέῃς scripsi : μὴ ἐκχέεις **MV** : ἔκχεον **BA** : ἐγχέῃς BeRu : μὴ ἐγχέῃς Zur¹ ‖ **11-2** βαλὼν — δύο **M** **V** : λαβὼν τὸν αὐτὸν ζωμόν· καὶ βαλὼν λίτρας β′ **BA** ‖ **12** ὡς **MV** : ὥστε **BA** ‖ **12-3** τὴν πρώτην ἀναλογίαν **BA** : τὴν πρώτην ἀναλογίαν **MBV** : ἔκχε **A** ‖ **13** ἔχε **MBV** : ἔκχε **A** ‖ **14** ποιησάτω **BA** : ποιήσει **M** : ποιήτω **V** : ποιήσῃ BeRu : πίῃ σοι dubit. Zur¹ ‖ **15** alt. εἶτα **MV** : ἔπειτα **BA** ‖ **16** λακχάν **MV** : -ὰν **BA** ut semper ‖ τρῖψον scripsi : ⑤ **MBVA** : τρίβε (sic) BeRu

Democritus, *On the Making of Purple and Gold: Natural and Secret Questions (=PM)*

1. For a pound of purple put (?) … iron slag into seven drachmas of urine: set it on a fire to bring it to the boil. Then take the decoction from the fire and pour it into a vessel—put the purple in first and pour the decoction over the purple: let it soak for a day and a night.[1] Then take seaweed,[2] two pounds, and pour in water so that it is four fingers above it. Keep it in this state until [the decoction] thickens. Then strain and warm the strained decoction; plunge the wool you have collected into it. Use the spongier sort so that the wash [i.e. the dyeing solution] soaks in deeply; leave it to stand for two days and nights. Afterwards take the wool out and dry it in the shade; but do not pour the wash away. Then put two pounds of seaweed into the same wash and pour water into the wash so that it has the previous ratio: keep it in this state until it thickens. Then strain and put in the wool as before and let [the solution?] work for a day and night. Then take the wool out, wash it with urine and dry it in the shade. Next take and crush *lakcha*;[3] then take four

---

[1]   om. **SyrL**.; aliqua praecepta de purpureis tincturis, Democriteis comparanda, servat **SyrC** (vide *3SyrC* §§ 5–7).

ἔκζεσον μετὰ οὔρου, ὡς λυθῆναι τὸ λάπαθον· καὶ ὑλίσας τὸ
ὕδωρ θαλάσσιον, βάλε τὸν λακχὰν καὶ ἕψει ἕως παχυνθῆ· καὶ
διυλίσας πάλιν τὸν λακχάν, βάλε τὴν ἐρέαν. Εἶτα μετὰ ταῦτα
20   πλῦνον οὔρῳ, εἶτα πάλιν ὕδατι· καὶ μετὰ ταῦτα ξηράνας ὁμοίως
ἐν σκιᾷ, θυμία ὄνυξι θαλασσίοις ἐναποβεβρεγμένην ἐν οὔρῳ
ἡμέρας δύο.

2. Εἰς δὲ τὴν κατασκευὴν τῆς πορφύρας τὰ εἰσερχόμενά
εἰσιν τάδε· φῦκος ὃ καλοῦσι ψευδοκογχύλιον, καὶ κόκκος, καὶ
25   ἄνθος θαλάσσιον, ἄγχουσα Λαοδικηνή, κρημνός, ἐρυθρόδανον
τὸ Ἰταλικόν, φυλάνθιον τὸ δυτικόν, σκώληξ ὁ πορφύριος, ῥόδιον
τὸ Ἰταλικόν. Ταῦτα τὰ ἄνθη προτετίμηται παρὰ τῶν προγε-
νεστέρων, καί εἰσι φευκτά, οὐ τίμια. Ἔστι δὲ ὁ τῆς Γαλατίας
σκώληξ, καὶ τὸ τῆς Ἀχαΐας ἄνθος ὃ καλοῦσιν λακχάν, καὶ τὸ
30   τῆς Συρίας ὃ καλοῦσιν ῥίζιον, καὶ τὸ κογχύλιον, καὶ τὸ κοχλιοκογχύλιον τὸ Λιβυκόν, καὶ ὁ Αἰγύπτιος κόγχος ὁ τῆς παραλίου,
ὃς καλεῖται πίννα, καὶ ἡ ἰσάτις βοτάνη, καὶ τὸ τῆς ἀνωτέρας
[καὶ τὸ τῆς] Συρίας ὃ καλοῦσιν κόγχον. Ταῦτά ἐστιν ἀκίνητα,
οὔτε τιμητὰ παρ' ἡμῖν, πλὴν τῆς ἰσάτεως.

---

18 ≈ [i. e. ὕδ(ωρ) θαλάσσι(ον)] **MBVA** : ὕδωρ BeRu ‖ τὸν om. **BA** ‖ ἕψει **MBV** : -η
A : ἕψε BeRu Zur[1] ‖ ἕως **MBA** : ἄχρις ἂν **V** : ἕως οὗ Zur[1] ‖ **19-20** μετὰ — ξηράνας
MV : πλύνον οὔρῳ· μετὰ ταῦτα ὕδατι· ἔπειτα ξηράνας **BA** ‖ **21** ἐναποβεβρεγμένην
MV : -οις **BA** ‖ **22** ἡμέρας **M** : **66 BVA** ‖ **23-4** τὰ εἰσερχόμενά εἰσι τάδε Zur[1] :
τὰ ἀπερχόμενά εἰσι τάδε **M** : εἰσι τάδε τὰ εἰσερχόμενα **V** : τὰ εἰσερχόμενά εἰσι
ταῦτα **BA** ‖ **24** ὃ **MBA** : ὅπερ **V** ‖ κόκκος Lag : κόκκον **MBV** : κόκον **A** ‖ **25**
ἄγχουσα Λαοδικηνή scripsi, praeeunte Lag (-κινή) : -αν λαδικήνη **MV** : -αν λαδικίνην
BA ‖ κρημνός **MBA** : om. **V** : ἡ κρ. BeRu ‖ **26** φυλάνθιον **MBVA**: φυλλάν- BeRu ‖
πορφύριος **MVA** : -ειος **B** ‖ post πορφύριος add. ἐκ τοῦ ἐρώᵘ γενόμενος **MV** : ἐκ τοῦ
ζώου γ. dubit. prop. BeRu : ἐκ τοῦ φαρίου γ. Lag : ἐκ τοῦ δρύου γ. prop. Colinet
(«Byzantion» 84) : glossema videtur, fort. ἐκ τοῦ ἐρίου γ., coll. *e lana confectus* Zuber (**Z**
63ʳ10) ‖ **27** προτετίμηται **MV** : -ηνται **BA** et Moyses Alch. ‖ **28** οὐ **MBVA** : om.
Moyses Alch. ‖ **29** τὸ τῆς Ἀχ. ἄνθος ὃ **MBVA** : τι τῆς Ἀχ. ἄν. ὃ BeRu : ὁ τῆς Ἀχ. ὃν
Moyses Alch. ‖ λακχάν **MBVA** : λαχάν Moyses Alch. ‖ **29-30** τὸ τῆς Σ. ὃ **MBVA** : ὁ
τῆς Σ. ὃν Moyses Alch. ‖ **30** ῥίζιον **MBVA** : ῥιζάριον Moyses Alch. ‖ καὶ τὸ
κοχλιοκογχύλιον **MBA** : κ. τὸ χλιοκογχύλιον **V** : om. Moyses Alch. ‖ **32** ὃς **BA** : ὃ
MV et Moyses Alch. ‖ πίννα **MBVA** : γλίνα (sic) Moyses Alch. ‖ καὶ τὸ om. **BA** ‖ **33**
καὶ τὸ τῆς **MBVA** : om. Moyses Alch. : καὶ τὸ del. Zur[1] ‖ κόγχον **MBVA** : -ος Moyses
Alch ‖ ante ἀκίνητα add. οὐκ BeRu : οὔτε Lag ‖ **34** ἰσάτεως **MV** : -τιδος **BA** :
διηγεῖσατέως (sic) Moyses Alch. : fort. διηγ<ηθείσης> ἰσάτεως leg.

pounds of monk's rhubarb [dock][4] and boil with urine to dissolve the dock. Then filter seawater, put *lakcha* in it and boil until it thickens. Strain the *lackha* again and plunge the wool in. Afterwards wash it with urine, and again with water: then dry in the shade as before and fumigate with seashells[5] the wool which has been left to soak in urine for two days.

2. These are the substances that enter into the preparation of purple dye:[6] the seaweed[7] that is called 'fake little shell,'[8] and kermes,[9] and sea flower,[10] and alkanet from Laodicea,[11] and *krēmnos*,[12] and Italian madder,[13] and western *phylanthion* (?),[14] and the purple worm, and Italian pomegranate [? or rose extract].[15] These flowers [i.e. dyes] were preferred by our predecessors, but they are unstable, not valuable. On the other hand there are the Galatian worm,[16] and the Achaean flower that is called *lakcha*,[17] and the Syrian [flower] that is called 'little root,'[18] and the little shell, and the small Libyan mollusc with a spiral shell, and the Egyptian shell from the coast, which is called *pinna*,[19] and the woad plant,[20] and the [flower] from the upper Syria, which is called 'shell.' These dyes are stable, but they are not considered valuable among us, except the woad plant.

§ 2   om. SyrL SyrC.
    Test. 27-34 Moyses Alch. (A, fol. 273ʳ19–273ᵛ1), ut monuerunt BeRu (*CAAG* II 307 n. 15), qui textum non ed.

35   3. Ταῦτα οὖν παρὰ τοῦ προειρημένου διδασκάλου μεμαθη-
κώς, καὶ τῆς ὕλης τὴν διαφορὰν ἐγνωκώς, ἠσκούμην ὅπως ἁρμόσω τὰς φύσεις. Εἰ γὰρ καὶ τέθνηκεν ἡμῶν ὁ διδάσκαλος, μηδέ-
πω ἡμῶν τελειωθέντων, ἀλλ᾽ ἔτι περὶ τὴν ἐπίγνωσιν τῆς ὕλης
ἀπασχολουμένων, ἐξ Ἅιδου τοῦτον φέρειν ἐπειρώμην· ὡς δὲ εἰς
40   τοῦτο ὥρμησα, εὐθὺς παρεκάλεσα λέγων· «παρέχεις δωρεὰς
ἐμοί, ἀνθ᾽ ὧν ἀπείργασμαι εἰς σέ»; Καὶ τοῦτο εἰπών, ἐσιώπα.
Ὡς δὲ πολλὰ παρεκάλουν ἠρώτων θ᾽ ὅπως ἁρμόσω τὰς φύσεις,
ἔφησέ μοι δύσκολον λέγειν, οὐκ ἐπιτρέποντος αὐτῷ τοῦ δαίμο-
νος. Μόνον δὲ εἶπεν· «αἱ βίβλοι ἐν τῷ ἱερῷ εἰσιν». Ἀναστρέψας
45   εἰς τὸ ἱερὸν ἐγενόμην ἐρευνήσων, εἴπερ δυνηθείην εὐπορῆσαι
τῶν βιβλίων· οὔτε γὰρ περιὼν τῷ βίῳ ταῦτα εἰρήκει· ἀδιάθετος
γὰρ ὢν ἐτελεύτα, ὡς μέν τινές φασιν, δηλητηρίῳ χρησάμενος
διὰ ἀπαλλαγὴν ψυχῆς ἐκ τοῦ σώματος, ὡς δὲ ὁ υἱός φησιν,
ἀπροσδοκήτως ἑστιώμενος. Ἦν δὲ πρὸ τῆς τελευτῆς ἀσφαλισά-
50   μενος μόνον τῷ υἱῷ φανήσεσθαι τὰς βίβλους, εἰ τὴν πρώτην
ὑπερβῇ ἡλικίαν· τούτων δὲ οὐδεὶς οὐδὲν ἡμῶν ἠπίστατο. Ὡς οὖν
ἐρευνήσαντες εὕρομεν οὐδέν, δεινὸν ὑπέστημεν κάματον ἔστε
ἂν συνουσιωθῶσι καὶ συνεισκριθῶσιν αἱ οὐσίαι καὶ αἱ φύσεις.
Ὡς δὲ ἐτελειώσαμεν τὰς συνθέσεις τῆς ὕλης, χρόνου τινὸς
55   ἐνστάντος καὶ πανηγύρεως οὔσης ἐν τῷ ἱερῷ πάντες ἡμεῖς
εἰστιώμεθα· ὡς οὖν ἦμεν ἐν τῷ ναῷ, ἐξ αὐτομάτου στήλη τις
[κίων ἦν· ἢ] διαρρήγνυται, ἣν ἡμεῖς ἑωρῶμεν ἔνδον οὐδὲν

---

35 οὖν **MBA** : τοίνυν **V** ‖ **36-7** ἁρμόσω **MBVA** : ἀκούσω Moyses Alch. ‖ **39** post
Ἅιδου add. φησίν **MV** : φασίν BiCu ‖ ὡς δὲ **MV** : καὶ ὡς **BA** ‖ **40** παρεκάλεσα **MV** :
-εσε **BA** ‖ **41** ἐσιώπα **MBA** : ἐσιώπησα **V** ‖ **42** ἠρώτων θ᾽ ὅπως BiCu : ἠ. τὸ πῶς **M** :
ἠ. πῶς **V** : ἠ. ὅπως **BA** ‖ **43** post ἔφησε add. δὲ **V** ‖ post δύσκολον add. ἔχω **BA**, τοῦτο
V ‖ **44** post εἶπεν add. ὅτι **V** ‖ post ἀναστρέψας add. οὖν **V** ‖ **45** εἰς τὸν ἱερὸν **MBA** :
ἐν τῷ ἱερῷ **V** ‖ ἐγενόμην **BAV** : ἐγιν- **M** ‖ **46** τῷ βίῳ ταῦτα **B** : τοῦτο **MV** : τῷ
βιβλίῳ ταῦτα **A** : τῷ βιβλίῳ τοῦτο BeRu : τῷ βίῳ τοῦτο BiCu ‖ **47** post ἐτελεύτα add.
οὔτε προεῖπε τοῦτο τινὰ τῶν συνήθων **V** ‖ **48** ψυχῆς **MV** : -ὴν **BA** ‖ τοῦ om. **MV** ‖
50 μόνον **M** et **V** a.c. : μόνῳ **BA** et **V** p.c. ‖ φανήσεσθαι **MBV** : φων- **A** ‖ **51** οὐδὲν
ἡμῶν **MV** : οὐδόλως ἐξ ἡμῶν **BA** ‖ **56** εἰστιώμεθα **MBVA** : εἰσθι- BeRu ‖ ὡς — ναῷ
MV : καὶ **BA** ‖ στήλη τίς **BA** : στηλήτις **M** : στηλίτις **V** ‖ **57** κίων ἦν· ἢ (sic **M**)
tamquam ex glossemate seclusi: post ἢ add. καὶ **V** : ἢν κίωνι· ἢ **BA** : κίων ἦν in parenth.
inser., nec non ἢ del. BiCu : <ἢ> κίων ἦν, ἢ BeRu : ἢ κιόνιον D-K ‖ διαρρήγνυται
MBA : -ηται **V** ‖ **57-60** ἢν ἡμεῖς — ἐγχύψαντες **MV** : καὶ ἐγκύψαντες ἔνδον, ὁρῶμεν
ἐν αὐτῇ τὰς πατρῴας βίβλους. καὶ προκομίσαντες εἰς μέσον **BA**

3. Having learned these things from the abovementioned master and known the differences of the matter, I strove to combine natures. Since our master died before our initiation was completed, while we were still devoted to investigating the matter, I tried to conjure him from Hades. As soon as I was ready to do it, I immediately conjured him by saying: "Are you giving me any gift in return for what I did for you?" So I spoke, but he kept silence. Since I conjured him several times asking how to combine natures, he replied to me that it was difficult to speak, because he was not allowed to do so by his daemon. He told me only: "The books are in the temple." I came back to the temple and prepared myself for exploring it in the hope of finding the books. He did not speak about them when he was alive, and he died intestate: some people claim that he swallowed a poison for separating soul from body; according to his son, he died suddenly during a banquet. But before dying he made sure that the books would have been shown only to his son after he had passed his first age [i.e. his childhood]: so none of us knew anything about them. Since we did not find anything despite our searches, we worked very hard to make substances and natures mix together and to bring them into aggregation. When we accomplished the combinations of the matter, after a little while a feast took place in the temple and all of us joined the banquet. We were in the *sancta sanctorum* when a column broke up by itself, which at first sight did not

⁵ ³   om. SyrL SyrC.
Test. 35–7 + 61–3 Moyses Alch. (A, fol. 273ᵛ1–4) CAAG II 307,15–7 ‖ 38–9 Olymp. Alch. CAAG II 100,6–8.

ἔχουσαν. †Ὁ δὲ οὔτ᾽ ἄν τις† ἔφασκεν ἐν αὐτῇ τὰς πατρῴας τε-
θησαυρίσθαι βίβλους, καὶ προκομίσας εἰς μέσον ἤγαγεν. Ἐγκύ-
60 ψαντες δὲ ἐθαυμάζομεν ὅτι μηδὲν ἦμεν παραλείψαντες· πλὴν
τοῦτον τὸν λόγον εὕρομεν ἐκεῖ πάνυ χρήσιμον· ἡ φύσις τῇ
φύσει τέρπεται, καὶ ἡ φύσις τὴν φύσιν νικᾷ, καὶ ἡ φύσις τὴν
φύσιν κρατεῖ. Ἐθαυμάσαμεν πάνυ ὅτι ἐν ὀλίγῳ λόγῳ πᾶσαν
συνήγαγε τὴν γραφήν.

58 ὁ δὲ οὔτ᾽ἄντις (sic) **M** : οὐδ᾽ ἄντις **V** : ὁ δ᾽ Ὀστάνης D-K : ὃ δὲ οὔτ᾽ ἄν τις
ἔφασκεν (= quod nemo dixerit) Zur¹ ‖ **59** καὶ **M** : πλὴν **V** ‖ **60** ἐθαυμάζομεν **MBV** :
-ζωμεν **A** ‖ παραλείψαντες **MBVA** : -λήψαντες Zur¹ ‖ **63** post κρατεῖ add. τὰ
δὲ ἕτερα προεγράφησαν et des. **V** 35ᵛ

contain anything inside. But < ... > said that the books of his father had been preserved within this column, and he took them out and showed them publicly.
Peering [into the books] we were surprised [to find] that we had not neglected anything, except this very helpful saying that we found there: "Nature delights in
nature, nature conquers nature, nature masters nature." We marvelled greatly at
how he had summarised all his work in such a short saying.[21]

---

§ 3   TEST. 61–3 Fragm. alch. *CAAG* II 20,5–6 et 22,4–7 et 359,3–4; Isis Alch. *CAAG* II 30,17–8; Zos. Alch. *CAAG* II
171,5–6; Sin. Alch. § 1, ll. 15–7; Moyses Alch. *CAAG* II 307,16f.; Steph. Alch. II 200,36–7 et 215,6–7 et 240,6–7
Ideler; Philos. Christ. Alch. *CAAG* II 395,9–10 et 416,20–1.

65   4. Ἥκω δὲ κἀγὼ ἐν Αἰγύπτῳ φέρων τὰ φυσικά, ὅπως τῆς πολλῆς περιεργίας καὶ συγκεχυμένης ὕλης καταφρονήσητε.

    5. Λαβὼν ὑδράργυρον, πῆξον τῷ τῆς μαγνησίας σώματι, ἢ τῷ τοῦ Ἰταλικοῦ στίμεως σώματι, ἢ θείῳ ἀπύρῳ, ἢ ἀφροσελήνῳ, ἢ τιτάνῳ ὀπτῷ, ἢ στυπτηρίᾳ τῇ ἀπὸ Μήλου, ἢ ἀρσενικῷ, ἢ ὡς

70   ἐπινοεῖς. Καὶ ἐπίβαλλε λευκὴν γενομένην χαλκῷ, καὶ ἕξεις χαλκὸν ἀσκίαστον. Ξανθὴν δὲ ἐπίβαλλε ἀργύρῳ, καὶ ἕξεις χρυσόν· χρυσῷ, καὶ ἔσται χρυσοκόραλλος σωματωθεῖσα. Τὸ δ᾽ αὐτὸ ποιεῖ καὶ ἀρσενικὸν ξανθὸν καὶ σανδαράχη οἰκονομηθεῖσα καὶ κιννάβαρις πάνυ ἡ ἐκστραφεῖσα. Τὸν δὲ χαλκὸν

75   ἀσκίαστον μόνη ἡ ὑδράργυρος ποιεῖ. Ἡ φύσις τὴν φύσιν νικᾷ.

    6. Πυρίτην ἀργυρίτην, ὃν καὶ σιδηρίτην καλοῦσιν, οἰκονόμει ὡς ἔθος, ἵνα ῥεῦσαι δυνηθῇ· ῥεύσει δὲ διὰ †νίθεως† ἢ λευκῆς λιθαργύρου, ἢ τῷ Ἰταλικῷ στίμει. Καὶ σκόρπισον μολύβδῳ· οὐχ

---

65 hic fort. compendium libri de auri confectione incipit ‖ ante ἥκω add. titulum ἐκ τῶν δημοκρίτου φυσικῶν καὶ μυστικῶν atque ll. 61-64 (ἡ φύσις — τὴν γραφήν) V 1ʳ ‖ **66** περιεργίας scripsi: -είας **MV** : περιερ[.....] **B** : -εργασίας **A** ‖ ante συγκεχυμένης add. οὐ **M** ‖ **67** 𐩠𐩠 [i.e. χρυσοποιία] **MBVA** in mg. : ܡܢ ܡܚܟܡܐ ܘܦܝܣܩܘܣ ܣܡܟܗܐ ܡܪܦ ܡܪܡ ܡܝ ܠܝܘܟܪܐ܂ ܕܘܝ ܠܩܩܘܣܐ, ܟܝܘܠ ܕܚܠ ܡܪܡ ܚܙܐ [ex philosophi Democriti doctrina, pars prima ex primo tractatu de auri confectione] **SyrL** : ܟܚܠ ܠܝܟܐ ܡܚܡܐ, ܠܩܩܘܣܐ, ܟܝܘܠ ܚܙܐ܂ܟܝ [Democriti liber. De solis (i.e. auri) splendentis confectione] **SyrC** ‖ ὑδράργυρον **BeRu** : ☽ **MBVA** ‖ **68** στίμεως **MV** : στίμμεος **B** : στίμμεως **A** ‖ ἀπύρῳ **MBV** : ἀπεί- **A** ‖ ἀφρο☾ω [i.e. ἀφροσελήνῳ **BeRu**] **MV** : ἀφρο☾ **BA** ‖ **70** ἐπίβαλλε **MV** : -βαλε **BA** ‖ γενομένην scripsi, coll. ܙܝܡܪ ܚܡ ܢܘܗ [et cum (argentum vivum) album fiat, destille] **SyrL** : γέαν **MV** : γαίαν **BA** ‖ ♀ω [i.e. χαλκῷ **BeRu**] **BA** et ܠܟ ܠܚܙܝܘܪ [ad Venerem (i.e. aes)] **SyrL** : ♀ᵘ [i.e. χαλκοῦ] **MV** ‖ **71** ♀ [i.e. χαλκὸν **BeRu**] **M** et ♀‟ **BA** : ♀ V ‖ ἐπίβαλλε **MV** : -βαλε **BA** ‖ ἀργύρῳ scripsi post Hoefer, *Histoire de la chimie*, vol. 1, 278 et ܠܚܙ ܡܠܝܙܘ, [ad plumbum] **SyrL** : σελήνην **M** : ☾ᴺ **B** : ☾ⁿˠ **A** : ☾ V : σελήνη Ruska, *Turba*, 189 n. 6 ‖ **72** ᾿◊ [i.e. χρυσὸν **BeRu**] **MV** : ◊ **B** : ◊ **A** ‖ **73** ἀρσένικον **MV** : β♀ **BA** ‖ σανδαράχη **MV** : ℓ♀ **BA** ‖ **74** κιννάβαρις **MV** : κινάβαρις⊙ **BA** ‖ χαλκὸν **MV** et ♀‟ **BA** ‖ **75** ὑδράργυρος **MV** : ☽ **BA** ‖ ante φύσις add. γὰρ **BeRu** ‖ **76** ante πυρίτην add. λαβὼν Moyses Alch. ‖ πυρίτην ἀργυρίτην **MVA** : -ης -ης **B** : σιδηρίτην **MBVA** : ἀργυροσιδηρίτην Moyses Alch. ‖ **77** διανίθεως **MV** : διὰ νύθου **A** et Moyses Alch. : διὰ νυθ **B** : ܠ̈ܝܪ܂ ܚܡ [fort. διὰ ῥητίνη] **SyrL SyrC** ‖ **78** ante τῷ add. ἐν **BA** et Moyses Alch. ‖ στίμει **BeRu** : στήμει **MV** : στίμμει **B** : στίμμι **A** ‖ μολύβδῳ **BeRu** : μολί- **MBVA** : ♄ Moyses alch. : ܚܚܘܣ [plumbo?] **SyrL** : (i.e. ܚܚܚܝܗ) ܗܚ [hoc (i.e. stibio)] **SyrC**

4. I too have come to Egypt to deal with natural substances, so that you may disregard many captious questions and the confused matter.

5. Take mercury and make it solid with the body of *magnēsia*, or with the body of Italian stibnite, or with unburnt sulphur, or with moon foam,[22] or with roasted lime, or with alum from Milos, or with orpiment, or according to your knowledge.[23] If it [i.e. mercury] turns white, lay it on copper, and you will have 'shadowless' copper. [If the mercury turns] yellow, lay it on silver and you will have gold;[24] on gold, and it will be solid gold coral.[25] Yellow orpiment produces the same effect, and processed realgar and the cinnabar that has been completely turned inside out. But mercury alone makes copper 'shadowless.'[26] Nature conquers nature.

6. Process silver pyrite that is also called *sidēritēs* as is customary, so that it may be melted.[27] It will melt with *nitheōs* (?),[28] or white litharge,[29] or Italian stibnite. So reduce it to powder, mixing with lead [lit. sprinkle it on

§ 4   om. SyrL SyrC.
   TEST. 65–6 Syn. Alch. § 5, ll. 61–2.
§ 5   SyrL: CMA II 10,3–11 (textus); 19 praec. 1 (translatio); SyrC : 1SyrC § 1.
   TEST. 67–70 Philos. Christ. Alch. CAAG II 397,2–5 ‖ 67 Zos. Alch. CAAG II 172,4–5 et 192,22–193,6; Syn. Alch.
   § 11, l. 184; Olymp. Alch. CAAG II 74,10; Steph. Alch. II 218,10 Ideler; Philos. Anon. Alch. (Zos. Alch. BeRu)
   CAAG II 123,20–1; Philos. Christ. Alch. CAAG II 397,13 ‖ 74–5 Zos. Alch. CAAG II 172,2–3 et 193,9.
§ 6   SyrL: CMA II 10, 12–6 (textus); 19 praec. 2 (translatio); SyrC: 1SyrC § 2.

ἁπλῶς λέγω, ἵνα μὴ πλανηθῇς, ἀλλὰ τῷ ἀπὸ Κοπτικοῦ καὶ
80 λιθαργύρου μέλανι τῷ ἡμῶν, ἢ ὡς ἐπινοεῖς· καὶ ὄπτησον, καὶ
ἐπίβαλλε ὕλῃ ξανθὸν γενόμενον, καὶ βάψεις. Ἡ γὰρ φύσις τῇ
φύσει τέρπεται.

7. Πυρίτην οἰκονόμει ἕως οὗ γένηται ἄκαυστος, ἀποβαλὼν
τὴν μελανίαν· οἰκονόμει δὲ ὀξάλμῃ, ἢ οὔρῳ ἀφθόρῳ, ἢ θαλάσσῃ,
85 ἢ ὀξυμέλιτι, ἢ ὡς ἐπινοεῖς, καὶ ὄπτησον ἕως οὗ γένηται ὡς
ψῆγμα χρυσοῦ ἄκαυστον. Καὶ ἐὰν γένηται, πρόσμιξον αὐτῷ
θεῖον ἄπυρον, ἢ στυπτηρίαν ξανθήν, ἢ ὤχραν Ἀττικήν, ἢ ὡς
ἐπινοεῖς. Καὶ ἐπίβαλλε ἀργύρῳ διὰ τὸν χρυσόν, καὶ χρυσῷ διὰ
τὸ χρυσοκογχύλιον. Ἡ γὰρ φύσις τὴν φύσιν κρατεῖ.

90 8. Τὸ κλαυδιανὸν λαβών, ποίει μάρμαρον καὶ οἰκονόμει ὡς
ἔθος, ἕως ξανθὸν γένηται. Ξάνθωσον οὖν, οὐ τὸν λίθον λέγω,
ἀλλὰ τὸ τοῦ λίθου χρήσιμον· ξανθώσεις δὲ μετὰ στυπτηρίας
ἐξιπωθείσης, <ἢ> θείου, ἢ ἀρσενικοῦ, ἢ σανδαράχης, ἢ τιτάνου,
ἢ ὡς ἐπινοεῖς. Καὶ ἐὰν ἐπιβάλλῃς ἀργύρῳ, ποιεῖς χρυσόν· ἐὰν
95 δὲ χρυσῷ, ποιεῖς χρυσοκογχύλιον. Ἡ γὰρ φύσις τὴν φύσιν
νικῶσα κρατεῖ.

79 τῷ **BA** : τὸ **MV** et Moyses Alch. ‖ ἀπὸ del. **Zur**[1] ‖ κοπτικοῦ **MBVA** : -ῷ **Zur**[1] :
ܢܩܦ ܣܘܡܣܡ ܘܡܚ ܘܚ [illud (plumbum vel stibium) quod ex Samo exit] **SyrL SyrC** ‖ **80**
λιθαργύρου **BA** : -ω **MV** ‖ **81** ἐπίβαλλε **MV** : -βαλε **BA** ‖ βάψεις Moyses Alch. et
ܓܚܚ [inficis] **SyrL**: -ει **MBVA** : -η **BeRu** ‖ **81-2** τῇ φύσει **MBV** : φύσι **A** ‖ **83** ante
πυρίτην add. ἐνταῦθα νοεῖ **BA** ‖ ἕως **MBA** : ἄχρις **V** ‖ **85** ὀξυμέλιτι **MV** : ὀξο- **BA** ‖
καὶ ὄπτησον om. **MV** et Moyses Alch. ‖ **86** ἄκαυστον **MBVA** : -καύστου **Zur**[1] et
Moyses Alch. ‖ **88** καὶ ἐπίβαλλε **MV** : ἐπίβαλε **BA** ‖ ☾ω [i.e. ἀργύρῳ **BeRu**] **MBV** et
ܙܘܦܠ ܠܘܙܐ [adice ad lunam (i.e. argentum)] **SyrC** : ☾ **A** : ܙܘܦܠ ܠܘܙܐ [adice lunam (i.e.
argentum)] **SyrL** ‖ χρυσόν **BeRu** : ☉ **MBVA** ‖ ω☉ [i.e. χρυσῷ **BeRu**] **MBVA** ‖ **89** τὸ
A : τὸν **MBV** ‖ χρυσοκογχύλιον **MBVA** : ☉κόλλιον Moyses Alch. ‖ **90** ante κλαυδιανὸν add. in mg. περὶ ξανθ(ώσεως) ☉ଅ [i.e. χρυσοποιία] **MVA** ‖ μάρμαρον **MBA** et **V**
p.c. : -γάρον **V** a.c. ‖ **92** ξανθώσεις **MBV** : -σις **A** ‖ **93** ἐξιπωθείσης **MV** : ἐκσηπτωθεί-
σης **BA** ‖ pr. ἢ addidi ‖ θείου **MV** : -ω **BA** ‖ ἀρσενίκου **MV** : -ω **BA** ‖ σανδαράχης
scripsi : σανδαραⁿ **MV** : -άχη **BA** ‖ τιτάνου scripsi : τετάνω **M** : τετάνου **V** : τιτάνω
BA : ܚܡܚܚܣ ܘܡ ܚܠܡܦܩܚܣ. ܘܐ ܚܠܐܙܡܣܡܚ. ܘܐ ܚܪܝܙܚܢܚ. ܘܐ ܚܚܠܚܐ [coque eum cum alumine
vel arrhenico vel sandaracā vel calce] **SyrC** ‖ **94** ἐπιβάλλῃς **MV** : βάλλῃς **BA** ‖ ☾ω
[i.e. ἀργύρῳ **BeRu**] **B** et ܙܘܦܠ ܣܘܡ ܚܡܙ ܘ/ [et si adicis partim ad lunam (i.e. argentum)]
SyrC : ☾ **MVA** ‖ χρυσόν **A** : ☉ **BV** : ⁰☉ [i.e. χρυσός] **M** ‖ **95** ω☉ [i.e. χρυσῷ **BeRu**]
MB et ܚܚܣܚ [ad solem (i.e. aurum)] **SyrC** : ☉ **VA**

lead]: I do not mean 'lead' in its general sense—do not be misled—but with our black |lead] composed from Coptic stibnite and litharge,[30] or according to your knowledge. So roast it and, when it turns yellow, lay it on the matter: you will dye it. For nature delights in nature.

7. Process pyrite until it becomes incombustible, losing its blackness: process it with vinegar and brine, or pure urine, or seawater, or honey and vinegar, or according to your knowledge; so roast it until it becomes like the incombustible gold-dust. And as it becomes so, mix with it unburnt sulphur, or yellow alum, or Attic ochre, or according to your knowledge.[31] So lay it on silver in order to have gold [lit. for the sake of gold] and on gold in order to have gold shell [lit. for the sake of gold shell].[32] For nature masters nature.

8. Take *klaudianon*,[33] make it gleaming and process it as is customary, until it turns yellow. You must make it yellow; I do not mean the stone, but the useful part of the stone: you will make it yellow with dried [lit. squeezed out] alum, <or> sulphur, or orpiment, or realgar, or lime, or according to your knowledge. If you lay it on silver, you will have gold; if you [lay it] on gold, you will have gold shell. For nature masters nature by conquering it.

§ 6   TEST. 78–81 Zos. Alch. *CAAG* II 177,20–178,2 et 193,7–8 et 194,13 et 198,15; Olymp. Alch. II 94,1–2; Philos. Anon. Alch. (Zos. Alch. BeRu) *CAAG* II 137,6.
§ 7   **SyrL**: *CMA* II 10,17–20 (textus); 20 praec. 3 (translatio); **SyrC**: *1SyrC* § 3.
      TEST. 83–9 Moyses Alch. (A, fol. 273ᵛ5–12), ut monuerunt BeRu (*CAAG* II 307 n. 18), qui textum non ed. ‖ 84–5 Zos. Alch. *CAAG* II 185,6–7; Pelag. Alch. *CAAG* II 255,12; Philos. Anon. Alch. (Zos. Alch. BeRu) *CAAG* II 126,11–2.
§ 8   om. **SyrL**; **SyrC**: *1SyrC* § 4.

9. Τὴν κιννάβαριν λευκὴν ποίει δι᾽ ἐλαίου, ἢ ὄξους, ἢ μέλιτος, ἢ ἅλμης, ἢ στυπτηρίας· εἶτα ξανθὴν διὰ μίσυος, ἢ σώρεως, ἢ χαλκάνθου, ἢ θείου ἀπύρου, ἢ ὡς ἐπινοεῖς. Καὶ ἐπίβαλλε
100 ἀργύρῳ· καὶ χρυσὸς ἔσται, ἐὰν χρυσὸν καταβάπτῃς· ἐὰν χαλκόν, ἤλεκτρον. Ἡ φύσις τῇ φύσει τέρπεται.

10. Τὴν δὲ Κυπρίαν καδμίαν, τὴν ἐξωσμένην λέγω, λεύκαινε ὡς ἔθος. Εἶτα ποίει ξανθήν· ξανθώσεις δὲ χολῇ μοσχείᾳ, ἢ τερεβινθίνῃ, ἢ κικίνῳ, ἢ ῥαφανίνῳ, ἢ ᾠῶν λεκίθοις, ξανθῶσαι
105 αὐτὴν δυναμένοις, καὶ ἐπίβαλλε ἀργύρῳ· χρυσὸς γὰρ ἔσται διὰ τὸν χρυσὸν καὶ διὰ τὸ χρυσοζώμιον. Ἡ γὰρ φύσις τὴν φύσιν νικᾷ.

11. Τὸν ἀνδροδάμαντα οἰκονόμει οἴνῳ αὐστηρῷ, ἢ θαλάσσῃ, ἢ οὔρῳ, ἢ ὀξάλμῃ, τοῖς δυναμένοις σβέσαι αὐτοῦ τὴν φύσιν.
110 Λείου μετὰ στίμεως Χαλκηδονίου. Οἰκονόμει δὲ πάλιν θαλασσίῳ ὕδατι, ἢ ἅλμῃ, ἢ ὀξάλμῃ· ἀπόπλυνον, ἕως ἂν φύγῃ τοῦ στίμεως ἡ μελανία. Φρῦξον ἢ ὄπτησον, ἕως ξανθίσῃ, καὶ ἕψει ὕδατι θείῳ ἀθίκτῳ. Ἐπίβαλλε δὲ ἀργύρῳ, <καὶ ἔσται χρυσός·> καὶ ὅταν θεῖον ἄπυρον προσβάλῃς, ποιεῖς χρυσοζώμιον. Ἡ γὰρ
115 φύσις τὴν φύσιν κρατεῖ. [Οὗτός ἐστιν ὁ λίθος ὁ λεγόμενος χρυσίτης].

97 κιννάβαριν MV : κινά- BA ‖ 98 σώρεως MV : -ρῦος B : -ριος A ‖ 99 χαλκάνθου MV : -θης BA ‖ θείου ἀπύρου BVA : -ω -ρω M ‖ ἐπίβαλλε MV : -βαλε BA ‖ 100 ℭᵚ [i.e. ἀργύρῳ BeRu] M : ℭ BA et V p.c. : ◊ V a.c. ‖ ˢ◊ [i.e. χρυσός BeRu] M : ◊ BA : ῞◊ [i.e. ἥλιος (?)] V ‖ καταβάπτῃς scripsi : -η MBVA ‖ 101 ante φύσις add. γὰρ BeRu ‖ 102 supra καδμίαν add. τουτίαν B ‖ ἐξωσμένην MBVA : fort. ἐξυ- BeRu : ἐξιοσμένην (sic) Zur¹ ‖ 104 τερεβινθίνη MBA : τερεκιν- V ‖ κικίνω BA : κη- MV ‖ ᾠῶν λεκήθοις V et M in ras. : ὠ. λεκύνθοις BA ‖ 105 ἐπίβαλλε MV : -βαλε BA ‖ ℭᵚ [i.e. ἀργύρῳ] MBV : ℭ A : χρυσῷ BeRu ‖ 106 διὰ om. BA ‖ τὸ BA : τὸν MV ‖ 108 supra ἀνδροδάμαντα add. ◊ MV ‖ 110 ante λείου add. καὶ V ‖ στίμεως MV : στίμμῦος BA ‖ χαλκηδονίου BA : χαλκι- MV ‖ 111 φύγῃ MBA : ἐκφύγη V : φυγῆ BeRu ‖ 112 στίμεως MV : στίμμῦος BA ‖ ante φρῦξον add. καὶ V ‖ ξανθίση MV : -θωθῆ BA ‖ 113 ὕδατι om. BA : ܠܡ̈ܝܐ ܕܟܒܪܝܬܐ [cum sulfurum aquā] SyrL et ܕܟܒܪܝܬ̈ܐ ܠܡ̈ܝܐ [idem] SyrC ‖ ἐπίβαλλε MV : -βαλε BA ‖ ℭᵚ [i.e. ἀργύρῳ BeRu] MV et ܪܦܣܐ ܡܢ ܠܣܗܪܐ [adice partim ad lunam (i.e. argentum)] SyrL SyrC : ℭ BA ‖ καὶ ἔσται χρυσός addidi, coll. ܕܗܒܐ ܗܘܐ [et aurum fit] SyrL et ܗܘܐ ܕܗܒܐ [et sol (i.e. aurum) fit] SyrC ‖ 114 ποιεῖς Zur¹ : ποίει MBVA ‖ 115 κρατεῖ MBA et V p.c. : τέρπεται V a.c. ‖ Οὗτος — χρυσίτης tamquam ex glossemate seclusi : om. SyrL SyrC

9. Make cinnabar white[34] with oil, or vinegar, or honey, or brine, or alum; then make it yellow with *misy*, or *sōri*, or copper flower, or unburnt sulphur, or according to your knowledge. So lay it on silver: and it will be gold, if you dip gold [into the solution?];[35] if you dip copper, [it will be] electrum. For nature conquers nature.

10. Whiten Cyprian cadmia as is customary; I mean the cadmia that has been forced out [of its ores].[36] Then make it yellow: you shall yellow it with the bile of a calf, or terebinth resin, or castor oil, or radish oil, or egg yolks, which substances can make it yellow.[37] Then lay it on silver; it will be gold by means of the gold and of the ferment [lit. sauce/wash] of gold. For nature conquers nature.

11. Process *androdamas*[38] with rough wine, or seawater, or urine, or vinegar and brine, substances that can quench its nature. Grind it with Chalcedonian stibnite. Process it again with seawater, or brine, or vinegar and brine: wash until the blackness of the stibnite disappears. Parch and roast until it turns yellow, and you shall boil it with untouched divine water. Lay it on silver, <and it will be gold >, and if you add unburnt sulphur, you make ferment [lit. sauce/wash] of gold. For nature conquers nature. (This is the stone called *chrysitēs*.)

§ 9   om. SyrL; SyrC: *1SyrC* § 5.
    Test. 97–98 Zos. Alch. *CAAG* II 185,8–9.
§ 10   om. SyrL; SyrC: *1SyrC* § 6.
§ 11   SyrL.: *CMA* II 10,21–11,4 (textus); 20 praec. 4 (translatio); SyrC: *1SyrC* § 7.
    Test. 111 Zos. Alch. *CAAG* II 185,10; Philos. Anon. Alch. (Zos. Alch. BeRu) *CAAG* II 134,3–4 ‖ 111–2 Olymp.
    Alch. *CAAG* II 99,17–8; Philos. Christ. Alch. *CAAG* II 410,5–6 ‖ 112–3 Zos. Alch. *CAAG* II 185,11 et 14–5 ‖
    114–5 Zos. Alch. *CAAG* 157,21–5.

12. Λαβὼν γῆν λευκήν, λέγω τὴν ἀπὸ ψιμυθίου καὶ ἑλκύσματος ἢ στίμεως Ἰταλικοῦ καὶ μαγνησίας ἢ καὶ λευκῆς λιθαργύρου, λευκάνης· λευκανεῖς δὲ αὐτὴν θαλάσσῃ ἢ ἅλμῃ
120 τεθρυμμένην ἢ ὕδατι ἀερίῳ, ἐν δρόσῳ λέγω καὶ ἡλίῳ, ὥστε αὐτὴν λειουμένην γενέσθαι λευκὴν ὡς ψιμύθιον. Χώνευσον οὖν τοῦτο καὶ ἐπίβαλλε αὐτῷ χαλκοῦ ἄνθος, ἢ ἰὸν ξυστόν, οἰκονομηθέντα λέγω, ἢ χαλκὸν κεκαυμένον λίαν φθαρέντα, ἢ χαλκίτην· καὶ κυανὸν ἐπίβαλλε, ἕως γένηται ἄρρευστος καὶ
125 ἄτρητος· εὐχερῶς δὲ γενήσεται. Τοῦτό ἐστιν τὸ μολυβδόχαλκον. Δοκίμαζε οὖν εἰ γέγονεν ἄσκιον, καὶ ἐὰν μὴ γέγονε, τὸν χαλκὸν μὴ μέμψῃ, μᾶλλον δὲ σαυτόν, ἐπεὶ μὴ καλῶς ᾠκονόμησας. Ποίει οὖν ἀσκίαστον καὶ λείου καὶ πρόσβαλλε τὰ ξανθῶσαι δυνάμενα καὶ ὄπτα, ἕως ξανθὸν γένηται· καὶ ἐπίβαλλε πᾶσι
130 τοῖς σώμασιν. Ὁ γὰρ χαλκὸς ἀσκίαστος ξανθὸς γενόμενος πᾶν σῶμα βάπτει. Ἡ γὰρ φύσις τὴν φύσιν νικᾷ.

13. Τῷ θείῳ τῷ ἀπύρῳ συλλείου σῶρι καὶ χάλκανθον· τὸ δὲ σῶρί ἐστιν ὡς κυανὸς ψωρώδης εὑρισκόμενος ἀεὶ ἐν τῷ μίσυι· τοῦτο καὶ χλωρὸν χάλκανθον καλοῦσιν. Ὄπτησον οὖν αὐτὸ
135 μέσοις φωσὶν ἡμέρας τρεῖς, ἕως γένηται ξανθὸν φάρμακον·

117 supra λαβὼν add. ⳹ MV ‖ λ. γῆν λευκήν ... τὴν BVA : λ. γῆν λευκόν ... τὸν M : fort. λ. μόλυβδον λευκόν, coll. مه احر سوزا [cape plumbum album] SyrL SyrC ‖ ψιμυθίου BeRu : ψιμυθ MV : ψιμμι- BA ‖ 118 στίμεως MV : στίμμυος BA ‖ 119 λιℂᵛ-- [i.e. λιθαργύρου BeRu] MV : λιℂ B : λιθαρℂ A ‖ λευκάνης om. BA ‖ λευκάνεις (sic) BA : -ης MV ‖ 120 τεθρυμμένην scripsi post -μένη BeRu et سوزا هسوا حفحت الخنه زبهٖ : و oههّ مهلحنم [sic (scil. plumbum) album fit cum maris aquā et conteritur] SyrL SyrC : τεθρεωμένη MBVA ‖ ἡλίῳ MV : ⳹ BA ‖ 121 ψιμύθιον MV : ψιμμ - BA ‖ 122 τοῦτο scripsi : -ον MBVA ‖ ἐπίβαλλε MV : -βαλε BA ‖ ♀ᵛ [i.e. χαλκοῦ BeRu] MV : ♀ BA ‖ 122-3 οἰκονομηθέντα MVA : ὀκο- B ‖ 123 χαλκὸν A et ♀' M : ♀ B : ♀ V ‖ 124 κυανὸν ἐπίβαλλε MV : κ. -βαλε BA : سوا بو زهنا حه احرٖ بو زهٖاٖل مهسه [adice ad hoc (i.e. plumbum) venus (i.e. aes) caeruleumque] SyrL SyrC ‖ post ἕως add. ἂν V ‖ 125 μολυβδόχαλκον BeRu : μολιβδό- MBVA ‖ supra μολυβδόχαλκον add. ♍ ♄ M ‖ 126 οὖν MV : γοῦν BA ‖ ἄσκιον BA : ἀσκίαστον MV ‖ χαλκὸν A et ♀' MV et ♀" B ‖ 127 ἐπεὶ MVA : ἐπὲ (sic) B ‖ ᾠκονόμησας BA : οἰκο- MV ‖ 128 πρόσβαλλε V : βάλε MBA ‖ 129 ἐπίβαλλε MV : -βαλε BA ‖ 130 ♀ˢ [i.e. χαλκὸς BeRu] MV : ♀ᵈ B : ♀A A ‖ ξανθὸς om. V ‖ post ξανθὸς add. ὢν MBA ‖ 134 χλωρὸν MVB : χλο- A ‖ supra χλωρὸν add. ☽ M ‖ 135 ἡμέρας MV : 66 BA

12. Take white earth[39]—I mean the earth composed by white lead[40] and *helkysma*,[41] or by Italian stibnite and *magnēsia*, or by white litharge[42] as well—and whiten it. You shall whiten it by crushing it in seawater, or brine, or rainwater—I mean under the dew and in the sun—so that after being ground it turns as white as white lead. Then melt this [product] and add to it copper flower, or the rust that has been scraped off—I mean the one that has been processed—or burnt copper in an advanced state of decay, or *chalkitēs*; then add azurite, until it becomes solid [lit. loses its liquidity] and unperforable; it will easily become so. This is the *molybdochalkon* [i.e. lead–copper alloy]. Then test whether it has become 'shadowless'; if not, do not blame the copper, but rather yourself, since you did not process it in the right way. Then make it 'shadowless,' grind it, add the substances that can make it yellow and roast until it turns yellow. Then lay it on any [metallic] body.[43] In fact, the 'shadowless' copper, after turning yellow, dyes any [metallic] body. For nature conquers nature.

13. Grind *sōri* and copper flower together with unburnt sulphur; *sōri* looks like azurite that easily peels off and is always found in *misy*. It is also called green copper flower. Roast it at moderate heat for three days, until

§ 12 **SyrL:** *CMA* II 11,5–12 (textus); 20–1 praec. 5 (translatio); **SyrC:** *1SyrC* § 8.
   TEST. 124–5 Steph. Alch. II 232,32 Ideler ‖ 126–7 Pelag. Alch. *CAAG* II 254,9–11; Philos. Anon. Alch. (Zos. Alch. BeRu) *CAAG* II 126,12–4 ‖ 130–1 Pelag. Alch. *CAAG* II 257,11–2; Philos. Anon. Alch. (Zos. Alch. BeRu) *CAAG* II 126,14–5.
§ 13 **132–7 SyrL:** *CMA* II 11,13–5 (textus); 21 praec. 6 (translatio); **SyrC:** *1SyrC* § 9.
   TEST. 133–4 Zos. Alch. *CAAG* II 146,19–20.

ἐπίβαλλε χαλκῷ ἢ ἀργύρῳ τῷ ἐξ ἡμῶν γενομένῳ, καὶ ἔσται χρυσός.

Τοῦτο κατάθες γενόμενον πέταλον εἰς ὄξος καὶ χάλκανθον καὶ μίσυ καὶ στυπτηρίαν καὶ ἅλας Καππαδοκικὸν καὶ νίτρον
140 πυρρόν, ἢ καὶ ὡς ἐπινοεῖς, ἐπὶ ἡμέρας τρεῖς ἢ πέντε ἢ ἕξ, ἕως γένηται ἰός, καὶ καταβάψεις· τὸν γὰρ χρυσὸν ποιεῖ ἡ χάλκανθος ἰὸν χρυσοῦ. Ἡ φύσις τῇ φύσει τέρπεται.

14. Χρυσόκολλαν τὴν τῶν Μακεδόνων τὴν ἰῷ χαλκοῦ παρεμφέρουσαν οἰκονόμει, λειῶν οὔρῳ δαμάλεως ἕως ἐκστραφῇ· ἡ
145 γὰρ φύσις ἔσω κρύπτεται. Ἐὰν οὖν ἐκστραφῇ, κατάβαψον αὐτὴν εἰς ἔλαιον κίκινον πολλάκις πυρῶν καὶ βάπτων· εἶτα δὸς ὀπτᾶσθαι, σὺν στυπτηρίᾳ προλειώσας, μίσυι ἢ θείῳ ἀπύρῳ· ποίει ξανθὴν καὶ ἐπίβαπτε πᾶν σῶμα <χαλκοῦ, ἀργύρου>, χρυσοῦ.

150 15. Ὦ φύσεις φύσεων δημιουργοί, ὦ φύσεις παμμεγέθεις ταῖς μεταβολαῖς νικῶσαι τὰς φύσεις, ὦ φύσεις ὑπὲρ φύσιν τέρπουσαι τὰς φύσεις. Ταῦτα δὴ οὖν εἰσι τὰ μεγάλην ἔχοντα τὴν φύσιν· τούτων τῶν φύσεων οὐκ εἰσὶν ἄλλαι μείζους ἐν βαφαῖς, οὐκ ἴσαι, οὐχ ὑποβεβηκυῖαι· ταῦτα ἀναλυόμενα πάντα
155 ἐργάζεται. Ὑμᾶς μὲν οὖν, ὦ συμπροφῆται, οἶδ᾽ οὐκ ἀπιστήσαντας, ἀλλὰ γὰρ καὶ θαυμάσαντας· ἴστε γὰρ τῆς ὕλης τὴν

---

136 ante ἐπίβαλλε add. καὶ **V** ‖ ἐπίβαλλε **MV** : -βαλε **BA** ‖ ♀ω [i.e. χαλκῷ BeRu] **BA** et ♀ω **MV** ‖ ☾ω [i.e. ἀργύρῳ BeRu] **M** : ☾ **BVA** ‖ **137** σ₴ [i.e. χρυσός BeRu] **M** et χρυσός cum ₴ s.l. **A** : ₴ **BV** ‖ **138-42** om. SyrL SyrC ‖ **139** νίτρον BeRu : Ν̅ **MBVA** ‖ **140** καὶ om. **BA** ‖ ἡμέρας **MV** : 66 **BA** ‖ **141** καταβάψεις **BA** : -ης **MV** ‖ ᾽₴ [i.e. χρυσὸν BeRu] **MV** : ₴ **BA** ‖ **142** ἰὸν χρυσοῦ scripsi : ἰὸν cum ₴ s.l. **MB** : ἰὸν₴ **A** : ἰὸν **V** ‖ ante φύσις add. γὰρ BeRu ‖ **143** ante χρυσόκολλαν add. ₴ **V** ‖ supra χρυσόκολλαν add. ₴ **M** ‖ ♀υ [i.e. χαλκοῦ BeRu] **MV** : ♀ **BA** ‖ **144** οὔρῳ **MBV** : -α **A** ‖ **145** κρύπτεται **MV** : κέκρυπται **BA** ‖ **146** αὐτὴν **BA** : -ὸν **MV** ‖ **148** ποίει **MVA** : πύει **B** ‖ ξανθὴν scripsi : ξανθ **MV** : -ὸν **BA** ‖ χαλκοῦ, ἀργύρου addidi, coll. Zos. Alch. *CAAG* II 195,11-2 : ܐܕܝܩ ܘܩܘ̈ ܘܐܫܟܚ ܐܝܠܝܢ [adice ac coque atque invenies quae quaeris] SyrL SyrC ‖ **149** χρυσοῦ **MV** : -όν **BA** ‖ **150** παμμεγέθεις **MBV** : -ης **A** ‖ **151** alt. φύσεις **MVA** : -ις **B** ‖ **152** οὖν om. **BA** ‖ **153** τὴν om. **BVA** ‖ **154** ὑποβεβηκυῖαι **MV** : ὑπερπεριβαίνουσαι **B** : ὑπερβαίνουσαι **A** ‖ **155** ἐργάζεται **MV** : -ζονται **BA** ‖ ante οἶδ᾽ οὐκ add. οὐκ **BA** ‖ **155-6** ἀπιστήσαντας **MBVA** : -οντας BeRu

it becomes a yellow drug.[44] Lay it on the copper or on the silver that is made by us and it will be gold.[45]

Make this into leaves and put them into vinegar, and copper flower, and *misy*, and alum, and Cappadocian salt, and red soda, or according to your knowledge, for three or five or six days until rust appears, and you will dye them: for copper flower makes gold into rust of gold.[46] For nature delights in nature.

14. Process the Macedonian malachite that looks like copper rust by grinding it with the urine of a heifer, until the malachite is turned inside out: for nature is hidden inside. When it is turned inside out, dip it into castor oil: warm and dip it several times; then let it be roasted after grinding it with alum, *misy* or unburnt sulphur: make it yellow and dip any [metallic] body [into it]: <copper, silver>, gold.[47]

15. O natures, artificers of natures! O greatest natures that conquer natures with your transformations! O natures above nature, which delight in natures! These are the substances that have a great nature; no other natures are greater than these natures in dyeing, no others are equivalent, no others are subordinated: these substances when dissolved produce everything. O you who are prophets with me, I know that you are not incredulous, but rather open to wonder; you know in fact the power of the matter. And [you

§ 13 **138–42** om. SyrL SyrC.
   Test. **141–2** Zos. Alch. *CAAG* II 161,20–1; Steph. Alch. II 244,13–4 Ideler.
§ 14 **SyrL:** *CMA* II 11,15–9 (textus); 21 praec. 7 (translatio); **SyrC:** *1SyrC* § 10.
   Test. **144–5** Syn. Alch. *CAAG* II 60,7–8; [Zos. Alch.] *CAAG.* II 202,20–1; Philos. Anon. Alch. (Zos. Alch. BeRu) *CAAG* II 129,12–3 ‖ **148–9** Zos. Alch. *CAAG* II 195,10–1
§ 15 om. SyrL; **SyrC:** *1SyrC* § 11.
   Test. **150–1** Syn. Alch. *CAAG* II 63,18–9; Pelag. Alch. *CAAG* II 260,14–5; Steph. Alch. II 215,16 et 215,26 et 247,13 Ideler; Philos. Anon. Alch. (Zos. Alch. BeRu) *CAAG* II 18–9; Philos. Christ. Alch. *CAAG* II 277,4–6.

δύναμιν. Τοὺς δὲ νέους πάνυ βλαβησομένους καὶ ἀπιστήσοντας
τῇ γραφῇ διὰ τὸ ἐν ἀγνοίᾳ τῆς ὕλης ὑπάρχειν αὐτούς, οὐκ
εἰδότας ὅτι ἰατρῶν μὲν παῖδες, ὁπηνίκα ὑγιεινὸν φάρμακον
160   βούλοιντο κατασκευάσαι, οὐκ ἀκρίτῳ ὁρμῇ τοῦτο πράττειν
ἐπιχειροῦσιν, ἀλλὰ γὰρ πρῶτον δοκιμάσαντες ποῖόν ἐστιν
θερμόν, ποῖον δὲ τούτῳ συνερχόμενον μέσην ἀποτελεῖ κρᾶσιν,
ψυχρὸν ἢ ὑγρὸν ἢ ὁποῖον τὸ πάθος, εἰ κατάλληλον τῇ μέσῃ
κράσει· καὶ οὕτως προσφέρουσιν τὸ πρὸς ὑγίειαν κριθὲν αὐτοῖς
165   φάρμακον.

16. Οὗτοι δὲ ἀκρίτῳ καὶ ἀλόγῳ ὁρμῇ τὸ τῆς ψυχῆς ἴαμα καὶ
παντὸς μόχθου λύτρον κατασκευάσαι βουλόμενοι, οὐκ αἰσθήσονται βλαβησόμενοι. Δοκοῦντες γὰρ ἡμᾶς μυθικόν, ἀλλ᾽ οὐ
μυστικὸν ἀπαγγέλλειν λόγον, οὐδεμίαν ἐξέτασιν ποιοῦνται τῶν
170   εἰδῶν· οἷον εἰ τόδε μέν ἐστι σμηκτικόν, τόδε δὲ ἐπιβλητέον, καὶ
εἰ τόδε μέν ἐστιν βαπτικόν, τόδε δὲ ἁρμοστέον, καὶ εἰ τόδε τὴν
ἐπιφάνειαν ποιεῖ, καὶ εἰ κατὰ τὴν ἐπιφάνειαν ἔσται φευκτὸν
καὶ ἐκ τοῦ βάθους φεύξεται, καὶ εἰ τόδε μέν ἐστι πυρίμαχον,
τόδε <δὲ> προσπλακὲν πυρίμαχον ποιεῖ· οἷον εἰ τὸ ἄλας σμήχει
175   τὰ ἐπάνω τοῦ χαλκοῦ καὶ τὰ ἐντὸς ἐξ ἅπαντος σμήχει, καὶ εἰ
ἰοῖ τὰ ἔξω μετὰ τὴν σμῆξιν καὶ τὰ ἐντὸς ἰοῖ· καὶ εἰ τὰ ἔξω τοῦ
χρυσοχάλκου λευκαίνει καὶ σμήχει ἡ ὑδράργυρος, καὶ τὰ ἐντὸς
λευκαίνει· καὶ εἰ φεύγει ἔξωθεν καὶ ἐκ τῶν ἐντὸς φεύξεται. Εἰ
ἐν τούτοις ὑπῆρχον ἀσκούμενοι οἱ νέοι, οὐκ ἂν ἐδυστύχουν,
180   κρίσει ἐπὶ τὰς πράξεις ὁρμῶντες· οὐ γὰρ ἐπίστανται τὰ τῶν
φύσεων ἀντιπαθῆ, ὡς ἓν εἶδος δέκα ἀνατρέπει. Ῥανὶς γὰρ
ἐλαίου οἶδε πολλὴν ἀφανίσαι πορφύραν, καὶ ὀλίγον θεῖον εἴδη

---

157 ἀπιστήσοντας Zur¹ : -αντας **MBVA** ‖ 159 εἰδότας scripsi : -ες **MBVA** ‖ ὁπηνίκα **BVA**: ὅπηνικᾶ (sic) **M** ‖ 160 ante ὁρμῇ add. δὲ **V** ‖ 161 γὰρ **M** s.l. et **V**: om. **BA** ‖
162 μέσην **BA** : -ον **MV** ‖ κρᾶσιν **MBA** : κράσει **V** ‖ 163 ὁποῖον **BA** : ὅπη ὂν **MV** ‖
164 ὑγίειαν **MV** : ὑγείαν **AB** ‖ αὐτοῖς **BA** et **V** p.c. : -ὴν **M** et **V** a.c. ‖ 166 ὁρμῇ
MVB : -ὴν **A** ‖ 167 λύτρον **MV** : λυτήριον **B** : λητήριον **A** ‖ 167-8 αἰσθήσονται
MVB : ἐσθή- **A** ‖ 170 δὲ om. **MV** ‖ 171 μέν om. **MV** ‖ δὲ om. **MV** ‖ εἰ τόδε τὴν
scripsi : τόδε εἰ τὴν **BA** : τῷδε εἰ τὴν **MV** ‖ 174 δὲ addidi ‖ ante προσπλακὲν add.
μᾶλλον **BA** ‖ 175 pr. τὰ **MV** : τὸ **BA** ‖ ♀ᵘ [i.e. χαλκοῦ BeRu] **M** : ♀̂ **BA** : ♀ **V** ‖ εἰ
om. **BA** ‖ 177 χρυσοχάλκου BeRu : ♀ᵘᵈ **MV** : ♀ᵈ **BA** ‖ ☽ᵟ [i.e. ὑδράργυρος BeRu]
MV : ☽ **BA** ‖ 179 ἀσκούμενοι **MBV** : διδασκού- (διδ- in ras.) **A** ‖ 180 κρίσει
MV : ἀκρίτως **BA** : ante κρίσει add. οὐ Zur²

know] that because of their ignorance of the matter young men will be misled and distrustful of this writing, since they do not know that students of physicians who want to prepare a beneficial drug do not set about making it on a rash impulse, but first of all they test which kind of [drug] is hot; which kind, when joined to it, produces a balanced mixture; which kind is cold or wet; and of which kind is the affection, whether it [the drug?] is appropriate for the balanced mixture. In this way they administer the drug that has been judged suitable for good health.

16. On the contrary, these novices who rashly and without reasoning want to prepare a medicine for the soul and for relieving any distress are not aware that they will fail. In fact, they believe that we are presenting a legendary rather than a secret discourse, so they do not carry out any close examination of the species: for example, where one species can cleanse, another can be applied; where one species can dye, another can combine; and whether one species can make things bright and, with respect to this brightness, whether it is vanishing and vanishes from inside; and whether one species can resist fire, and another, when mixed, can make things fire-resisting; for instance, whether salt cleanses the surface of the copper and whether it properly cleanses its inner part; and, after this cleansing process, whether it rusts the surface and whether it rusts the inner part; and whether mercury cleanses and whitens the surface of a gold–copper alloy, and whether it makes its inner part white; and whether it [i.e. the whitening produced by mercury] vanishes from the surface and whether it will vanish from inside. If these novices had practised these kinds of investigations, they would not be in trouble, since they could set to work with good judgement. But they do not know the antipathies of natures, how one species upsets ten: a drop of oil can remove much purple, and a pinch of

κατακαῦσαι πολλά. Ταῦτα μὲν οὖν περὶ τῶν ξηρίων καὶ ὅπως
δεῖ προσέχειν τῇ γραφῇ εἰρήσθω. Φέρε δὴ καὶ τοὺς ζωμοὺς
185 καθεξῆς εἴπωμεν.

17. Λαβὼν τὸ Πόντιον ῥᾶ, λείου ἐν οἴνῳ Ἀμιναίῳ αὐστηρῷ
καὶ ποίει πάχος κηρωτῆς. Δέξαι πέταλον τὸ μήνης, ἵνα ποιήσῃς
τὸν χρυσόν· κατεργάζου ὀνυχοπαχὲς καὶ τούτου πάλιν ἰσχνό-
τερον. Χρῖσον τοῦ φαρμάκου <τὸ ἥμισυ> καὶ θὲς εἰς καινὸν ἀγ-
190 γεῖον περιφιμῶν πάντοθεν· ὑπόκαιε ἠρέμα ἕως μεσασθῇ. Εἶτα θὲς
τὸ πέταλον εἰς τὸ τοῦ φαρμάκου λείψανον καὶ ἄνες οἴνῳ τῷ
τεταγμένῳ ἕως χυλώδης ζωμός σοι φανῇ· εἰς τοῦτον κάθες εὐθὺς
τὸ πέταλον, μήπω ψυγέν· ἔα συμπιεῖν. Εἶτα λαβὼν χώνευσον,
καὶ εὑρήσεις χρυσόν. Ἐὰν δὲ τὸ ῥᾶ ᾖ παλαιὸν τῷ χρόνῳ, πρόσμι-
195 ξον αὐτῷ ἐλυδρίου τὸ ἴσον προταριχεύσας ὡς ἔθος. Τὸ γὰρ ἐλύδριον ἔχει συγγένειαν πρὸς τὸ ῥᾶ. Ἡ φύσις τῇ φύσει τέρπεται.

18. Δέξαι κρόκον Κιλίκιον· ἄνες ἅμα ἄνθη τοῦ κρόκου τῷ
προταγέντι χυλῷ τῆς ἀμπέλου· ποίει ζωμὸν ὡς ἔθος. Βάπτε
ἄργυρον ἐκ πετάλων ἕως ἀρέσῃ τὸ χρῶμα· ἐὰν δὲ χάλκεον τὸ

---

183 πολλά **MBA** : πλεῖστα **V** ‖ ξηρίων **MV** : -ρῶν **BA** ‖ 184 δεῖ **MBA** : χρὴ **V** ‖
εἰρήσθω **MV** : εἴρηται **BA** ‖ 185 εἴπωμεν **MBVA** : -ομεν **Vᵃ** ‖ 186 supra Πόντιον add.
Ⲙ **BA** : Ποντικὸν dubit. prop. BeRu ‖ ῥᾶ **MBVA** : ράκος **Vᵃ** ‖ ἀμιναίῳ
MVVᵃ : ἀμοινέω et s.l. ◉ **BA** ‖ 187 πέταλον **MBVAVᵃ** : -α Phil. Anon. Alch. *CAAG* II
264,14 et ـمه ـܪܝ [cape folia] **SyrC** ‖ μήνης **MVVᵃ** et B qui add. ☽ s.l. : μίνης☽ **A** ‖
ποιήσῃς **MBV** : -εις **A** : ποιή (sic) **Vᵃ** ‖ 188 ⲟ [i.e. χρυσόν BeRu] **MV** : ⲟ **BAVᵃ** ‖
ὀνυχοπαχὲς scripsi : ὀνυχοπαχ **MV** : -χόπαχον **BA** : -χοπαχῆ **Vᵃ** : ὄνυχος πάχη
Heiberg in *CMAG* II 333 ‖ τούτου **MVVᵃ** : -ω **BA** ‖ 189 χρῖσον **V** et ܣܘܩ [illine] **SyrC** :
χρήσου **M** : χρήσι **B** : χρῆσι **A** : χρῖσαι **Vᵃ** : χρίση prop. BeRu ‖ τοῦ φαρμάκου **MV** :
τῶ -ω **BA** : ἐκ τοῦ -ου **Vᵃ** ‖ τὸ ἥμισυ addidi, coll. *AP* § 7, l. 64 et Phil. Anon. Alch.
CAAG II 264,15 ‖ θὲς **MBVA** : ἐπίθες **Vᵃ** ‖ 190 περιφιμῶν **B** : περίφιμον **MVVᵃ** :
περιφ' ἡμῶν **A** ‖ post ἕως add. οὗ **BA** ‖ 191 καὶ om. **M** ‖ τῷ om. **VVᵃ** ‖ 192 χυλώδης
MVVᵃ : χυλώδ **B** : χυλώδεις **A** ‖ φανῇ **MBVAVᵃ** : φάνη (incuria?) BeRu ‖ κάθες
MVVᵃ: κατάθες **BA** ‖ εὐθὺς **A** : εὐθὺ **MBVVᵃ** ‖ 193 ante ἔα add. εἶτα **Vᵃ** ‖ 194 ⲟ
[i.e. χρυσόν BeRu] **M** : ⲟ **BAVᵃ** : ⲟ **V** ‖ 196 ante φύσις add. γὰρ **VVᵃ** ‖ 197 κιλίκιον
MV et B qui ⲃ s.l. add. : κιλίκιονⲃ **A** ‖ ἄνθη BeRu : -ει **MBVA** ‖ 199 post ἄργυρον
add. ☾ **A** ‖ πετάλων **MV** : -ου **A** et B ut vid. (πε[...]ου) : ܣ[...] , ܣܩܝܠ ܚܘܣ [inice
veneris (i.e. aeris) folia] **SyrC** ‖ ἀρέσῃ **MV** et ܣܘܩ ܩܦܢ ܚܘܣ [donec eorum color
placeat] **SyrC** : ἀρεστόν σοι φανῇ [-εῖ **A**] **BA**

sulphur can burn many species. Let this suffice concerning dry substances and concerning how people must approach this writing. Let us deal with washes [i.e. dyeing liquids] in the following part.[48]

17. Take[49] Pontic rhubarb[50] and triturate it into rough Aminean wine and make it as thick as a salve. Take a silver leaf in order to make gold; make it as thick as a fingernail and even thicker than that. Rub <half> of the drug [on the leaf] and put it into a new vessel well closed all around; burn by applying a gentle fire below until [the drug] soaks in [lit. reaches the middle (of the leaf)]. Then put the leaf into the rest of the drug and dilute with the wine prescribed above, until the wash looks thick. Put the leaf again into this solution before it gets cold: let [the leaf] soak it up. Then take it out, melt it, and you will find gold. When the rhubarb is stale, mix it with the same amount of celandine that has been macerated as is customary. For celandine has affinity with rhubarb. Nature delights in nature.

18. Take Cilician saffron; dissolve the saffron flowers with the above prescribed grape juice; make a wash as is customary. Dip in silver leaves, until you like their colour. But, it is better if the leaf is made of copper;

§ 17 om. SyrL; SyrC: 1SyrC § 13.
    Test. 183–91 Philos. Anon. Alch. (Joan. Alch. BeRu) CAAG II 264,13–8.
§ 18 om. SyrL; SyrC: 1SyrC § 14.
    Test. 197–8 Steph. Alch. II 247,15–7 Ideler.

200 πέταλον ἔσται, βέλτιον. Προκάθαιρε δὲ τὸν χαλκὸν ὡς ἔθος. Εἶτα βαλὼν ἀριστολοχίας βοτάνης μέρη δύο, καὶ κρόκου καὶ ἐλυδρίου τὸ διπλοῦν, ποίει πάχος κηρωτῆς· καὶ χρίσας τὸ πέταλον,
ἀπεργάζου τῇ πρώτῃ ἀγωγῇ, καὶ θαυμάσεις. Ὁ γὰρ Κιλίκιος
κρόκος τὴν αὐτὴν τῇ ὑδραργύρῳ ἔχει ἐνέργειαν, ὡς ἡ κασία τῷ
205 κινναμώμῳ. Ἡ φύσις τὴν φύσιν νικᾷ.

19. Λαβὼν μόλυβδον τὸν ἡμῶν, τὸν γενόμενον ἄρρευστον διὰ
γῆς Χίας καὶ Πάρου καὶ στυπτηρίας, χώνευσον ἀχύροις καὶ
κατέρα εἰς πυρίτην καὶ κρόκον καὶ κνήκου καὶ οἰχομενίου
ἄνθος καὶ ἐλύδριον καὶ κροκόμαγμα καὶ ἀριστολοχίαν. Λείου
210 ὄξει δριμυτάτῳ, καὶ ποίει ζωμὸν ὡς ἔθος· καὶ τῇ ῥᾷ τὸν
μόλυβδον ἔα συμπιεῖν, καὶ εὑρήσεις χρυσόν. Ἐχέτω δὲ τὸ σύνθεμα καὶ θεῖον ἄπυρον ὀλίγον. Ἡ φύσις τὴν φύσιν κρατεῖ.

20. Αὕτη ἡ <ἀγωγὴ τοῦ> Παμμένους ἐστίν, ἣν ἐπεδείξατο
τοῖς ἐν Αἰγύπτῳ ἱερεῦσιν. Ἕως τῶν φυσικῶν τούτων ἐστὶν ἡ τῆς
215 χρυσοποιίας ὕλη. Μὴ θαυμάσητε δὲ εἰ ἓν εἶδος τὸ τοιοῦτον
ἀπεργάζεται μυστήριον· οὐχ ὁρᾶτε ὡς πολλὰ φάρμακα καὶ
μόλις χρόνῳ τὴν ἐκ σιδήρου κολλήσει τομήν, κόπρος δὲ
ἀνθρωπεία οὐ χρόνῳ τοῦτο ποιεῖ; Καὶ καυστῆρσι μὲν πολλὰ

200 post βέλτιον add. ἔστι **V** ‖ χαλκὸν BeRu : ♀ **MV** : ♀ **B** : χαλκὸν♀ **A** ‖ **201** μέρη
BA : ⏑̄ **MV** ‖ **203** τῇ **MBA** : τῶ **V** ‖ **205** κινναμώμω **MBV** : κινα- **A** ‖ ἡ φύσις —
νικᾷ om. **V** ‖ ante φύσις add. γὰρ BeRu ‖ **206** μόλυβδον BeRu : μόλιβ- **MBVA** ‖ supra
μόλυβδον add. ☽ **BA** ‖ **207** χίας **BA** : χείας **MV** ‖ supra Χίας et Πάρου et στυπτηρίας
add. ℳ **BA** ‖ **208** πυρίτην **MBVA** : ܘܐܝܘܐ [swṭws (?); ἄνθος prop. Berthelot-Duval
CMA II 12] SyrL : ܘܐܝܘܐ [swṭws (?)] SyrC ‖ κνήκου BeRu : κνίκου **MV** : κνίκον
BA ‖ οἰχομενίου **MBVA** : ܢܘܐܘܡܟ [kmwnyqwn (?)] SyrL : ܢܘܐܘܡܟ [kmwnywn (?)]
SyrC : ἠχουμενίου Fal ‖ **209** λείου **BA** : -οι **MV** ‖ **211** μόλυβδον BeRu : μόλι- **MB**
VA ‖ ⟨ [i.e. χρυσόν BeRu] **M** : ⟨ **BVA** ‖ **212** ante φύσις add. γὰρ BeRu ‖ **213** ἀγωγὴ
τοῦ addidi : μέθοδος τοῦ add. Hammer Jensen, *Die älteste Alchymie*, 88 ‖ Παμμένους
MBVA : Παμμεγέθους Preisendanz *RE* XVIII/2 (1942) 1633, post Hammer Jensen, *Die
älteste Alchymie*, 88 : ܢܘܐ ܐܝܘܡܟ ܘܪܐ [ista (scil. natura?) maxima (est), i.e. fort. αὕτη ἡ
παμμεγέθης] SyrL SyrC ‖ **213-4** ἣν ἐπεδείξατο — ἱερεῦσιν **MBVA** : ܠܝܘܡܟ ܐ̈ܚܕܡܚܐ̇ܘ
ܡ ܠܐ ܘ ...ܢܘܐܝܘܡܟ ܘܠܐ ܡ [(natura?) quam docui Aegyptios cum sacerdotes venisset...] SyrC : fort.
ἐπεδειξάμην leg. ‖ **214** ἱερεῦσιν. Ἕως distinxi : ἱερεῦσιν· ἔ. **MBVA** : ἱερεῦσιν, ἔ.
BeRu ‖ **216** ὁρᾶτε **MBV** : -αι **A** ‖ **217** κολλήσει **MBVA** : fort. οὐ κ. BeRu ‖ **218**
ἀνθρωπεία **MV** : ἀνθρώπου **BA**

clean first the copper as is customary. Then add two parts of the herb *Aristolochia*, the double amount of saffron and celandine, and make them as thick as a salve. Rub the leaf [with it], work following the foregoing method and you will marvel. In fact, Cilician saffron produces the same effect as mercury as well as cassia [produces the same effect as] cinnamon.[51] Nature conquers nature.

19. Take our lead,[52] the lead that has become harder [by losing its own moisture] by means of Chian earth, Parian earth, and alum; melt it with [a fire of] chaff and pour it over pyrite, and saffron, and flowers of safflower and *oichomenion*,[53] and celandine, and saffron-sauce, and *Aristolochia*. Triturate [these ingredients] with the sharpest vinegar, and make a wash as is customary. Let the lead soak up rhubarb, and you will find gold. Let the compound have also a pinch of unburnt sulphur.[54] Nature masters nature.

20. This is <the method> of Pammenes which he taught to the Egyptian priests.[55] The matter for the making of gold extends up to these natural substances. But do not marvel that just one species can perform such a mystery. Do you not see that many drugs can [only] with difficulty, even given time, close up a wound made by iron, while human excrement produces this effect immediately? And often, when many drugs are

§ 18 TEST. 200–1 Zos. Alch. *CAAG* II 183,3–5.
§ 19 SyrL: *CMA* II 11,19–12, 2 (textus); 21 praec. 8 (translatio); SyrC: *1SyrC* § 15.
    TEST. 206–7 Zos. Alch. *CAAG* II 162,7–9.
§ 20 SyrL: *CMA* II 12,2–5 (textus); 22 praec. 9 (translatio); SyrC: *1SyrC* § 16.
    TEST. 213 Zos. Alch. *CAAG* II 148,15–6; Steph. Alch. II 234,25ff. Ideler |ubi Παμμένης leg. (M, fol. 31ᵛ3–4) : παμμέγης Ideler|.

προσφερόμενα φάρμακα οὐδὲν ἀνύσει πολλάκις, μόνη δὲ
220   ἄσβεστος οἰκονομηθεῖσα ἰᾶται τὸ πάθος· καὶ ὀφθαλμίᾳ μὲν
πολλάκις ποικίλη προσφερομένη πραγματεία οἶδε καὶ βλάψαι,
ῥάμνος δὲ τὸ φυτὸν πρὸς πᾶν τὸ τοιοῦτον ποιοῦσα πάθος οὐκ
ἀποτυγχάνει. Δεῖ οὖν καταφρονεῖν τῆς ματαίας καὶ ἀκαίρου
ὕλης ἐκείνης, χρᾶσθαι δὲ μόνοις τοῖς φυσικοῖς. Νῦν δὲ καὶ ἐκ
225   τούτου κρίνατε· ἄνευ τῶν προειρημένων φύσεων, τίς ἀπείρ-
γασταί ποτε; Εἰ δὲ ἄνευ τούτων οὐδὲν ἔστιν ποιῆσαι, τί ἀγαπῶ-
μεν τὴν πολύυλον φαντασίαν; Τί ἡμῖν καὶ πολλῶν εἰδῶν ἐπὶ τὸ
αὐτὸ συνδρομή, μιᾶς φύσεως νικώσης τὸ πᾶν; Ἴδωμεν δηλαδὴ
καὶ τῶν εἰς ἀργυροποιίαν εἰδῶν τὴν σύνθεσιν.

223 ἀποτυγχάνει **BV** : -η **M** : ὁποτυγ- (sic) **A** ‖ δεῖ **MBA** : χρὴ **V** ‖ **224** χρᾶσθαι **MV**
χρῆ- **AB** ‖ καὶ om. **BA** ‖ **225** post κρίνατε add. ὅτι **BA** ‖ **225-6** τίς... ποτε **MV**
οὐδεὶς... ποτὲ τί **BA** ‖ **226** ποιῆσαι **MBV** : -εῖσαι **A** ‖ **227** ἡμῖν **MV** : ὑμῖν **BA** ‖ **22**
συνδρομή **MV** : -ῇ **BA** : fort. -εῖ BeRu ‖ ἴδωμεν **MBV** : εἶδ- **A** ‖ post ἴδωμεν ad(
οὖν **V**

administered for cauterising, they do not produce any effect, while treated quicklime alone cures the disease. And often, when a complex treatment is provided for *ophthalmia* it can be harmful, while the *Rhamnus* plant, which has an effect on any disease of this kind, does not fail. So we have to disregard such a worthless and unsuitable matter, and make use only of the natural substances. And now evaluate the question also on the basis of this point: who could ever work without the above mentioned natures? And, if working is not possible without these natures, why do we like the illusion that a plurality of matters exists? Why do we need many species to concur in order to do the same, when one nature conquers everything? Let us clearly examine also the composition of the species for the making of silver.

§ 20 TEST. 226–7 Syn. Alch. § 5, ll. 59–60; Steph. Alch. II 247,9–10 Ideler ‖ 227–8 Pelag. Alch. *CAAG* II 257,13–4; Steph. Alch. II 200,24–5 et 200,35 et 214,37–215,2 et 216,30 et 247,10 Ideler; Philos. Anon. Alch. *CAAG* II 437,5.

Περὶ ἀσήμου ποιήσεως

1   1. <Λαβὼν> ὑδράργυρον τὴν ἀπὸ τοῦ ἀρσενικοῦ, ἢ σανδαράχης, ἢ ὡς ἐπινοεῖς, πῆξον ὡς ἔθος, καὶ ἐπίβαλλε χαλκῷ <ἢ>
σιδήρῳ θειωθέντι, καὶ λευκανθήσεται. Τὸ δ' αὐτὸ ποιεῖ καὶ μαγνησία λευκανθεῖσα καὶ ἀρσενικὸν ἐκστραφὲν καὶ καδμία ὀπτὴ
5   καὶ σανδαράχη ἄπυρος καὶ πυρίτης λευκανθεὶς καὶ ψιμύθιον
ἅμα θείῳ ὀπτηθέν. Τὸν δὲ σίδηρον λύσεις, μαγνησίαν ἐπιβάλλων, ἢ θείου τὸ ἥμισυ, ἢ μάγνητος βραχύ. Ὁ γὰρ μάγνης ἔχει
συγγένειαν πρὸς τὸν σίδηρον. Ἡ φύσις τῇ φύσει τέρπεται.

2. Λαβὼν τὴν προγεγραμμένην νεφέλην, ἕψει ἐλαίῳ κικίνῳ ἢ
10   ῥαφανίνῳ, προσμίξας βραχὺ στυπτηρίας. Εἶτα λαβὼν κασσίτε-

M, fols. 71ʳ7-72ᵛ8
B, fols. 17ʳ16-20ʳ18
V, fols. 7ʳ17-10ᵛ8
A, fols. 29ᵛ4-31ʳ22

CAAG II 49,23-53,15

Tit. περὶ ἀσήμου ποιήσεως **M** : περὶ π. ἀσήμου **BA** : περὶ ἀργύρου **V** : حلدا ولذي
ومحمدهم ܣܡ/ܝܘܡܚܠܝ ܟܠܝܠܚܐ [Philosophi Democriti liber secundus] **SyrC** ‖ **1** ante λαβὼν add.
محم وذني/كحمك، دحم (**SyrL** ܚܡܙܐ ܣܡܠ) ܘܚܡܙܐ ܢܒܠܠ ܣܪܢ [vide vim medicamentorum
quae ad argentum pertinent, haec sunt] **SyrC SyrL** ‖ λαβὼν add. Moyses Alch. (λαβὼν
νεφέλην Zos. Alch. *CAAG* II 179,9 et λαβὼν ☽ Steph. Alch. II 235,37 Ideler) : om.
MBVA SyrL SyrC ‖ ὑδράργυρον **BVA** : -ος **M** ‖ supra ὑδράργυρον add ☽ **A** ‖ τὴν **V** :
ἢ **M** : om. **BA** ‖ **2** καὶ om. **BA** ‖ ἐπίβαλλε **MV** : -βαλε **BA** ‖ **2-3** ἢ add. Moyses Alch.
(coni. Zur²) : ♀ʷ [i.e. χαλκῷ BeRu] σιδήρῳ **M** : ♀‵ [i.e. χαλκόν] σιδήρῳ **BA** : ♀ σιδήρῳ
V :ܠܝܐܘܘܦܣܚܠܐ ܘ/ (**SyrL** ܚܘܙ) ܣܚܘܚܕܚܘ ܐܘܣ ܠܚܘܙ/ܘ [et adice partim ad mercurium
(liquorem? SyrL) vel ad alumen] **SyrC SyrL** ‖ **3-4** supra μαγνησία add. ☉ **M** ‖ **4** supra
ἀρσενικὸν add. ♂ **M** ‖ supra ἐκστραφὲν add. ☾ **M** ‖ **4-5** supra καδμία et σανδαράχη
add. ☿ **M** ‖ **5** supra ἄπυρος add. ἀληθ **M** ‖ supra λευκανθεὶς add. ☉ **M** ‖ ψιμύθιον **V**
et **M** qui add. ☽ s.l. : ψιμμίθιον **BA** ‖ **6** λύσεις **MBVA** : λευκαίνεις Moyses Alch. ‖ **7**
supra θείου add. ἀληθ **M** ‖ ἥμισυ **BA** : ⅂ **M** : ⅋ **V** ‖ alt. ἢ om. **V** ‖ μάγνητος **MBVA** :
μαγνίτου Moyses Alch. ‖ μάγνης **MBVA** : μαγνίτης Moyses Alch. ‖ **8** σίδηρον BeRu :
♂ **MV** : �ြ **BA** ‖ ante φύσις add. γὰρ BeRu ‖ **9** νεφέλην **V** et **M** qui add. ☽ s.l. : ☽
BA ‖ ἕψει **MBV** : -η **A** ‖ supra κικίνῳ add. ♋ **MB** ‖ **10** βραχὺ **MBA** : -χύτι **V** ‖
κασσίτερον **MV** : καττί- **A** et **B** qui add. ♀ [signum incertum] s.l.

On the Making of Silver (= *AP*)

1. \<Take> the mercury that comes from orpiment, or from realgar, or according to your knowledge, and make it solid as is customary;[1] lay it on the copper \<or> on the iron[2] which have been purified with sulphur,[3] and they will turn white. Whitened *magnēsia* produces the same effect as well, and orpiment that has been turned inside out, and roasted cadmia, and unburnt realgar, and whitened pyrite, and white lead roasted together with sulphur.[4] You will melt iron by adding *magnēsia*,[5] or the half of sulphur, or a pinch of magnetite. For magnetite has affinity with iron. Nature delights in nature.

2. Take the above mentioned volatile substance[6] and boil it with castor or radish oil, mixing in a pinch of *magnēsia*. Then take tin and purify it as is

§ 1   **SyrL:** *CMA* II 12, 5–11 (textus); 23 praec. 1 (translatio); **SyrC:** *2SyrC* § 1.
     TEST. 1–8 Moyses Alch. (A, fol. 272ᵛ7–15), ut monuerunt BeRu (*CAAG* II 306,14), qui textum non ed. ‖ 1–3 Zos. Alch. *CAAG* II 179,9–11 et 180,3; Steph. Alch. II 235,37–236,2 Ideler ‖ 7–8 Zos. Alch. *CAAG* II 197,12–3.
§ 2   **SyrL:** *CMA* II 12,11–5 (textus); 23 praec. 2 (translatio); **SyrC:** *2SyrC* § 2.
     TEST. 9–10 Zos. Alch. *CAAG* II 154,14 et 180,7–8; |Zos. Alch.| *CAAG* II 202,7–8.

ρον, κάθαιρε τῷ θείῳ ὡς ἔθος, ἢ τῷ πυρίτῃ, ἢ ὡς ἐπινοεῖς. Καὶ
κατέρα κατὰ τῆς νεφέλης, καὶ ποίει μῖγμα. Δὸς ὀπτᾶσθαι φωσὶν
εἰλικτοῖς, καὶ εὑρήσεις ψιμυθίῳ παρεμφερές· τὸ φάρμακον τοῦτο λευκαίνει πᾶν σῶμα. Πρόσμισγε δὲ αὐτῷ ἐν ταῖς ἐπιβολαῖς
15 γῆν Χίαν, ἢ ἀστερίτην, ἢ ἀφροσέληνον, ἢ ὡς ἐπινοεῖς· τὸ
γὰρ ἀφροσέληνον τῇ ὑδραργύρῳ μιγὲν πᾶν σῶμα λευκαίνει. Ἡ
φύσις τὴν φύσιν νικᾷ.

3. <Λαβὼν> μαγνησίαν λευκήν, λευκάνῃς δὲ αὐτὴν ἅλμῃ
καὶ στυπτηρίᾳ σχιστῇ ἐν ὕδατι θαλασσίῳ, ἢ χυλῷ, κίτρων λέ-
20 γω, ἢ θείου αἰθάλῃ. Ὁ γὰρ καπνὸς τοῦ θείου λευκὸς ὢν πάντα
λευκαίνει· ἔνιοι δέ φασιν καὶ τὸν καπνὸν τῶν κοβαθίων
λευκαίνειν αὐτήν. Πρόσμιξον αὐτῷ μετὰ τὴν λεύκωσιν καὶ
σφέκλης τὸ ἴσον, ἵνα λίαν γένηται λευκή· καὶ δεξάμενος
χαλκοῦ ὑπολεύκου, ὀρειχάλκου λέγω, οὐγγίας τέσσαρας,
25 χώνευε, ἐπιβάλλων κατ᾽ ὀλίγον κασσιτέρου προκαθαρισθέντος
οὐγγίαν μίαν, καθύπο χεῖρα κινῶν ἕως συγγαμήσωσιν αἱ
οὐσίαι· ἔσται ῥηγνύμενον. Ἐπίβαλλε οὖν τοῦ λευκοῦ φαρμάκου
τὸ ἥμισυ, καὶ ἔσται πρῶτον· ἡ γὰρ μαγνησία λευκανθεῖσα οὐκ

---

11 supra πυρίτῃ add. ◊ M ‖ **12** κατὰ MBVA : μετὰ BeRu ‖ supra μῖγμα add. ṁ M ‖
13 εἰλικτοῖς MBVA : fort. μειλικτοῖς sive ἀλήκτοις BeRu : حفوـ [furno] SyrL SyrC ‖
post εὑρήσεις add. τί BeRu ‖ ψιμυθίῳ V et M qui add. ṁ s.l. : ψιμμϊ- BA ‖ **14-6**
πρόσμισγε — λευκαίνει om. SyrL SyrC ‖ **14** αὐτῷ MBV : -ὸ A ‖ ἐν MV : ἐπὶ BA ‖
15 supra γῆν add. ☉ M et ☉ A ‖ χίαν BA : χεί- MV ‖ ἀστερίτην A : -ιν MB : -εν V
‖ ἀφροσέληνον — τὸ γὰρ om. BA ‖ **16** ἀφροσέληνον MV : ἀφροℭ B : ἀφροℭ^ω
[i.e. ἀφροσελήνῳ] A ‖ ὑδραργύρῳ BeRu : ☽ MBVA ‖ **17** ante φύσις add. γὰρ BeRu ‖
18 λαβὼν addidi, coll. محني [ⲥⲓ]) (SyrC ـ) ـصد [cape magnesiam albam] SyrL SyrC ‖
supra μαγνησίαν add. ☉ MV ‖ λευκάνῃς MV : -εις B : -αίνεις A ‖ **19** στυπτηρίᾳ
σχιστῇ BeRu : στυπτηρία✳ MV : στυπτηρία✱ BA ‖ supra ὕδατι add. ☽ MB ‖ κίτρων
BA et حنبا, ومص [citrorum aceto] SyrC : κίτρω M : κίτρου V ‖ λέγω om. BA ‖ **20** supra
αἰθάλῃ add. ☿ MV ‖ **21** τὸν MBV: τῶν A ‖ **23** σφέκλης MV : φέ- BA ‖ **24** χαλκοῦ V
et ♂^υ M : ♀ B : ♀̈ [i.e. χαλκόν] A ‖ ὀρειχάλκου BA : ὀρι- MV : om. SyrL SyrC ‖
οὐγγίας scripsi : 𐆄 MBA : ουϒᴷ V ‖ **25** κατ᾽ ὀλίγον prop. BeRu : κάτω ὀλίγον BA :
κάτω ὀλίγου MV : om. SyrL SyrC ‖ κασσιτέρου M : κασσιᵗᵖ BV : κασιτήρου A ‖ **26**
οὐγγίαν scripsi : 𐆄 MBA : ουϒᴷ V ‖ **27** ῥηγνύμενον MBV : ῥηγνή- A ‖ ἐπίβαλλε MV :
-βαλε BA ‖ supra φαρμάκου add. ṁ MBA : محـ علـ ljojo [et adice ad hoc
magnesiam] SyrL (similiter SyrC) ‖ **28** ἥμισυ BeRu: ↳ M : ⌇ BVA : om. SyrL SyrC ‖
ante ἔσται lacunam indicavit Zur[1] ipse καὶ οὐκ ῥηγνύμενον ἔσται vel καὶ οὐ ῥήγνυται
dubit. supplens

customary, with sulphur, or with pyrites, or according to your knowledge.[7] Pour it over the volatile substance and make a mixture. Let it be roasted in an enveloping fire, and you will find it similar to white lead. This drug makes any [metallic] body white. When you lay it on, mix Chian earth in it, or *asteritēs*, or moon foam, or according to your knowledge; for moon foam, when mixed with mercury, makes any [metallic] body white. Nature conquers nature.

3. <Take> white *magnēsia*: whiten it with brine and scissile alum in sea water, or with juice – I mean citron juice – or with sulphur vapour. For sulphur fumes whiten everything since they are white. Some people claim that *kobathia* fumes[8] whiten it [i.e. *magnēsia*] as well. After whitening, also mix an equal amount of wine dregs with it [i.e. the sulphur], so that [the *magnēsia*] turns extraordinarily white. Then take four ounces of whitish copper – I mean of brass – and melt it by adding gradually one ounce of previously purified tin; stir with your hands until the substances join together:[9] it [i.e. the metallic alloy] will be breakable. So, add half of the white drug and it [i.e. the alloy] will be of first quality. For whitened *magnēsia*

§ 2   TEST. 12 Zos. Alch. *CAAG* II 180,9 et 197,15 ‖ 12–3 Zos. Alch. *CAAG* II 147,25 ‖ 13–4 Zos. Alch. *CAAG* II 197,15–6 ‖ 14–5 Zos. Alch. *CAAG* II 185,17–8 et 186,4–5.
§ 3   SyrL: *CMA* II 12,15–21 (textus); 24 praec. 3 (translatio); SyrC: 2SyrC § 3.
       TEST. 20–1 Zos. Alch. *CAAG* II 189,2–3 et 9–11; Olymp. Alch. *CAAG* II 85,1–2 ‖ 26–7 Zos. Alch. *CAAG* II 162,4–6 et 178,7 et 179,17–8; [Zos. Alch.] *CAAG* II 201,17–8; Steph. Alch. II 236,23–4 Ideler ‖ 27–8 Zos. Alch. *CAAG* II 153,10–2; Steph. Alch. II 240,14–6 Ideler.

ἐᾷ ῥήγνυσθαι τὰ σώματα, οὐδὲ τὴν σκιὰν τοῦ χαλκοῦ ἐπιφέ-
30 ρεσθαι. Ἡ φύσις τὴν φύσιν κρατεῖ.

4. Λαβὼν θεῖον τὸ λευκόν, λευκάνῃς δὲ αὐτὸ οὔρῳ λειῶν ἐν
ἡλίῳ ἢ στυπτηρίᾳ καὶ ἄλμῃ τῇ τοῦ ἁλός· ἀνθήσει πάνυ λευκότατον. Λείου αὐτὸ σὺν σανδαράχῃ ἢ οὔρῳ δαμάλεως ἡμέρας
ἕξ, ἕως γένηται τὸ φάρμακον μαρμάρῳ παρεμφερές· καὶ ἐὰν
35 γένηται, μέγα ἐστὶ μυστήριον· τὸν γὰρ χαλκὸν λευκαίνει,
μαλάσσει τὸν σίδηρον, ἄτριστον ποιεῖ τὸν κασσίτερον, τὸν μόλυβδον ἄρρευστον, ἀρρήκτους ποιεῖ τὰς οὐσίας, ἀφεύκτους τὰς
βαφάς· τὸ γὰρ θεῖον θείῳ μιγὲν θείας ποιεῖ τὰς οὐσίας, πολλὴν
ἔχοντα τὴν πρὸς ἄλληλα συγγένειαν. Τέρπονται γὰρ αἱ φύσεις
40 ταῖς φύσεσιν.

5. Τὴν δὲ λευκανθεῖσαν λιθάργυρον τρῖψον σὺν θείῳ, ἢ
καδμίᾳ, ἢ ἀρσενικῷ, ἢ πυρίτῃ, ἢ ὀξυμέλιτι, ἵνα μηκέτι ῥεύ-
σῃ. Ὄπτησον οὖν αὐτὸ λαμπροτέροις φωσίν, ἀσφαλισάμενος τὸ
σκεῦος. Ἐχέτω δὲ τὸ σύνθεμα καὶ τιτάνου ὀπτοῦ βραχέντος ὄξει
45 ἡμέρας τρεῖς, ἵνα γένηται σμηκτικώτερον. Ἐπίβαλλε οὖν αὐτὸ
λευκὸν γενόμενον μᾶλλον ἢ τὴν ψιμύθιον. Γίνεται δὲ πολλάκις

29 ῥήγνυσθαι **MBV** : ῥήγνη- **A** ‖ χαλκοῦ **A** : ♀ **MV** : ♀ **B** ‖ **30** ante φύσις add. γὰρ
BeRu ‖ **31** λευκόν **MV** : λευκανθέν **BA** : ܠܝܐ ܐܘܗܪܘ ܐܝܠ [i.e. θεῖον ἄπυρον λευκόν]
SyrC : om. **SyrL** ‖ λευκάνῃς **M** : -άνεις **B** : -αίνεις **A** : -κανθήσεται **V** ‖ αὐτὸ om.
BeRu ‖ **31-40** λευκάνῃς — ταῖς φύσεσιν om. **SyrL SyrC** ‖ **32** ἡλίῳ **BeRu** : ♄ **MA** : ˆ♄
BV ‖ τῇ om. **BA** ‖ ἀνθήσει scripsi post Pizzim f. 10ʳ4 (florebit) : ἀνθίγῃ **MV** : ἀθιϻ
B : ἀθϻ **A** : ἄθικτον θεῖον **BeRu** ‖ **33** αὐτὸ **BA** et **V** p.c. : -ῷ **M** et **V** a.c. ‖ **35** ♀` [i.e.
χαλκὸν **BeRu**] **M** et ♀` **BA** : ♀ **V** ‖ **36** σίδηρον **BeRu**: ♂ **MV** : ♇ **A** et **B** qui add.
σίδηρος s.l. ‖ ἄτριστον scripsi : ἄτρηστον **MV** et **BeRu** : ἄτρυτον **B** : ἄτριτον **A** ‖
κασσίτερον **MV** : -τυρον **B** : κασίτηρον **A** ‖ **36-7** μόλυβδον **BeRu** : μόλιβ- **A** : ♄ **M** et
B qui add. μόλιβδος s.l. : ♄ **V** ‖ **39** ἔχοντα **MVA** et **B** ut vid. ([.....]νᵗᵃ) : ἐχούσας prop.
BeRu ‖ ἄλληλα **MBA** : ἀλλήλας **V** ‖ συγγένειαν **MBV** : -γένιαν **A** ‖ αἱ φύσεις om.
A ‖ **41** supra λιθάργυρον add. ϻ **MBA** ‖ τρῖψον scripsi : ⑥ **MVA** et **B** qui add. λείου
s.l. : λείου **BeRu** : ܣܚܩ [contere] **SyrC** ‖ **42** ἀρσενικῷ **MV** : ἀρσ\᷍ ̔ **B** : -νίου **A** ‖
ὀξυμέλιτι **MV** : ὄξο- **BA** ‖ **43** λαμπροτέροις **MBVA** : -τάτοις prop. **BeRu** ‖ **44** ἐχέτω
MBV : -ο **A** ‖ σύνθεμα **MV** : -θημα **BA** ‖ τιτάνου **BVA** : τίτανος **M** ‖ **45** ἡμέρας
MV : 66 **BA** ‖ σμηκτικώτερον **BA** : σμι- **MV** ‖ ἐπίβαλλε **MV** : -βαλε **BA** ‖ αὐτὸ
BA : -ῷ **MV** ‖ **46** τὴν ψιμύθιον **BeRu** : τὴν ψιμυ\θ̔ **MV** : τὸ ψιμμΐθιον **BA**

does not let the [metallic] bodies break, nor the blackness of copper [lit. 'the shadow of copper'] come out. Nature masters nature.

4. Take white sulphur,[10] whiten it by grinding it in the sun with urine, or alum and salt brine; it will shine completely white. Grind it with realgar or with the urine of a heifer for six days, until it becomes a drug similar to marble.[11] If it does, it is a great mystery: for it whitens copper, softens iron, takes away the cry of tin, takes away the fluidity of lead, makes substances unbreakable, and makes dyes stable. For sulphur, when mixed with sulphur, makes substances divine, since [sulphurs] have great affinity with each other. For natures delight in natures.

5. Crush whitened litharge with sulphur, or cadmia, or orpiment, or pyrite, or vinegar and honey, so that it loses its fusibility. Roast this [compound] over quite high flames, sealing the vessel. The compound must also have a bit of roasted lime that has been moistened with vinegar for three days, so that it acquires a stronger cleansing property. When it has turned whiter than white lead, lay it [on metallic bodies].[12] However, it often also turns yellow if you make too

---

§ 4   om. **SyrL**; **SyrC**: *2SyrC* § 4, l. 1 tantum λαβὼν θεῖον τὸ λευκόν praebet.
    Test. 34–8 Zos. Alch. *CAAG* II 161,6–9 et 162,12–7 ‖ **38–9** |Zos. Alch.| *CAAG* II 199,17–8; Philos. Christ. Alch. *CAAG* II 399,5–6.
§ 5   om. **SyrL**; **SyrC**: *2SyrC* § 4, ll. 2ff.
    Test. 46 Zos. Alch. *CAAG* II 147,10.

καὶ ξανθή, ἐὰν πλεονάσῃ τὰ φῶτα· ἀλλ᾽ ἐὰν γένηται ξανθή, οὐ
χρησιμεύσει σοι νῦν· λευκᾶναι γὰρ βούλει τὰ σώματα. Καῦ-
σον οὖν αὐτὸ τῇ συμμετρίᾳ καὶ ἐπίβαλλε παντὶ σώματι χρείαν
50 ἔχοντι λευκώσεως· ἡ γὰρ λιθάργυρος, ἐὰν γένηται ἄρρευστος,
οὐκέτι ἔσται μόλυβδος· γίνεται δὲ εὐκόπως· ταχὺ γὰρ εἰς πολλὰ
μετατρέπεται ἡ τοῦ μολύβδου φύσις. Αἱ γὰρ φύσεις νικῶσι τὰς
φύσεις. <...>

6. Λαβὼν κρόκον Κιλίκιον, τρῖψον θαλάσσῃ ἢ ἄλμῃ, καὶ
55 ποίησον ζωμόν, εἰς ὃν πυρῶν κατάβαπτε πέταλα χαλκοῦ,
μολύβδου, σιδήρου, ἕως σοι ἀρέσῃ· γίνονται δὲ λευκά. Εἶτα
λαβὲ τοῦ φαρμάκου τὸ ἥμισυ, καὶ συλλείου σανδαράχῃ, ἢ
ἀρσενικῷ λευκῷ, ἢ θείῳ ἀπύρῳ, ἢ ὡς ἐπινοεῖς· καὶ ποίησον
κηρωτῆς πάχος. Χρῖσον τὸ πέταλον καὶ θὲς εἰς καινὸν ἀγγεῖον·
60 περιφίμωσον ὡς ἔθος, θεὶς εἰς πρισματοκαύστην ἡμέραν ὅλην·
εἶτα ἐξενέγκας κάθες εἰς καθαρὸν ζωμόν, καὶ ἔσται λευκός,
λευκότατος ὁ χαλκός. Κατεργάζου λοιπὸν ὡς τεχνίτης· ὁ γὰρ
Κιλίκιος κρόκος θαλάσσῃ μὲν λευκαίνει, οἴνῳ δὲ ξανθοῖ. Ἡ
φύσις τῇ φύσει τέρπεται.

65 7. Δέξαι λευκὴν τὴν λιθάργυρον, καὶ λείου αὐτὴν μετὰ
φύλλων δάφνης καὶ Κιμωλίας καὶ μέλιτος καὶ σανδαράχης

---

47 pr. ξανθή **MBA** : -εῖ **V** ‖ alt. ξανθή **BA** : -όν **MV** ‖ **49** ἐπίβαλλε **MBV** : -βαλε **A** ‖ χρείαν **MV** : χροί- **B** : χρή- **A** ‖ **50** λι₵ᵒˢ [i.e. λιθάργυρος BeRu] **BA** : λιθάρ₵ **M** : λιθάρᵞᴾ **V** ‖ **51** μόλυβδος **M** : μόλι- **BVA** ‖ εὐκόπως **MV** : -κόλως **A** et **B** ut vid. (εὐκόλ[..]) ‖ **52** ♄ᵛ [i.e. μολύβδου BeRu] **M** : ♄ **BV** : μολίβδου **A** ‖ νικῶσι **BVA** : -σαι **M** ‖ **53** post φύσεις lacunam conieci, coll. SyrC, fols. 95ʳ10-95ᵛ3 qui caput cum *PM* §§ 15-6 comparandum servat (vide *2SyrC* § 5) ‖ **54** λαβὼν **BVA** : ⚗ **M** : ante λ. titulum ἀργυροποιίας ζωμοί add. Moyses Alch. ‖ τρῖψον **BA** : ⊛ **MV** : λείωσον Moyses Alch. ‖ **55** εἰς ὃν πυρῶν **BA** : εἰσὸν πυρός **M** : εἰς ὃν πυρός **V** : om. SyrL SyrC ‖ ⬜ᵛ⬜ [i.e. πέταλα BeRu] **MBVA** ‖ ♀ᵛ [i.e. χαλκοῦ BeRu] **MV** : ♀ **BA** ‖ **56** μολύβδου BeRu : ♄ **MBVA** ‖ σιδήρου BeRu : ♂ **M** : ♂] **BVA** ‖ post ἕως add. ἄν **BA** ‖ **57** ἥμισυ BeRu : ⤵ **M** : ⤴ **BVA** ‖ συλλείου **MBVA** : συνλύει (sic) Moyses Alch. ‖ **59** ante χρῖσον add. καὶ **BA** ‖ πέταλον BeRu : ⬜ **MBVA** : πέταλα Moyses Alch. : ܠܛܦ̈ܐ [i.e. πέταλα] SyrL SyrC ‖ **60** περιφίμωσον **B** : περιφ᾽ἤμωσον (sic) **A** : περίφιμον **MV** : fort. περιφίμωτον BeRu ‖ **61** κάθες **MBVA** : θὲς Moyses Alch. : ܐܪܡܝ [adice] SyrL SyrC : κάτθες BeRu ‖ **62** ♀ˢ [i.e. χαλκός BeRu] **MV** : ♀ **BA** ‖ **64** ante φύσις add. γὰρ BeRu ‖ **65** λιθάργυρον BeRu : λιθάρ₵ **M** : λι₵ **BA** : λιθάρᵞᴾ **V** ‖ αὐτὴν **MBA** : ταύτην **V**

strong a fire, but if it turns yellow it will not be useful to you at present, since you want to whiten the [metallic] bodies. So heat it up on a suitable fire and lay it on any [metallic] body that must be whitened.[13] For litharge after losing its fusibility will no longer be lead; this happens easily: the nature of lead, in fact, quickly undergoes many transformations.[14] For natures conquer natures. <...>

6. Take Cilician saffron,[15] crush it with sea water or brine, and make a wash into which to dip heated leaves of copper, lead, and iron, until you like the result: they will turn white. Then take half of the drug and grind it up together with realgar, or white orpiment,[16] or unburnt sulphur, or according to your knowledge: make it as thick as a salve. Rub the leaf and put it in a newly-made vessel: seal [the vessel] as is customary, after setting it on a fire of sawdust for a whole day, then take [the leaf] out and dip it in a fresh wash: it will be white, the copper leaf whitest. Do the rest as a craftsman does: for Cilician saffron with sea water makes [metallic leaves] white, with wine yellow.[17] Nature delights in nature.

7. Take white litharge and grind it with bay leaves and Cimolian earth and honey and white realgar: make a glutinous mixture. Rub half of this

§ 5   TEST. 46–8 Zos. Alch. *CAAG* II 139,18–20 = VIII 43–4 Mertens; |Zos. Alch.| *CAAG* II 204,6–7 ‖ 51–2 Steph. Alch. II 232,34 et 36 Ideler.

§ 6   **SyrL:** *CMA* II 12, 22–13,5 (textus); 24 praec. 5 (translatio); **SyrC:** *2SyrC* § 6.
      TEST. 54–64 Moyses Alch. (A, fol. 275ᵛ2–14), ut monuerunt BeRu (*CAAG* II 310 n. 9), qui textum non ed.

§ 7   om. **SyrL**; **SyrC:** *2SyrC* § 7.

λευκῆς, καὶ ποίησον γλοιῶδες. Χρῖσον τοῦ φαρμάκου τὸ ἥμισυ, καὶ ὑπόκαιε ὡς ἔθος. Κατάβαπτε εἰς τὸ τοῦ φαρμάκου λείψανον, ἀναλύσας ὕδατι σποδοῦ λευκίνων ξύλων· τὰ γὰρ ἀνούσια
70   μίγματα καλῶς ἐνεργοῦσι χωρὶς πυρός. Ποίει αὐτὰ τοῖς ζωμοῖς πυρίμαχα. Ἡ γὰρ φύσις τὴν φύσιν νικᾷ.

8. Λαβὼν τὴν προγεγραμμένην νεφέλην, συλλείου αὐτῇ στυπτηρίαν καὶ μίσυ· ὄξει τε περιπλύνας, βάλε αὐτῇ καὶ ὀλίγην λευκὴν καδμίαν, ἢ μαγνησίαν, ἢ ἄσβεστον, ἵνα γένηται σῶμα ἡ
75   ἀσώματος. Τρῖψον σὺν μέλιτι λευκοτάτῳ· ποίει ζωμόν, εἰς ὃν πύρου καταβάπτων ὃ βούλει· ἔασον κάτω, καὶ γενήσεται. Ἐχέτω δὲ τὸ σύνθεμα καὶ ὀλίγον ἄπυρον θεῖον, ἵνα διαδύνη τὸ φάρμακον ἐντός. Ἡ φύσις τὴν φύσιν κρατεῖ.

9. Δέξαι ἀρσενικοῦ οὐγγίαν μίαν, καὶ νίτρου οὐγγίας τὸ
80   ἥμισυ, καὶ φλοιοῦ φύλλων Περσαίου ἁπαλῶν οὐγγίας δύο, καὶ ἅλατος ἥμισυ, καὶ συκαμίνου χυλοῦ οὐγγίαν μίαν, <στυπτηρίας> σχιστῆς τὸ ἴσον. Λείου ὁμοῦ ἐν ὄξει, ἢ οὔρῳ, ἢ ἀσβέστῳ

67 καὶ om. **MV** ‖ ποίησον **BA** : ℣ **MV** ‖ γλοιῶδες **BA** : γλυ- **MV** ‖ ante χρῖσον add. καὶ **BA** ‖ χρῖσον **BVA** : χρῆσον **M** : ܟܕ ܡܢ ܘܠ ܐܘ [et illine horum externam partem] **SyrC** : om. **SyrL** ‖ τοῦ **MVA** : τὸ **B** ‖ ἥμισυ **BeRu** : ܫ **M** : ܦ **BVA** ‖ **68** ὡς **MBV** : ἕως **A** ‖ ante κατάβαπτε add. καὶ **BA** ‖ **69** ὕδατι **V** et ܚܡܣ [aquā] **SyrC** : ὕδαᵗ **M** : ὕδατος **BA** ‖ σποδοῦ **MVA** : non legit. **B** : fort. καὶ σποδῷ **BeRu** ‖ λευκίνων **MBV** : -καίνων **A** : fort. πευκίνων **BeRu** ‖ **70** ποίει **A** : ℣ **MBV** ‖ **71** πυρίμαχα **BA** : πυρίμαχ **MV** ‖ **72** αὐτῇ **BeRu** : αὐᵗ **M** : αὐτὴν **BVA** ‖ **72-3** στυπτηρίαν καὶ μίσυ **M** : στυπτηρία καὶ μίσυι **BVA** ‖ **73** βάλε **MVA** : βάλλε **B** ‖ **74-5** ἡ ἀσώματος scripsi, coll. *CAAG* II 195,7 : ἀπὸ σώμαᵗ **MV** : ἀπὸ σώματος **BA** : ܠܫ ܘܗܢ ܗܘܣ ܩ [ut hoc fiat album corpus] **SyrC** ‖ **75** τρῖψον scripsi post **BeRu** (τρίψον sic) : ⑥ **MBVA** ‖ λευκοτάτῳ **BA** : λευκοᵗ **MV** ‖ ποίει **A** : iter. **B** : ℣ **MV** ‖ **76** πύρου **MBA** : πῦρ **V** ‖ βούλει **MV** : βουλ **BA** ‖ **77** σύνθεμα **MV** : -θημα **BA** ‖ **78** ante φύσις add. γὰρ **BeRu** ‖ **79** ℥ [i.e. οὐγγί(αν)] **MBA** : ουʸʸ **V** ‖ νίτρου **BeRu** : ℣ **MBVA** ‖ ℥ [i.e. οὐγγί(ας)] **MBA** : ουʸʸ **V** ‖ τὸ om. **V** ‖ **80** ἥμισυ **BeRu** : ܫ **M** : ܦ **BVA** ‖ φλοιοῦ **BeRu** : φλοός **MB** : φλοιο **V** : φλοιῶν **A** : ܠܝܟܠ ܘܠ.ܙܚܛܠ ܐܩܨ ܡܢ ܘ ܠܨܒܘ ܡܚܕܠ [Perseae cortex vel mollium foliorum duo drachmae] **SyrC** ‖ περσαίου **MV** : περσαί[...] **B** : περσέων **A** ‖ ℥ β′ [i.e. οὐγγί(ας) δύο] **MB** : ουʸʸ β′ **V** : om. **A** ‖ **81** ἥμισυ **BeRu** : ܫ **M** : ܦ **BVA** ‖ χυλοῦ **B** : -όν **MV** : χειλῶν **A** ‖ ℥ [i.e. οὐγγί(αν)] **MBA** : ουʸʸ **V** ‖ **81-2** στυπτηρίας σχιστῆς scripsi post Zur² (σχ. στ.) : σχιστῆς **MV** : σχιστῆς et ✗ s.l. **BA** : ܠܚܒܘܩܠܣ [i.e. στυπτηρία] **SyrC** ‖ **82** ante λείου add. καὶ **V** ‖ ἀσβέστῳ **MV** : -ου **BA**

drug [on a metallic leaf?] and heat from below as is customary.[18] Dip [the metallic leaf?] into the rest of the drug, by diluting with the water of ashes of white poplar: for mixtures that are not thick[19] are very effective without fire. Make them fire-resisting with washes. For nature masters nature.

8. Take the abovementioned volatile substance,[20] and grind alum and *misy* together with it; after washing with vinegar, also add to it a bit of white cadmia, or *magnēsia*, or quicklime, so that this incorporeal [vapour/substance] becomes a body [i.e. solid]. Crush with whitest honey: make a wash in which to dip and warm up whatever you want; keep it dipped and [what you are seeking for] will happen. Let the compound have a bit of unburnt sulphur as well, so that the drug may sink in. Nature conquers nature.

9. Take orpiment, one ounce, and soda, a half-ounce, and the skin of the tender leaves of *Persaion*, two ounces, and salt, a half-ounce, and mulberry juice, one ounce, and the same amount of scissile <alum>.[21] Grind together in vinegar, or urine, or filtrate of quicklime, until a wash is formed: dip into

§ 7   TEST. 67–70 Zos. Alch. *CAAG* II 168,12–4 ‖ 69–70 Zos. Alch. *CAAG* II 160,5–6 et 168,6–7 et 189,13–4.
§ 8   om. SyrL; SyrC: 2*SyrC* § 8.
         TEST. 72–5 Zos. Alch. *CAAG* II 195,5–7 ‖ 75–6 Zos. Alch. *CAAG* II 185,23–4 ‖ 76–7 Zos. Alch. *CAAG* II 174,16; Steph. Alch. II 247,21 Ideler; Philos. Christ. Alch. *CAAG* II 274,14 et 420,14–5 ‖ 77–8 Philos. Christ. Alch. *CAAG* II 277,13–4; Hieroth. Alch. *CAAG* II 451,5–6.
§ 9   om. SyrL; SyrC: 2*SyrC* § 9.
         TEST. 79–84 Zos. Alch. *CAAG* II 178,18–179,1 ‖ 79–81 Zos. Alch. *CAAG* II 153,1 et 163,22–164,2 et 217,14–6 ‖ 80 Zos. Alch. *CAAG* II 155,2–3 ‖ 81–4 Zos. Alch. *CAAG* II 155,17–9.

στακτῇ, ἕως γένηται ζωμός· εἰς τοῦτον τὰ ἔνσκια πυρὶ κατάβαπτε πέταλα, καὶ ἀποσκιώσεις. Ἡ φύσις τὴν φύσιν κρατεῖ.

85   10. Ἀπέχετε πάντα τὰ χρυσῷ καὶ ἀργύρῳ χρήσιμα. Οὐδὲν
ὑπολείπεται, οὐδὲν ὑστερεῖ, πλὴν τῆς νεφέλης καὶ τοῦ ὕδατος ἡ
ἄρσις· ἀλλὰ ταῦτα ἑκὼν παρεσιώπησα, διὰ τὸ ἀφθόνως αὐτὰ
ἐγκεῖσθαι καὶ ἐν ταῖς ἄλλαις μου γραφαῖς. Ἔρρωσθε [ἐν ταύτῃ
τῇ γραφῇ].

84 □ᵃ□ [i.e. πέταλα BeRu] **MBVA** ‖ ἀποσκιώσεις **MBA** : -άσεις **V** ‖ ante φύσις add.
γὰρ BeRu ‖ **85** ἀπέχετε **MBV** : -ται **A** ‖ τὰ om. **BA** ‖ **88** ἐγκεῖσθαι **MBV** : ἐγγ- **A** ‖
ἔρρωσθε **MBV** : -σθαι **A** ‖ **88-9** ἐν ταύτῃ τῇ γραφῇ secl. BeRu : ἔρρωσθε — γραφῇ om.
SyrL SyrC

this wash the dark metallic leaves while heating them, and you will make them 'shadowless.' Nature masters nature.

10. You have received everything useful for gold and silver. Nothing has been left out; nothing is missing, except how to sublime volatile substances and to distil waters. But I have intentionally passed these topics over in silence, since they are covered enough in my other writings. Farewell [with this writing].

§ 10 **SyrL:** *CMA* II 13,6–8 (textus); 24 praec. 6 (translatio); **SyrC:** *2SyrC* § 10.
Test. 85–6 Zos. Alch. *CAAG* II 148,10–1 et 152,16–7 et 181,3–6; Pelag. Alch. *CAAG* II 260,19–21; Steph. Alch. II 205,30–5 et 217,11–5 et 233,28–9 et 244,10–1 Ideler; Philos. Anon. Alch. (Zos. Alch. BeRu) *CAAG* II 125,6–7 et 136,12–3 ‖ 87–8 Zos. Alch. *CAAG* II 225,10–3 = IV (versio 2) 24–30 Mertens et *CAAG* II 234,5–7 = I 194–5 Mertens et *CAAG* II 237,9–18 = IV 10–34 Mertens.

Excerpta ex *Moysis Chymica*

<Democriti catalogi>

1   1. Ὕλη χρυσοποιίας·

[Λαβὼν] ὑδράργυρος ἡ ἀπὸ κινναβάρεως, σῶμα μαγνησίας,
χρυσόκολλα, ὅ ἐστι βατράχιον – ἐν τοῖς χλωροῖς λίθοις εὑρίσκεται – κλαυδιανόν, ἀρσενικὸν τὸ ξανθόν, καδμία, ἀνδροδάμας,
5   στυπτηρία ταπεινωθεῖσα, θεῖον ἄπυρον, ὅ ἔστι ἄκαυστον, πυρίτης, ὤχρα Ἀττική, σινωπὶς Ποντική, θεῖον ὕδωρ ἄθικτον, ἐὰν
ἀκούσῃς τοῦ ἀπὸ μόνου θείου· ἐὰν δὲ ἀπολελυμένως, τὸ διὰ
ἀσβέστου· θείου αἰθάλη, σῶρι ξανθόν, χάλκανθος ξανθὴ καὶ κιννάβαρις.

10   2. Ὕλη ζωμῶν

Ζωμοί· τὰ δὲ ἐν ζωμοῖς εἰσιν ταῦτα· κρόκος Κιλίκιος, ἀριστολοχία, κνήκου ἄνθος, ἐλύδριον, ἄνθος ἀναγαλλίδος τῆς τῶν
κυανέων, <ῥᾶ Πόντιον>, κυανός, χάλκανθος, κόμμι ἀκάνθης Αἰ-
γυπτίας, ὄξος, οὖρον ἄφθορον, ὕδωρ θαλάσσιον, ὕδωρ ἀσβέστου,

A, fols. 272ᵛ16-273ʳ19
CAAG II 306,15-307,14

---

1 ꓔποιίας [i.e. χρυσοποιίας BeRu] A ‖ **2** λαβὼν seclusi : λαβὲ prop. BeRu ‖
ὑδράργυρος ἡ scripsi, coll. Syn. Alch., ll. 43-4, 125, 168, 202-3 et 306-7 : ꝅ τὴν A :
ὑδράργυρον τὴν BeRu ‖ κινναβάρεως BeRu : κῦνα- A ‖ μαγνησίας BeRu : μ̄ A ‖ **3**
χρυσόκολλα scripsi : ꝅκόλλην [i.e. χρυσοκόλλην BeRu] A ‖ post βατράχιον add. καὶ
BeRu ‖ χλωροῖς BeRu : χλο- A ‖ **4** καδμία scripsi : καθμίαν A ‖ ἀνδροδάμας A :
-δάμαντα BeRu ‖ **5** στυπτηρία scripsi : ⚹ A : -αν σχιστὴν BeRu ‖ ἄκαυστον BeRu :
αὔκαστον A ‖ πυρίτης scripsi : -ην A ‖ **6** ὤχρα ἀττική A : -αν -ήν BeRu : om. Syn.
Alch. ‖ σινωπὶς scripsi : -ώπη A : -ὴν BeRu ‖ θεῖον A : θείου Syn. Alch., l. 224, fort.
mel. ‖ ὕδωρ BeRu : ꭎ A ‖ **7** ἀπολελυμένως, τὸ scripsi, coll. Syn. Alch., l. 226 : -ος τῷ
A ‖ **8** θείου αἰθάλη scripsi, coll. Syn. Alch., l. 230 : θεῖον· αἰθάλην A ‖ σῶρι scripsi :
-ιν A ‖ χάλκανθος scripsi : ⚸ A : -ον BeRu ‖ ξανθὴ A : -ὴν BeRu ‖ **8-9** κιννάβαρις
scripsi : κῦνάβαρης A : -ιν BeRu ‖ **12** κνήκου BeRu : κνϊ- A ‖ ἐλύδριον om. Syn.
Alch. ‖ τῆς BeRu : τοῖς A ‖ **13** ῥᾶ Πόντιον addidi, coll. Syn. Alch., ll. 253-8 ‖
χάλκανθος BeRu : ⚸ A ‖ **14** ὄξος BeRu : ☌ A : om. Syn. Alch. ‖ οὖρον BeRu : οὖρος
A ‖ ἄφθορον scripsi : ἀφθόριον A ‖ ὕδωρ BeRu : ꭎ A ut semper

Excerpts from *The Chemistry of Moses* ⟨The *Catalogues by Democritus*⟩ (= Cat.)

1. Substances for the making of gold:

Mercury that comes from cinnabar,[1] *magnēsia*'s body, malachite, that is 'ranunculus' [lit. little frog] – it is found among the green stones – *klaudianon*, yellow orpiment, cadmia, *androdamas*, 'humbled' alum, unburnt sulphur, that is, the incombustible one, pyrite, Attic ochre, earth of Sinope from Pontus [i.e. the Black Sea], untouched divine water – if you understand the water produced only from sulphur, if you [understand water] in a general sense [lit. without further qualification],[2] it is [the water] produced with quicklime – sulphur's vapour, yellow *sōri*, yellow copper flower and cinnabar.

2. Substances for [the making of] washes [i.e. dyeing liquids]

Washes: these are the substances that fall within washes: Cilician saffron, *Aristolochia*, Carthamus flower, celandine, flower of the Anagallis that has blue blossoms, <rhubarb from Pontus [i.e. the Black Sea]>, azurite, copper flower, Egyptian acacia gum, vinegar, pure urine, sea water, quicklime

---

§ 1 TEST. ET LOCI SIMILES 1–9 Zos. Alch. (?) *CAAG* II 159,5–10; Syn. Alch. §§ 13–4, ll. 202–33 ‖ 1–3 Syn. Alch. § 5, ll. 43–4; § 8, l. 125; § 11, l. 168; § 19, ll. 306–7; *De lapide philosophiae CAAG* II 199,19–20 ‖ 3 Syn. Alch. § 12, ll. 188–91 ‖ 5–6 Zos. Alch. *CAAG* II 227,8–9 = IV 81–3 Mertens; Zos. Alch. (?) *CAAG* II 159,19–20; *De ovo CAAG* II 18,18–9 (≈ *CAAG* II 21,16–8).

§ 2 TEST. ET LOCI SIMILES 10–7 Syn. Alch. §§ 15–7, ll. 242–75 ‖ 11–2 Zos. Alch. *CAAG* II 227,9–11 = IV 83–4 Mertens; Zos. Alch. (?) *CAAG* II 160,1–3.

15   ὕδωρ σποδοκράμβης, ὕδωρ φέκλης, ὕδωρ στυπτηρίας, ὕδωρ νίτρου,
ὕδωρ ἀρσενικοῦ, ὕδωρ θείου ἀθίκτου, οὐρὸς γάλακτος ὀνείου,
ἀπὸ κυνὸς γάλα.

Αὕτη ἡ ὕλη τῆς χρυσοποιίας· ταῦτά ἐστιν τὰ ἀλλοιοῦντα τὴν
ὕλην, ταῦτα πυρίμαχά εἰσιν· ἐκτὸς τούτων οὐδέν ἐστι ἀσφαλές.
20   Ἐὰν ἦς νοήμων καὶ ποιήσῃς ὡς γέγραπται, ἔσῃ μακάριος. Ἐπίβαλλε χαλκόν, χρυσὸν <...> διὰ ταῦτα· <ποτὲ χρυσὸν> διὰ τὸ
χρυσοκοράλλιον, ποτὲ ἄργυρον διὰ τὸν χρυσόν, ποτὲ χαλκὸν διὰ
τὸ ἤλεκτρον, ποτὲ μόλυβδον διὰ τὸ μολυβδόχαλκον. Αὕτη ἡ ὕλη
εἰς τὴν χρυσοποιίαν εἰρήσθω.

15 σποδοκράμβης BeRu : σποδω- **A** ‖ στυπτηρίας BeRu : ⚹ **A** ‖ νίτρου BeRu : Ñ **A** ‖
16 θείου ἀθίκτου scripsi : ꟽ **A** : θείου (✕) Syn. Alch., l. 267 ‖ οὐρὸς scripsi : οὖρος
(sic) **A** : οὖρον BeRu ‖ **18** ἀλλοιοῦντα **A** : μεταλλ- Syn. Alch., l. 74 ‖ **19** post ὕλην
add. καὶ μεταλλεύοντα Syn. Alch., l. 75 ‖ εἰσὶν **A** : ποιεῖ Syn. Alch., l. 279 ‖ **20**
ποιήσῃς scripsi : -εις **A** ‖ post μακάριος add. νικήσεις γὰρ μεθόδῳ πενίαν,
τὴν ἀνίατον νόσον Syn. Alch., l. 58 ‖ **20-1** ἐπίβαλλε prop. BeRu : -ει **A** ‖ **21** χαλκόν
BeRu : ♀ **A** ‖ χρυσὸν scripsi : ♂ **A** : χρυσῷ BeRu ‖ post χρυσὸν lacunam indicavi ‖
ποτὲ χρυσὸν addidi, coll. Syn. Alch., l. 293 ‖ **22** ἄργυρον BeRu : ☾ **A** ‖ τὸν χρυσόν
BeRu : τὸ ♂ **A** ‖ ♀` [i.e. χαλκόν BeRu] **A** ‖ **23** ἤλεκτρον **A** : χρυσόν Syn. Alch.,
l. 294 ‖ μόλυβδον BeRu : ♄ **A** ‖ post μόλυβδον add. ἢ κασσίτερον Syn. Alch., l. 295 ‖
μολυβδόχαλκον scripsi, coll. Syn. Alch., l. 295 : ♄ **A** : μόλυβδον BeRu ‖ **24** ♂ποιΐαν
[i.e. χρυσοποιίαν BeRu] **A**

water, water of cabbage ashes, water of lees, alum water, soda water, orpiment water, pure sulphur water, the serous part of jenny's milk, dog's milk.

These are the substances for the making of gold; these are the substances that change the matter; these substances are fire-resisting. Nothing is certain beyond these substances. If you are intelligent and operate according to what has been written, you will be happy. Throw upon copper, gold, <...> for [making] these metals: <now [throw upon] gold> for [making] gold coral, now silver for [making] gold, now copper for [making] electrum [i.e. gold-silver alloy], now lead for [making] *molybdochalkon* [i.e. lead-copper alloy].[3] These are the substances for the making of gold; let this suffice.

§ 2   Test. et loci similes 14–7 Zos. Alch. *CAAG* II 226,9–14 = IV 47–53 Mertens et *CAAG* II 184,6–14; Syn. Alch. § 6, ll. 69–72; *De ovo CAAG* II 19,7–12 (vide etiam *CAAG* II 21,18–9) ‖ **18–20** Syn. Alch. § 6, ll. 74–7 et § 17, ll. 282–4 ‖ **21–3** Syn. Alch. § 18, ll. 293–5 ‖ **23–4** Syn. Alch. § 18, l. 299.

25   3. Ὕλη ἀργυροποιίας ἐστὶ δέ·

Ὑδράργυρος ἡ ἀπὸ ἀρσενικοῦ, ἢ σανδαράχης, ἢ ψιμυθίου, ἢ
μαγνησίας, ἢ στίμεως Ἰταλικοῦ· ποιήσεις εἴς <τι> τοιοῦτον ὃ ἐὰν
βούλη ἐκ<σ>τρέψας. Ἐὰν χαλκὸν οἰκονομήσῃς ὡς δέον, φέρεις
ἔξω τὴν φύσιν· γῆ Χία, καδμία λευκή, γῆ ἀστερίτης, Κιμωλία,
30   ἀρσενικὸν τὸ λευκόν, μίσυ ὀπτόν, μίσυ ὠμόν, λιθάργυρος λευκή,
ψιμύθιον, νίτρον πυρρὸν ὅ ἐστι ρίθεον, ἅλας Καππαδοκικόν,
μαγνησία λευκή, ἀφροσέληνον ὑαλοῦν, κυανός, τίτανος ὀπτή.

---

25 Ⅽποιίας [i.e. ἀργυροποιίας] **A** ‖ **26** ὑδράργυρος BeRu : ☽ **A** ‖ σανδαράχης **A** :
θείου (✕) Syn. Alch., l. 305 ‖ ψιμυθίου scripsi : ψιμμί̈θεως **A** ‖ **27** ποιήσεις **A** : -σει
BeRu ‖ τι addidi ‖ τοιοῦτον BeRu : τοιούτων **A** ‖ **28** ἐκστρέψας BeRu : ἐκτρε- **A** ‖
♀‟ [i.e. χαλκὸν BeRu] **A** : οὖν Syn. Alch., l. 87 ‖ οἰκονομήσῃς BeRu : -εις **A** ‖ δέον
BeRu : δέων **A** ‖ **29** καδμία BeRu : κατμία **A** ‖ γῆ ἀστερίτης BeRu : -ῆν -ην **A** ‖
κιμωλία BeRu : -αν **A** ‖ **30** ἀρσενικὸν scripsi : -οῦ **A** ‖ μίσυ BeRu : μυσι **A** ut
semper ‖ λιθάργυρος BeRu : λἵℂ **A** ‖ **31** ψιμύθιον scripsi : ψιμμΐ- **A** ‖ νίτρον πυρρὸν
BeRu : Ν̄ πυρὸν **A** ‖ **32** μαγνησία λευκή BeRu : μ̄ λευκῆς **A** ‖ ἀφροσέληνον
BeRu : ἀℂνον **A** ‖ ὑαλοῦν scripsi : -οῦ **A** ‖ ὀπτή BeRu : -ῆς **A**

3. **Here are the substances for the making of silver:**

Mercury that comes from orpiment, or from realgar,[4] or from white lead, or from *magnēsia*, or from Italian stibnite:[5] you will work towards [producing] something similar by turning whatever substance you want inside out. If you process copper properly, you lead its nature outside: Chian earth, white cadmia, *asteritēs* earth, Cimolian earth, white orpiment, roasted *misy*, raw *misy*, white litharge, white lead, red soda that is *ritheon* (?),[6] salt of Cappadocia, white *magnēsia*, moon foam of glass, azurite, roasted lime.

5 3 TEST. ET LOCI SIMILES **26–7** Syn. Alch. § 8, ll. 125–6 et § 19, ll. 304–5 ‖ **28–30** Syn. Alch. § 7, ll. 87–9 ‖ **29–30** Zos. Alch. *CAAG* II 226,23–5 = IV 68–70 Mertens et *CAAG* II 185,17–8 (= 186,4–5); Zos. Alch. (?) *CAAG* II 159,12–4; Syn. Alch. § 6, ll. 88–9.

Συνεσίου φιλοσόφου πρὸς Διόσκορον
εἰς τὴν βίβλον Δημοκρίτου ὡς ἐν σχολίοις

1   Διοσκόρῳ ἱερεῖ τοῦ μεγάλου Σαράπιδος τοῦ ἐν ᾿Αλεξανδρείᾳ
θεοῦ τε συνευδοκοῦντος Συνέσιος φιλόσοφος χαίρειν.

1. Τῆς πεμφθείσης μοι ἐπιστολῆς παρὰ σοῦ περὶ τῆς τοῦ θείου
Δημοκρίτου βίβλου οὐκ ἀμελέστερον ἔσχον, ἀλλὰ σπουδῇ πολλῇ
5   καὶ πόνῳ ἐμαυτὸν βασανίσας, ἔδραμον πρός σέ. ᾿Εν ᾧ οὖν
πρόκειται ἡμῖν εἰπεῖν τίς ἂν εἴη ὁ ἀνὴρ ἐκεῖνος, ὁ φιλόσοφος
Δημόκριτος, ἐλθὼν ἀπὸ ᾿Αβδήρων, φυσικὸς ὢν καὶ πάντα τὰ
φυσικὰ ἐρευνήσας καὶ συγγραψάμενος τὰ ὄντα κατὰ φύσιν.
῎Αβδηρα δέ ἐστι πόλις Θρᾴκης· ἐγένετο δὲ ὁ ἀνὴρ λογιώτατος, ὃς
10   ἐλθὼν ἐν Αἰγύπτῳ ἐμυσταγωγήθη ὑπὸ τοῦ μεγάλου ᾿Οστάνου ἐν
τῷ ἱερῷ τῆς Μέμφεως σύν τε πᾶσι τοῖς ἱερεῦσι Αἰγύπτου. ᾿Εκ
τούτου λαβὼν ἀφορμάς, συνεγράψατο βίβλους τέσσαρας βαφικάς,
περὶ χρυσοῦ καὶ ἀργύρου καὶ λίθων καὶ πορφύρας. Λέγω δὴ· τὰς
ἀφορμὰς λαβών, συνεγράψατο παρὰ τοῦ μεγάλου ᾿Οστάνου.
15   ᾿Εκεῖνος γὰρ ἦν πρῶτος ὁ γράψας ὅτι ἡ φύσις τῇ φύσει τέρπεται
καὶ ἡ φύσις τὴν φύσιν κρατεῖ καὶ ἡ φύσις τὴν φύσιν νικᾷ καὶ τὰ
ἑξῆς.

M, fol. 2ʳ (ubi tit. legitur) + fols. 72ᵛ9-78ʳ4
V, fols. 79ʳ4-91ʳ5. Manus recentior (Vᵃ) fols. 82ʳ-83ᵛ (nunc deleta sunt) necnon fol. 87ʳ⁻ᵛ
scripsit
B, fols. 20ʳ19-31ᵛ13
A, fols. 31ʳ23-37ᵛ15

CAAG II 56-69

Tit. διόσκορον **M** (72ᵛ et 2ʳ) **VA** : -σκουρον **B** ‖ εἰς τὴν β. Δ. ὡς ἐν σχολ. **M** 72ᵛ **BVA** :
διάλεξις περὶ τῆς τοῦ θείου Δημοκρίτου βίβλου **M** 2ʳ ‖ σχολίοις **BVA** : -είοις **M** ‖ **1**
διοσκόρῳ **MBV** : -ώρῳ **A** ‖ alt. τοῦ om. BeRu (qui in *CAAG* II 471 add.) ‖ ἀλεξανδρείᾳ
MBV : -έου **A** ‖ **2** χαίρειν **MBV** : -ιν **A** ‖ **8** ἐρευνήσας καὶ συγ- non legit. **B** ‖ **8-9**
-σιν ῎Αβδ- non legit. **B**, sed manus secunda in mg. ἀβδ- add. ‖ **9** θρᾴκης **BVA** : -ις **M** ‖ ὁ
ἀνὴρ non legit. **B** ‖ **10** ὑπὸ **MV** : παρὰ **BA** ‖ ὀστάνου **BA** : -ους **MV** ut semper ‖ **11**
σύν τε πᾶσι prop. Scardino («BMCR» 2013.02.55) : σὺν καὶ πᾶσι **MV** : σὺν καὶ παισὶ
BA ‖ **13** χρυσοῦ BeRu : ♎ **MBVA** ‖ ἀργύρου BeRu : ☾ **MBVA** ‖ **14** παρὰ **BA** : περὶ
MV ‖ **15** ὁ om. **MV**

The philosopher Synesius to Dioscorus: Notes on Democritus' Book (= *Syn. Alch.*)

With God's approval, the philosopher Synesius greets Dioscorus, priest of the great Serapis in Alexandria.

1. I did not neglect the letter you sent me about the book by the divine Democritus; rather, after having zealously and laboriously questioned myself, I hastened to you. Therefore we set forth right now to say who was that famous man, the philosopher Democritus: he came from Abdera and as a natural philosopher he investigated all natural questions and composed writings about all natural phenomena. Abdera is a Thracian city, but he became a very wise man when he went to Egypt, and was initiated in the temple of Memphis along with all the Egyptian priests by the great Ostanes. He took his basic principles from him and composed four books on dyeing, on gold, silver, [precious] stones and purple. I stress this point: he wrote by taking his basic principles from the great Ostanes. For he was the first to write that nature delights in nature, and nature masters nature, and nature conquers nature, and so on [Ps.-Dem. Alch. *PM* § 3, ll. 61-3].[1]

---

§ 1 TEST. ET LOCI SIMILES 7 Philos. Anon. Alch. *CAAG* II 425,1–2; Philos. Christ. Alch. *CAAG* II 395,6; *CAAG* II 447,12 ‖ 9–14 Olymp. Alch. *CAAG* II 102,17–8 (et *CAAG* II 78,11); Syncell. 297,24–298,1 Mosshammer (= 68 B 300,16 D-K; Bidez-Cumont, *Mages*, vol. 2, 311 fr. A3).

2. Ἀλλ' ἡμῖν ἀναγκαῖόν ἐστι τὰ τοῦ φιλοσόφου ἀνιχνεῦσαι καὶ μαθεῖν τίς ἡ γνώμη καὶ ποία ἡ τάξις τῆς ἐν αὐτῷ ἀκο-
20 λουθίας. Ὅτι μὲν οὖν δύο καταλόγους ἐποιήσατο δῆλον ἡμῖν γέγονεν, λευκοῦ καὶ ξανθοῦ· καὶ πρῶτον μὲν τὰ στερεὰ κατέ-λεξεν, ἔπειτα δὲ τοὺς ζωμούς, τουτέστι τὰ ὑγρά, καίτοι μηδενὸς τούτων προσλαμβανομένου ἐπὶ τῆς τέχνης. Αὐτὸς γὰρ μαρτυρεῖ λέγων περὶ τοῦ μεγάλου Ὀστάνου ὅτι οὗτος ὁ ἀνὴρ οὐκ ἐκέχρητο
25 ταῖς τῶν Αἰγυπτίων ἐπιβολαῖς οὐδὲ ὀπτήσεσιν, ἀλλὰ ἔξωθεν διέχριε τὰς οὐσίας καὶ πυρῶν εἰσέκρινε τὸ φάρμακον. Εἶπε δὲ ὅτι ἔθος ἐστὶν Πέρσαις τοῦτο ποιεῖν· ὃ δὲ λέγει, τοῦτό ἐστιν· ὅτι εἰ μὴ ἐκλεπτύνῃς τὰς οὐσίας καὶ ἀναλύσῃς καὶ ἐξυδατώσῃς οὐδὲν ποιήσεις.

30 3. Ἔλθωμεν οὖν ἐπὶ τὴν τοῦ ἀνδρὸς ῥῆσιν καὶ ἀκούσωμεν αὐτοῦ λέγοντος· λέγεται δὲ καὶ τὸ Πόντιον ῥᾶ. Βλέπε τοσαύτην παρατήρησιν τοῦ ἀνδρός· ἀπὸ βοτανῶν ᾐνίξατο, ἵνα μηνύσῃ τὸ ἄνθος· αἱ γὰρ βοτάναι ἀνθοφόροι εἰσίν. Εἶπε δὲ καὶ τὸ Πόντιον ῥᾶ, ὡς ὅτι ὁ Πόντος καταρρέοιτο ὑπὸ τῶν ποταμῶν· καὶ <γὰρ>
35 πάντες οἱ ποταμοὶ εἰς αὐτὸν καταρρέουσιν. Κατάδηλον οὖν ἡμῖν ποιούμενος σημαίνει τὴν ἐξυδάτωσιν καὶ ἀχλύωσιν καὶ λεπτυ-σμὸν τῶν σωμάτων ἤτοι οὐσιῶν.

4. Διόσκορος λέγει. Καὶ πῶς εἶπεν ὅτι ὅρκια ἡμῖν ἔθετο μηδενὶ σαφῶς ἐκδοῦναι;

40 - Καλῶς εἶπε μηδενί, τουτέστι μηδενὶ τῶν ἀμυήτων· τὸ γὰρ μηδενὶ οὐ κατὰ παντὸς κατηγορεῖται. Αὐτὸς γὰρ περὶ τῶν <μὴ> μεμυημένων καὶ γεγυμνασμένον τὸν νοῦν ἐχόντων εἶπεν.

18 ἀρχ in mg. **MV** : ἀρχὴ in mg. **BA** ‖ 21 post λευκοῦ add. γὰρ **BA** ‖ post ξανθοῦ add. καταλόγους ἐποιήσατο **BA** ‖ 26 πυρῶν **B** : -ὸν **MVA** ‖ 27 post ἐστὶν add. οὕτω **BA** ‖ 28 ἀναλύσῃς **MBV** : -εις **A** ‖ ἐξυδατώσῃς **MB** et V s.l. : -σεις **V** : -όσῃς **A** ‖ 29 ποιήσεις **A** : -ῃς **MBV** ‖ 31 post δὲ add. πρῶτον **A** ‖ πόντιον **MBVA** : Ποντικὸν prop. BeRu ‖ τοσαύτην **BA** : -η **M** : ὅση ἡ **V** ‖ 32 παρατήρησιν **B** :-ρεισιν **A** : -ρησις **MV** ‖ ᾐνίξατο **MBVA** : ἤρξατο Fabr post Pizzim 12ʳ22 (*exorsus est*) ‖ 34 καταρρέοιτο ὑπὸ τῶν ποταμῶν **MV** : ὑπὸ τ. π. καταρρέεται **B** : ἀπὸ τοῦ κατωρρέων (καταρρεῖν Fabr) τὸ ὑπὸ τῶν ποταμῶν **A** et Fabr : *a ponto defluunt flumina* Pizzim 12ʳ24 ‖ 34-5 καὶ — καταρρέουσιν om. **B** ‖ 34 γὰρ addidi ‖ 35 κατάδηλον **MBV** : κατάλληλον **A** ‖ 36 ποιούμενος **MV** : ποιησάμενος **BA** ‖ post σημαίνει add. δὲ **A** ‖ ἀχλύωσιν **BA** : ἄχλυσιν **MV** ‖ post ἀχλ. add. καὶ κατάλυσιν **A** ‖ 36-7 λεπτυσμὸν **MBV** : λελεπτυσ-μένων **A** ‖ 38 λέγει **MV** : φησίν **BA** ‖ 41 μὴ addidi post Pizzim 12ᵛ4-5 (*de imperitis et rudibus dixit*) ‖ 42 ἐχόντων τὸν νοῦν **BA**

2. But it is necessary that we follow in the footsteps of the philosopher and learn what his doctrine is, and in what order his arguments follow one another. It is clear to us that he composed two catalogues, on the white [i.e. on silver] and on the yellow [i.e. on gold]. First he listed the solid substances, afterwards the washes – that is, the liquid substances – although none of them is taken into account in our Art.[2] For the philosopher himself, in speaking about the great Ostanes, testifies that this man did not make use of the Egyptian methods for applying and roasting [the substances], but he used to smear the substances on the outside and heat them to make the drug sink in. He said that the Persians were in the habit of working thus.[3] Here is what he said: "If you do not make the substances very thin and do not dissolve and make them watery,[4] you will not achieve any result."

3. Let us move on to the words of this man [i.e. Democritus] and listen to him speaking. He said: "and rhubarb from Pontus [i.e. the Black Sea]" [Ps.-Dem. Alch. *PM* § 17, l. 186; *Cat.* § 2, l. 13]. Look at how perceptive are the remarks of this man. By speaking of plants he spoke in riddles, in order to indicate their flower: for plants produce flowers.[5] But he said "and rhubarb from Pontus," just because Pontus is fed by rivers; <in fact,> all the rivers flow into it.[6] So he made clear to us and showed how to make the bodies or the substances watery, and darken and make them thin.

4. Dioscorus says: And in what sense did he say that he [i.e. Ostanes?] made us swear not make any clear disclosures to anybody?

– He was right when he said "to nobody," that is, to no one among uninitiated people. For "to nobody" is not asserted with a general meaning. He was speaking about those who have <not> been initiated and who do <not> have a well-trained mind.[7]

§ 2   TEST. ET LOCI SIMILES 20–3 Olymp. Alch. *CAAG* II 99,20–100,5 ≈ *De lapide philosophiae*, *CAAG* II 199,25–200,6; vide etiam infra, Syn. Alch. § 8, ll. 110–2 ‖ 23–7 Philos. Anon. Alch. (Joan. Alch. BeRu) *CAAG* II 264,19–265,6 ‖ 28–9 Philos. Anon. Alch. (Zos. Alch. BeRu) *CAAG* II 131,5–16 ≈ *De lapide philosophiae CAAG* II 203,2–12.

§ 3   TEST. ET LOCI SIMILES 33–5 Steph. Alch. II 234,4–27 Ideler; vide etiam imag. in **M**, fol. 10ʳ (*CAAG* I 141–2); Philos. Christ. Alch. *CAAG* II 417,8–11.

§ 4   TEST. ET LOCI SIMILES 38–9 Isis Alch. *CAAG* II 29–30; Philos. Christ. Alch. (Pappus Alch. BeRu) *CAAG* II 27,19–28,4; vide etiam **SyrC**, fol. 144ᵛ (translatio in *CMA* II 326–7) et **SyrC**, fol. 137ʳ (translatio in *CMA* II 320) ‖ 40–2 Steph. Alch. II 234,35–235,3 Ideler.

5. Βλέπε γὰρ ἐν τῇ εἰσβολῇ τῆς χρυσοποιίας τί εἶπεν· ὑδράργυρος ἡ ἀπὸ κινναβάρεως, χρυσόκολλα.

45   Διόσκορος. Καὶ τοιούτων χρεία ἐστίν;

Συνέσιος. Οὐχί, Διόσκορε.

Διόσκορος. Ἀλλὰ τίνος ἐστὶ χρεία;

– Ἤκουσας καὶ πάλιν ἄκουσον. Ἡ ἀνάλυσίς ἐστι τῶν σωμάτων, ἵνα ἀναλύσῃς αὐτὰ καὶ ὕδατα αὐτὰ ποιήσῃς, ἵνα
50   ῥεύσωσι καὶ ἀχλυωθῶσι καὶ λεπτυνθῶσι· τοῦτο δὲ καλεῖται ὕδωρ
θεῖον καὶ ὑδράργυρος καὶ χρυσόκολλα καὶ θεῖον ἄπυρον καὶ ὅσα
ἄλλα ὀνόματά εἰσιν. Ἡ γὰρ λεύκωσις καῦσίς ἐστι, καὶ ἡ
ξάνθωσις ἀναζωοπύρησις· αὐτὰ γὰρ ἑαυτὰ καίουσι, καὶ αὐτὰ
ἑαυτὰ ἀναζωοπυροῦσιν. Ὁ δὲ φιλόσοφος πολλοῖς ὀνόμασιν ἐκά-
55   λεσεν αὐτά, ποτὲ μὲν ἑνικῶς, ποτὲ δὲ πληθυντικῶς, ἵνα γυμνάσῃ
ἡμᾶς καὶ ἴδῃ εἴ ἐσμεν νοήμονες. Εἴρηκε γὰρ ὑποκατιὼν οὕτως·
ἐὰν ᾖς νοήμων καὶ ποιήσῃς ὡς γέγραπται, ἔσῃ μακάριος· νικήσεις
γὰρ μεθόδῳ πενίαν, τὴν ἀνίατον νόσον. Ἀποδιαπεμπόμενος οὖν
καὶ ἀποπερισπῶν ἡμᾶς τῆς ματαίας πλάνης, ὥστε ἀπαλλαγῆναι
60   ἡμᾶς τῆς πολυύλου φαντασίας· πρόσεχε δὲ ἐν τῇ εἰσβολῇ τῆς
βίβλου τί εἶπεν· ἥκω δὴ κἀγὼ ἐν Αἰγύπτῳ φέρων τὰ φυσικά, ὅπως
τῆς πολλῆς ὕλης καταφρονήσητε. Φυσικὰ δὲ εἴρηκε τὰ στερεὰ
σώματα· εἰ μὴ γὰρ αὐτὰ ἀναλυθῶσι, καὶ πάλιν παγῶσιν, οὐδὲν
<ἂν> εἰς πέρας προέλθοι τοῦ πράγματος.

65   6. Καὶ ἵνα νοήσωμεν ὅτι ἐκ τῶν στερεῶν λαμβάνεται τὰ
ὕδατα, τουτέστι τὸ ἄνθος, ὅρα πῶς εἶπε· τὰ δὲ ἐν ζωμοῖς· κρόκος
Κιλίκιος καὶ ἀριστολοχία καὶ τὰ ἑξῆς. Τὰ ἄνθη εἰπών, ἐδήλωσεν

---

43 ϑποιίας [i.e. χρυσοποιίας] **MBVA** ‖ ὑδράργυρος BeRu : ☽ **MBVA** ‖ **44** κινναβάρεως BeRu : ☉ **MBV** : ☉ʸ **A** ‖ χρυσόχολλα BeRu : ♆ **MBVA** ‖ **45** ἐρωτ. ἀπρ. [i.e.
fort. ἐρώτησις· ἀπόκρισις·] in mg. **A** ‖ **45-7** χρεία **MBV** : χρήα **A** ‖ **49** ἀναλύσῃς
MBV : -εις **A** ‖ pr. αὐτὰ om. **BA** ‖ ποιήσῃς **MBV** : -εις **A** ‖ alt. ἵνα **MV** : καὶ **BA** ‖
50-1 ὕδωρ θεῖον BeRu: ὖΧ **MBV** et A qui s.l. θεῖον add. ‖ **51** ὑδράργυρος BeRu : ☽
MBVA : σελήνη Fabr ‖ χρυσόκολλα BeRu : ♆ **MBVA** ‖ alt. θεῖον BeRu : Χ **MB**
VA ‖ **51-2** καὶ ὅσα — εἰσιν **MV** : ἀλλὰ δὴ καὶ ὅσα λοιπὰ ὀνόματά εἰσιν **BA** ‖
53 ἀναζωοπύρησις **MBV** : -ρισης **A** ‖ **54** ἀναζωοπυροῦσιν **MVA** : -ζωπυροῦσιν **B** ‖
55 ὅρα πονηρίαν φιλοσόφου (-ων **B**) in mg. **BA** ‖ **56** ἴδῃ **MBV** : εἶ- **A** ‖ νοήμονες
MBV : νοεί- **A** ut plerumque ‖ **57** ποιήσῃς **MBV** : -εις **A** ‖ **58** post οὖν add. ἦν **BA** ‖
62 ἀπ᾽ ὧδε in mg. **M** ‖ **62-3** supra τὰ στερεὰ σώματα add. ⸗⸗⸗⸗ [signum incertum] **B** ‖
64 ἂν addidi ‖ προέλθοι **MV** : προσέλθη **BA** : προσέλθοι BeRu ‖ **66-7** κρόκος
Κιλίκιος scripsi : -ον -ον **MBVA** ‖ supra κρόκος et ἀριστολοχία add ☽ **BA** ‖ **67**
ἀριστολοχία scripsi : -λοΧ **MV** : -χίαν **BA**

5. Look at what he said in his introduction to the making of gold: "Mercury that comes from cinnabar, malachite ..." [Ps.-Dem. Alch. *Cat.* § 1, ll. 1-2].

Dioscorus. And do we really need such substances?
Synesius. We do not, Dioscorus.
Dioscorus. But, what do we need?

– You have listened to me, but listen again. The solution concerns solid bodies, in order to dissolve them and make them liquid [lit. waters]: in this way they flow and are made dark and thin. This is called divine water and mercury and malachite and unburnt sulphur and all the other names that are suitable. Whitening, in fact, consists in a burning process, and yellowing in a further burning [which regenerates bodies].[8] The bodies burn each other and kindle each other again [by regenerating one another]. The philosopher called them by many names both singular and plural, in order to train us and to see whether we are intelligent. He spoke these words further on in the text: "If you are intelligent and will operate according to what has been written, you will be happy" [Ps.-Dem. Alch. *Cat.* § 2, l. 20]; "for through this method you will overcome poverty, the incurable disease." Therefore he prevented and drew us away from being vainly led astray, so that we dismiss the illusion that a plurality of matters exist [Ps.-Dem. Alch. *PM* § 20, l. 227]. Turn your attention to what he said in the introduction of his book: "I too came to Egypt to deal with natural substances, so that you may turn your mind away from the plurality of matter" [Ps.-Dem. Alch. *PM* § 4]. He gave to solid bodies the name of natural substances: if they had not been dissolved and made solid again, we would not advance at all towards the end of our practice."

6. And in order to understand that the waters [i.e. the liquid substances] – that is, the flowers – are obtained from solid substances, look at how he spoke: "These are the substances that fall among the washes: Cilician saffron and

§ 5   Test. et loci similes 48–50 vide supra, Syn. Alch. § 2, ll. 28–9 ‖ 50–2 Zos. Alch. *CAAG* II 184,5–6; [Zos. Alch.] *CAAG* II 144,20–1; Olymp. Alch. *CAAG* II 79,11–2; Philos. Christ. Alch. *CAAG* II 416,5–8 ; *De lapide philosophiae*, *CAAG* II 200,13–4 et 201,10 ‖ 52–4 Zos. Alch. *CAAG* II 179,12–3 et 183,17–20; Philos. Anon. Alch. *CAAG* II 441,10–5; Steph. Alch. II 210,6–20 Ideler; vide etiam infra, *De blancatione* § 2, ll. 4–7 ‖ 56–7 Zos. Alch. *CAAG* II 233,25–7 = I 187–8 Mertens et *CAAG* II 212,21; Steph. Alch. II 233,23–4 et 235,4 Ideler; Philos. Christ. Alch. *CAAG* II 414,9–10; *CAAG* II 285,3; vide etiam infra, Syn. Alch. § 6, ll. 76–7 ‖ 60 Steph. Alch. II 247,9 Ideler.

ἡμῖν ὅτι ἐκ τῶν στερεῶν τὰ ὕδατα λαμβάνεται. Καὶ ἵνα ἡμᾶς
πείσῃ ὅτι ταῦτα οὕτως ἔχει, μετὰ τὸ εἰπεῖν οὖρον ἄφθορον, εἶπε·
70 καὶ ὕδωρ ἀσβέστου καὶ ὕδωρ σποδοκράμβης καὶ ὕδωρ φέκλης καὶ
ὕδωρ στυπτηρίας, καὶ ἐπὶ τέλει εἶπε κυνὸς γάλα. Καὶ δῆλον ἡμῖν
ἐστιν ὅτι τὸ ἐκ τοῦ κοινοῦ ἀναφερόμενον. Τὰ γὰρ λυτικὰ τῶν
σωμάτων προεισήνεγκεν ὕδωρ νίτρου καὶ ὕδωρ φέκλης. Καὶ ὅρα
πῶς εἶπεν· αὕτη ἡ ὕλη τῆς χρυσοποιίας· ταῦτά εἰσι τὰ μεταλ-
75 λοιοῦντα τὴν ὕλην καὶ μεταλλεύοντα καὶ πυρίμαχα ποιοῦντα.
Ἐκτὸς γὰρ τούτων οὐδέν ἐστιν ἀσφαλές· ἐὰν οὖν ᾖς νοήμων καὶ
ποιήσῃς ὡς γέγραπται, ἔσῃ μακάριος.

7. Διόσκορος. Καὶ πῶς ἔχω νοῆσαι, φιλόσοφε; Τὴν μέθοδον
παρὰ σοῦ βούλομαι μαθεῖν· ἐὰν γὰρ ἀκολουθήσω τοῖς εἰρημέ-
80 νοις, οὐδὲν ὀνήσομαί τι παρ' αὐτῶν.

– Ἄκουσον, Διόσκορε, αὐτοῦ λέγοντος καὶ ὄξυνόν σου τὸν
νοῦν, Διόσκορε, καὶ βλέπε πῶς λέγει· ἔκστρεψον αὐτῶν τὴν
φύσιν· ἡ γὰρ φύσις ἔνδον κέκρυπται.

– Ὦ Συνέσιε, τίνα ἐκστροφὴν λέγει;

85 – Τὴν τῶν σωμάτων λέγει

– Καὶ πῶς αὐτὴν ἐκστρέψω; Ἤ πῶς φέρω τὴν φύσιν ἔξω;

– Ὄξυνόν σου τὸν νοῦν, Διόσκορε, καὶ πρόσεχε πῶς λέγει· ἐὰν
οὖν οἰκονομήσῃς ὡς δεῖ, φέρεις τὴν φύσιν ἔξω· γῆ Χία καὶ
ἀστερίτης, καδμία λευκὴ καὶ τὰ ἑξῆς. Βλέπε πόση παρατήρησις
90 τοῦ ἀνδρός, πῶς πάντα τὰ λευκὰ ἠνίξατο, ἵνα δείξῃ τὴν λεύκωσιν. Ὁ λέγει οὖν, Διόσκορε, τοιοῦτόν ἐστι· βάλε τὰ σώματα
μετὰ τῆς ὑδραργύρου καὶ ῥίνισον εἰς λεπτόν, καὶ ἀναλάμβανε
ὑδράργυρον ἑτέραν· πάντα γὰρ ἡ ὑδράργυρος εἰς ἑαυτὴν ἕλκει.

69 ἄφθορον **MBVA** : ἀφθόρου Zur¹ ‖ **70** pr. καὶ om. **BA** ‖ post ἀσβέστου des. **V** et inc.
Vᵃ 82ʳ-83ᵛ, quae autem deleta sunt ‖ σποδοκράμβης **MV** : σπονδο- **BA** ‖ tert. ὕδωρ
BA : ὖ **M** ‖ **71** στυπτηρίας BeRu : ⸎ **MBA** ‖ **72** τὸ om. **BA** ‖ **73** προεισήνεγκεν
BA : προσήνεγκεν **M** ‖ pr. et alt. ὕδωρ **A** : ὖ **MB** ‖ **74** ϑποιίας [i.e. χρυσοποιίας
BeRu] **MBA** ‖ **74-5** μεταλλοιοῦντα **MB** : -λιοῦντα **A** ‖ **77** ποιήσῃς **MB** : -εις **A** ‖ **78**
Διόσκορος om. **BA** ‖ **79** βούλομαι **MB** : -ωμαι **A** ‖ **80** ὀνήσομαι **MB** : -ωμαι **A** ‖ **81**
καὶ om. **M** ‖ σου **MA** : σον **B** ‖ **82** Διόσκορε **M** : τοῖς ἐγκειμένοις **BA** ‖ βλέπε **M** :
πρόσχες **BA** ‖ **85** τὴν — λέγει **M** : in mg. **BA** ‖ **87** πῶς λέγει **M** : τοῖς εἰρημένοις
BA ‖ **88** οἰκονομήσῃς **MB** : -εις **A** ‖ χία **BA** : χεία **M** ‖ **89** ἀστερίτης BeRu : -ρίτις
MB : -ρίτˢ **A** ‖ **90** τοῦ om. **M** ‖ τὰ om. **B** : add. s.l. **A** ‖ **91** ση (?) in mg. **A** ‖ supra βάλε
add. ⸎ **BA** ‖ **92** ὑδραργύρου BeRu : ☽ **MBA** ‖ ῥίνισον **MBA** : ῥίνη- BeRu ‖ **93**
ὑδράργυρον BeRu : ☽ **MBA** ‖ φανερόν (litt. inversae) in mg. **MBA** ‖ ὑδράργυρος
BeRu : ☽ **MBA**

Aristolochia" and so on [Ps.-Dem. Alch. *Cat.* § 2, ll. 11-2]. By speaking about flowers he showed us that the waters are obtained from solid substances. And in order to persuade us that this is the way things are, after mentioning "pure urine" he said: "and quicklime water, and water of cabbage ashes, and water of lees, and alum water," and at the end he said "dog's milk" [Ps.-Dem. Alch. *Cat.* § 2, ll. 14-7]. It is clear for us that it [i.e. this milk] is what [i.e. the vapour that] rises up from what is common [i.e. from the substances of everyday use].[9] Before [the dog's milk] he introduced the substances that can dissolve bodies, i.e. "soda water and water of lees." And look at how he spoke: "These are the substances for the making of gold; these are the substances that change the matter, that make it possible to mine [metallic bodies][10] and make them fire-resisting. Nothing is certain beyond these substances: If you are intelligent and operate according to what has been written, you will be happy" [Ps.-Dem. Alch. *Cat.* § 2, ll. 18-20].

7. Dioscorus. How can I become intelligent, my philosopher? I want to learn the method from you; for if I try to follow what has been said, I will not have any benefit from that.

– Listen to him speaking, O Dioscorus, and sharpen your mind, Dioscorus. Look at how he says: "Turn their nature inside out: for nature is hidden inside" [Ps.-Dem. Alch. *PM* § 14, l. 145; *Cat.* § 3, ll. 27-8].
– O Synesius, what transformation [lit. turning inside out] is he speaking about?
– He speaks about the transformation of the bodies.
– And how can I turn it [i.e. the nature] inside out? How can I lead the nature outside?
– Sharpen your mind, O Dioscorus, and turn your attention to how he speaks: "Therefore if you perform the right treatments, you lead the nature outside: Chian earth and *asteritēs*, white cadmia," and so on [Ps.-Dem. Alch. *AP* § 2, ll. 14-5; *Cat.* § 3, ll. 29-30]. Look at how perceptive are the remarks of this man, how he hinted at all white substances in order to show the whitening process. Therefore this is what he said, Dioscorus: "Mix the bodies with mercury, file them finely and add any other mercury: for mercury attracts everything to itself.[11]

§ 6   Test. et loci similes 74–5 Steph. Alch. II 209,6–7 Ideler ‖ 76–7 vide supra, Syn. Alch. § 5, l. 57.

§ 7   Test. et loci similes 82–3 Zos. Alch. CAAG II 223,25–6; Philos. Anon. Alch. (Zos. Alch. et Joan. Alch. BeRu) CAAG II 129,12–3 et CAAG II 264,2–3; *De lapide philosophiae*, CAAG II 201,1 ‖ 88–9 CAAG II 18,8–9; Zos. Alch. CAAG II 226,23–5 = IV 68–9 Mertens; Zos. Alch. CAAG II 185,17–8 et 186,4–5; Zos. Alch. (?) CAAG II 159,12–4; *De quattuor elementis*, CAAG II 341,9–13.

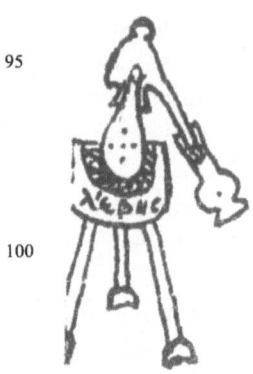

Καὶ ἔασον πεφθῆναι ἡμέρας τρεῖς ἢ τέσσαρας·
καὶ βάλε αὐτὴν εἰς βωτάριον ἐπὶ θερμοσποδιᾶς
μὴ ἐχούσης τὸ πῦρ διάπυρον, ἀλλὰ ἐπὶ θερμοσποδιᾶς πραείας [ὅ ἐστι κηροτακίς]. Ταύτῃ
οὖν τῇ ἀναδόσει τοῦ πυρός, συναρμόζεται τῷ
βωταρίῳ ὑάλινον ὄργανον ἔχον μαστάριον· ἐπὶ
τὰ ἄνω προσέχων, [καὶ] ἐπικέφαλα κείσθω· καὶ
τὸ ἀνερχόμενον ὕδωρ διὰ τοῦ μαζοῦ δέχου καὶ
ἔχε εἰς σῆψιν· τοῦτο λέγεται ὕδωρ θεῖον, αὕτη
ἐστὶν ἐκστροφή· ταύτῃ τῇ ἀγωγῇ φέρεις ἔξω τὴν
φύσιν τὴν ἔνδον κεκρυμμένην· αὕτη καλεῖται λύσις σωμάτων.
Τοῦτο ὅταν σαπῇ καλεῖται ὄξος καὶ οἶνος Ἀμιναῖος καὶ τὰ
ὅμοια.

8. Καὶ ἵνα θαυμάσῃς τὴν τοῦ ἀνδρὸς σοφίαν, βλέπε πῶς δύο
καταλόγους ἐποιήσατο, χρυσοποιίας καὶ ἀργυροποιίας, καὶ πάλιν δύο ζωμούς, τὸν μὲν ἕνα ἐν τῷ ξανθῷ, τὸν δὲ ἕτερον ἐν τῷ
λευκῷ, τουτέστιν χρυσῷ καὶ ἀργύρῳ· καὶ ἐκάλεσε τὸν τοῦ
χρυσοῦ κατάλογον χρυσοποιίαν, τὸν δὲ τοῦ ἀργύρου ἀργυροποιίαν.

94 imag. servat in mg. **A** ‖ ἡμέρας **M** : 66 **BA** ‖ 95 βωτάριον **M** : βο- **BA** : *in texta*
Pizzim 13ᵛ12 [i.e. μοτάρια] ‖ 95-6 ἐπὶ θερμοσποδίας (sic) — ἀλλὰ **M** : om. **BA**, sed μὴ
ἐχούσις (sic) τὸ διάπυρον· ἀλλ᾽ ἐπὶ θερμοποδίας add. in mg. **A** ‖ 96-7 θερμοσποδιᾶς
πραείας **BA** : -ίαν -εῖαν **M** ‖ 97 ὅ ἔστι κηροτακίς (sic **M**) tamquam ex glossemate
seclusi : ὁ δὴ βοτάριόν ἐστι κηρ. **BA** ‖ 99 βωταρίῳ **M** : βο- **BA** ‖ ὑάλινον **BA** : ὑέ-
M ‖ 100 προσέχων **M** : -ον **BA** ‖ καὶ seclusi ‖ ἐπικέφαλα κείσθω **M** : κατωκάρα
κείμενον **BA**, sed ἤγουν ἐπικέφαλα κ[εί]σθω add. in mg. **A** ‖ 102 εἰς σῆψιν **BA** : καὶ
σῆψον **M** ‖ ὕδωρ θεῖον **BA** : ὗ✕ **M** ‖ 103-4 ταύτῃ — κεκρυμμένην om. **BA** ‖ τὴν
φύσιν **M** : om. BeRu ‖ 105 ἀμιναῖος scripsi : ἀμη- **M** : ἀμηνέος **BA** ‖ imag. in ima
pagina servat **B** ‖ 108 ♂ποιίας [i.e. χρυσοποιίας BeRu] **MBA** ‖ ☿ποιίας [i.e. ἀργυροποιίας BeRu] **MBA** ‖ 109 τὸν δὲ ἕτερον **BA** : καὶ τὸν ἕνα **M** ‖ 110 χρυσῷ Zur¹ : ♂
MBA : -ὸν BeRu ‖ ἀργύρῳ Zur¹ : ☿ **MBA** : -ον BeRu ‖ 111 χρυσοῦ BeRu : ♂ **MBA** ‖
♂ποιίαν [i.e. χρυσοποιίαν BeRu] **MBA** ‖ τοῦ **M** : τῆς **BA** ‖ ἀργύρου BeRu : ☿ **MBA**

Let it macerate for three or four days and put it in a vessel [placed] not on hot ashes with a high flame, but in ashes at a milder temperature [this is the *kērotakis*]. During this application of heat, a glass instrument that has a breast-shaped protuberance is fitted onto the vessel. Put it on the top of the vessel and turn it upside down; collect the water going up through the breast and keep it for the fermentation."[12] This water is called divine water, and this is how to turn [the bodies] inside out. Through this method you lead the hidden interior nature outside; this is called the melting of bodies. This water after being macerated is called vinegar and Aminaios wine and similar names.[13]

8. In order to admire the wisdom of this man, look at how he composed two catalogues – for the making of gold (*chrysopoeia*) and the making of silver (*argyropoeia*) – and how he composed [two catalogues] of washes, the first in the yellow, that is, in [his book on] gold, the second in the white, that is, in [his book on] silver; and he gave to the catalogue of gold the name of "the making of gold," and to the catalogue of silver the name of "the making of silver."

§ 7   TEST. ET LOCI SIMILES **91–103** Zos. Alch. CAAG II 236,2–237,5 = III 1–27 Mertens (ll. 2–20 Mariae Ebraiae tribuendae sunt) et CAAG II 141,3–20 = IX 3–23 Mertens ‖ **98–102** |Zos. Alch.| CAAG II 210,11–2 et 250,18–251,2; Olymp. Alch. CAAG II 105,6; Steph. Alch. II 208,20–1 Ideler; Philos. Christ. Alch. CAAG II 275,10–4 et 278,11–2; Philos. Anon. Alch. CAAG II 439,23–440,4 ‖ **103–4** Zos. Alch. CAAG II 223,24–6; De quattruor elementis, CAAG II 338,17–8 et 340,2–3; De lapide philosophiae, CAAG II 200,26–7; vide etiam supra, Syn. Alch. § 6, ll. 82–3.
§ 8   TEST. ET LOCI SIMILES **107–12** vide supra, Syn. Alch. § 2, ll. 20–3.

– Πάνυ καλῶς ἔφης, ὦ φιλόσοφε Συνέσιε· καὶ ποῖον πρῶτόν ἐστι τῆς τέχνης, τὸ λευκᾶναι ἢ τὸ ξανθῶσαι;

115   Συνέσιος. Μᾶλλον τὸ λευκᾶναι.

Διόσκορος. Καὶ διὰ τί τὴν ξάνθωσιν εἶπε πρῶτον;

– Ἐπειδὴ προτετίμηται ὁ χρυσὸς τοῦ ἀργύρου.

– Καὶ οὕτως ὀφείλομεν ποιῆσαι, Συνέσιε;

– Οὔ, Διόσκορε, ἀλλὰ διὰ τὸ γυμνάσαι ἡμῶν τὸν νοῦν καὶ
120   τὰς φρένας, οὕτω συνετάγησαν. Ἄκουσον αὐτοῦ λέγοντος· ὡς νοήμοσιν ὑμῖν ὁμιλῶ, γυμνάζων ὑμῶν τὸν νοῦν. Ἐὰν δὲ βούλῃ τὸ ἀκριβὲς γνῶναι, πρόσεχε εἰς τοὺς δύο καταλόγους, ὅτι πρὸ πάντων ἡ ὑδράργυρος ἐτάγη, καὶ ἐν τῷ ξανθῷ, τουτέστιν χρυσῷ, καὶ ἐν τῷ λευκῷ, τουτέστιν ἀργύρῳ. Καὶ ἐν μὲν τῷ χρυσῷ εἶπεν·
125   ὑδράργυρος ἡ ἀπὸ κινναβάρεως· ἐν δὲ τῷ λευκῷ εἶπεν· ὑδράρ-γυρος ἡ ἀπὸ ἀρσενικοῦ ἢ σανδαράχης καὶ τὰ ἑξῆς.

9. Διόσκορος εἶπεν. Διάφορος οὖν ἐστιν ἡ ὑδράργυρος;

Συνέσιος. Ναί, διάφορός ἐστι μία οὖσα.

Διόσκορος. Καὶ εἰ μία ἐστὶ πῶς ἐστι διάφορος;

130   Συνέσιος. Ναί, διάφορος γίνεται καὶ μεγίστην δύναμιν ἔχει. Ἦ οὐκ ἤκουσας τοῦ Ἑρμοῦ λέγοντος· τὸ κηρίον τὸ λευκὸν καὶ τὸ κηρίον τὸ ξανθόν;

Διόσκορος. Ναί, ἤκουσα. Ὅπερ δὲ βούλομαι μαθεῖν, Συνέσιε, τοῦτο με δίδαξον τὸ ποίημα· πάντως αὕτη τὰ εἴδη πάντων δέχε-
135   ται;

Συνέσιος. Ἐνόησας, Διόσκορε· ὥσπερ γὰρ ὁ κηρὸς οἷον δ' ἂν προσλαμβάνῃ χρῶμα δέχεται, οὕτω καὶ ἡ ὑδράργυρος, φιλόσοφε·

---

113 Συνέσιε φιλόσοφε **BA** ‖ 117 ὁ om. **A** ‖ χρυσὸς BeRu : ☊ **MBA** ‖ ἀργύρου BeRu : ☽ **MBA** ‖ 118-9 καὶ σὺ κἂν δὲ οὐ δοκῇ ὦ διόσκορε add. in mg. **A** ‖ 119 διὰ om. **BA** : διὰ τὸ om. BeRu ‖ ἡμῶν **MA** : ὑ- **B** ‖ 121 νοήμοσιν **B** : -ωσιν **M** : νοεί- **A** ‖ 121-2 βούλῃ τὸ ἀκριβὲς **M** (idem in Olymp.): Βουλ^λ τὸ ἀ. **A** : βει (sic) τὸ ἀκριβῶς **B** ‖ 123 ὑδράργυρος BeRu : ☽ **MBA** ‖ χρυσῷ BeRu : ☊ **MBA** ‖ 124 ἀργύρῳ BeRu : ☽ **MBA** ‖ χρυσῷ BeRu : ☊ **MBA** ‖ 125 ὑδράργυρος BeRu : ☽ **MBA** ‖ κινναβάρεως BeRu : ☉ **MBA** ‖ 125-6 ὑδράργυρος BeRu : ☽ **MBA** ‖ 126 ☽^ου [i.e. ἀρσενίκου BeRu] **M** : ♌ **BA** ‖ σανδαράχης **BA** : σαν^δ **M** ‖ 127 διάφορος οὖν **M** : καὶ δ. **BA** ‖ ὑδράργυρος BeRu : ☽ **MBA** ‖ 131 ἦ scripsi : ἢ **BA** : om. **M** ‖ κηρίον **MB** : κῦρ- **A** ‖ 134 ποίημα non legit. **B** ‖ post ποίημα add. ὅτι σὺ ἐπίστασαι **A** ‖ 134-5 πάντως — δέχεται Synesio trib. Pizzim 14^r23-24, qui Συνέσιος. Ἐνόησας Δ. (l. 136) om. ‖ 136 Συνέσιος **A** : om. **MB** ‖ ἐνόησας **MB** : νόησον **A** (νόεισον in mg. add.) ‖ ante Διόσκορε add. ὦ **A** ‖ 137 supra προσλαμβάνῃ add. ὁμιλήσι (sic) **A** (vide *De lap. phil. CAAG* II 199,22 οἷον ἂν χρῶμα προσομιλήσῃ) ‖ ἡ om. **A** ‖ ὑδράργυρος BeRu : ☽ **MBA**

– You spoke wonderfully, O philosopher Synesius. What is the first operation of the
Art, whitening or yellowing?

Synesius. Whitening, to be sure.

Dioscorus. So why did he first speak about yellowing?

– Because gold is more valued than silver.

– So must we work following this order, Synesius?

– No, Dioscorus, but he chose this order with the intent of training our mind and
intellect. Listen to him speaking: "I deal with you as intelligent people, by training
your mind." If you want to obtain accurate knowledge, turn your attention to the
two catalogues: he listed mercury before all other substances, both in the yellow,
that is, in [his book on] gold, and in the white, that is, [in his book on] silver. He
said in [his book on] gold: "Mercury that comes from cinnabar" [Ps.-Dem. Alch.
Cat. § 1, l. 2], and he said in [his book on] silver: "Mercury that comes from orpiment'[4] or realgar," and so on [Ps.-Dem. Alch. *Cat.* § 3, l. 26].

9. Dioscorus said. So is mercury of different kinds?[15]

Synesius. Yes, it is of different kinds, although it is only one.

Dioscorus. But if it is only one, how can it be of different kinds?

Synesius. Yes, it becomes of different kinds and it has the greatest power. Did you
not listen to Hermes saying: "The white honeycomb and the yellow honeycomb"?

Dioscorus. Yes, I listened to him. Teach me, O Synesius, exactly this procedure
that I want to learn: does mercury take on entirely the forms of all substances?

Synesius. You did understand, Dioscorus: for as wax takes any colour it receives,
in the same way mercury does also, O philosopher; it whitens all substances, draws

§ 8   Test. et loci similes **119–21** Olymp. Alch. *CAAG* II 103,9–10; Christ. Alch. *CAAG* II 397,17–8; vide etiam infra,
Syn. Alch. § 7, ll. 289–90 ‖ **121–6** Olymp. Alch. ut monuerunt BeRu (*CAAG* II 90 n. 15), qui textum non ed. |vide **M**,
fol. 172ʳ14–22; **V**, fol. 21ᵛ19–22ʳ6; **A**, fol. 207ʳ21- 207ᵛ3| ‖ **124–6** *De lapide philosophiae*, *CAAG* II 199,19–21.

§ 9   Test. et loci similes **127–44** Olymp. Alch. ut monuerunt BeRu (*CAAG* II 90 n. 15), qui textum non ed. |vide **M**, fol.
172ʳ22–172ᵛ7; **V**, fol. 22ʳ7–22ᵛ3; **A**, fol. 207ᵛ4–19|; vide etiam **SyrL** in *CMA* II 45–7 (textus) et 82–5 (translatio);
SyrC, fols. 55ᵛ-58ᵛ (translatio in *CMA* II 242–5) ‖ **131–2** Philos. Christ. Alch. *CAAG* II 420,4–14 ‖ **136–7** *De
lapide philosophiae*, *CAAG* II 199,22–4.

αὕτη λευκαίνει πάντα καὶ πάντων τὰς ψυχὰς ἕλκει καὶ ἐφ᾽
ἑαυτὴν ἐπισπᾶται. Διοργανιζομένη οὖν καὶ ἔχουσα ἐν ἑαυτῇ τὰς
140 ὑγρότητας πάντως καὶ σῆψιν ὑφισταμένη, ἀμείβει πάντα τὰ
χρώματα καὶ ὑποστατικὴ γίνεται, ἀνυποστάτων αὐτῶν ὑπαρ-
χόντων. Μᾶλλον δέ, ἀνυποστάτου αὐτῆς ὑπαρχούσης, τότε καὶ
κατόχιμος γίνεται ταῖς οἰκονομίαις ταῖς διὰ τῶν σωμάτων καὶ
τῶν ὑλῶν αὐτῶν.
145   10. Διόσκορος. Καὶ ποῖά εἰσι ταῦτα τὰ σώματα καὶ αἱ ὗλαι
αὐτῶν;
      Συνέσιος. Ἡ τετρασωμία καὶ τούτων τὰ συγγενῆ.
      Διόσκορος. Καὶ ποῖά εἰσι τὰ τούτων συγγενῆ;
      Συνέσιος. Ἤκουσας ὅτι αἱ ὗλαι αὐτῶν ψυχαὶ αὐτῶν εἰσιν;
150   Διόσκορος. Καὶ αἱ ὗλαι οὖν αὐτῶν ψυχαὶ αὐτῶν εἰσιν;
      Συνέσιος. Ναί. Ὥσπερ γὰρ ὁ τέκτων, ἐὰν λάβῃ ξύλον <καὶ
ποιῇ καθέδραν> ἢ δίφρον ἢ ἄλλο τι, μόνον τὴν ὕλην ἐργάζεται,
οὕτω καὶ ἡ τέχνη αὕτη, ὦ φιλόσοφε, ἐπειδὴ ἔτεμεν αὐτήν. Ἄκουσον, ὦ Διόσκορε· ὁ λιθοξόος ξέει τὸν λίθον ἢ πρίζει,
155 ἵνα ἐπιτήδειος γένηται εἰς τὴν χρείαν αὐτοῦ. Ὁμοίως καὶ ὁ
τέκτων τὸ ξύλον πρίζει καὶ ξέει ὥστε γενέσθαι θρόνον ἢ δίφρον,
καὶ οὐδὲν χαρίζεται ὁ τεχνίτης εἰ μὴ μόνον τὸ εἶδος· οὐδὲν γάρ
ἐστιν, εἰ μὴ ξύλον. Ὁμοίως καὶ <ὁ> χαλκὸς γίνεται ἀνδριὰς ἢ
ἄλλο σκεῦος, τοῦ τεχνίτου αὐτῷ μόνον τὸ εἶδος χαριζομένου.
160 Οὕτως οὖν καὶ ἡ ὑδράργυρος φιλοτεχνουμένη ὑφ᾽ ἡμῶν πᾶν

138-9 ἐφ᾽ ἑαυτὴν Olymp. Alch. (cods. **MV**) : ἐψεῖ (-ῇ **A**) αὐτὰ καὶ **MBA** et Olymp.
Alch. (cod. **A**) ‖ **140** πάντα BeRu, coll. Olymp. Alch. (cods. **MV**) : παν^τ **M** : πάντως **BA**
et Olymp. Alch. (cod. **A**) ‖ **143** κατόχιμος **MB** : κατώχυμος **A** ‖ post οἰκ. ταῖς des. **V**ᵃ
et rursus inc. **V** ‖ **147** post τετρασωμία add. φησὶν Olymp. Alch. (cods. **MV**) ‖ τὰ
τούτων σ. **V** ‖ **149** post ὗλαι add. οὖν **A** ‖ pr. αὐτῶν om. BeRu ‖ post αὐτῶν εἰσι iter.
συνέσιος· ἡ τετρασωμία — ψυχαὶ αὐτῶν εἰσι (ll. 147-8) **B** ‖ **150-1** Διόσκορος — ναὶ
om. **V** ‖ **151-2** καὶ ποιῇ καθέδραν addidi, coll. ποιεῖ καθ. Olymp. Alch. : καὶ ποιῇ
θρόνον add. BeRu ‖ **152** post ἐργάζεται add. καὶ οὐδὲν ἄλλο αὐτῷ χαρίζεται ὁ
τεχνίτης εἰ μὴ μόνον τὸ εἶδος Olymp. Alch. (cods. **MV**) ‖ **153-4** αὐτὴν **V** : αὐ^τ **M** :
αὐτὰ **BA** ‖ **155** ἐπιτήδειος **MBV** : -διος **A** ‖ **157** οὐδὲν — εἶδος om. **A** ‖ post pr.
οὐδὲν add. ἄλλο BeRu, coll. Olymp. Alch. ‖ post χαρίζεται add. αὐτῷ Olymp. Alch.
(cods. **MV**) ‖ post alt. οὐδὲν (**VA**) add. ἄλλο BeRu, coll. Olymp. Alch. : οὐδὲ **MB** ‖ **158**
ὁ add. BeRu, coll. Olymp. Alch. ‖ χαλκὸς BeRu : ♀ **MV** : ☉ **BA** ‖ ἀνδριὰς **BA** : -ειὰς
MV ‖ post ἀνδριὰς add. ἢ κύκλος BeRu, coll. Olymp. Alch. ‖ **159** post ἄλλο add. τι
BeRu, coll. Olymp. Alch. ‖ αὐτῷ scripsi, coll. Olymp. Alch. (cods. **MV**): -ὸ **MBVA** ‖
160 ὑδράργυρος BeRu : ☽ **MVA** : non legit. **B**

to itself and absorbs their souls. Since it is processed with the appropriate tools[16] and holds in itself all the moistures and sustains the macerating process, it takes in exchange all the colours and becomes their support, because they are unsubstantial. Rather, exactly when mercury loses its substance [lit. becomes unsubstantial], then it may be retained [by the metallic bodies] by means of the treatments carried out using the bodies and their matters [i.e. constituents].[17]

10. Dioscorus. And what are these bodies and their matters?

Synesius. They are the *tetrasōmia* [lit. the four bodies] and their related substances.[18]

Dioscorus. And what are the substances related to these bodies?

Synesius. Did you hear that the matters [i.e. constituents] of these bodies are their souls?

Dioscorus. So are the matters of these bodies their souls?[19]

Synesius. Yes, they are. In fact, [just as] when a carpenter takes wood and <makes a chair> or a chariot or something else, he works only upon the matter, so it is with the Art itself, O philosopher, since it cuts the matter. Listen to me, Dioscorus: the stonemason polishes or saws the stone that has to become suitable for his needs. In the same way the carpenter also saws and smoothes down the wood so that it becomes a chair or a chariot, and the craftsman does not confer anything other than the form alone: in fact, there is nothing other than wood. In the same manner copper becomes a statue or a some other object, since the craftsman confers only the form upon it. And likewise, that is how also the mercury that we produce

εἶδος αὐτὴ ἀναδέχεται καὶ πεδηθεῖσα, ὡς εἴρηται, ἐν τῷ τετρα-
στοίχῳ σώματι ἰσχυρὰ καὶ ἀδίωκτος μένει, κρατοῦσα καὶ
κρατουμένη. Διὰ τοῦτο καὶ Πιβήχιος πολλὴν συγγένειαν ἔχειν
ἔλεγεν.

165   11. Διόσκορος. Καλῶς ἐπέλυσας, φιλόσοφε· ἐδίδαξάς με, φιλόσοφε. Βούλομαι οὖν ἐπὶ τὴν τοῦ ἀνδρὸς ἀναδραμεῖν ῥῆσιν καὶ ἐξ
ὑπαρχῆς ἀναλαβεῖν τὰ ὑπ᾽ αὐτοῦ λελοξευμένα ὡς εἰρημένα·
ὑδράργυρος ἡ ἀπὸ κινναβάρεως. Πᾶσα οὖν ὑδράργυρος ἀπὸ σωμάτων γίνεται. Οὗτος δὲ κιννάβαριν εἶπεν, ὡς δῆλον αὐτὴν ἀπὸ
170 κινναβάρεως οὖσαν. Καίτοι γε ἡ κιννάβαρις ὑδράργυρος ξανθή
ἐστιν, αὕτη δὲ λευκὴ ἡ ὑδράργυρος.

Συνέσιος. Ἐνεργείᾳ μὲν λευκὴ ὑπάρχει ἡ ὑδράργυρος, δυ-
νάμει δὲ ξανθὴ γίνεται.

Διόσκορος. Μὴ ἆρα τοῦτο ἔλεγεν ὁ φιλόσοφος· ὦ φύσεις
175 οὐράνιοι φύσεων δημιουργοί, ταῖς μεταβολαῖς νικῶσαι τὰς
φύσεις;

Συνέσιος. Ναί, διὰ τοῦτο εἴρηκεν· εἰ μὴ γὰρ ἐκστραφῇ,
ἀδύνατον γενέσθαι τὸ προσδοκώμενον, καὶ μάτην κάμνουσιν οἱ
τὰς ὕλας ἐξερευνῶντες καὶ μὴ φύσεις σωμάτων μαγνησίας ζη-
180 τοῦντες. Ἔξεστι γὰρ τοῖς ποιηταῖς καὶ συγγραφεῦσι τὰς αὐτὰς
λέξεις ἄλλως τε καὶ ἄλλως σχηματίζειν. Σῶμα οὖν μαγνησίας
εἴρηκε, τουτέστιν τὴν μίξιν τῶν οὐσιῶν· καὶ διὰ τοῦτο ὑποκατιὼν ἔφη ἐν τῇ εἰσβολῇ τῆς ποιήσεως τοῦ χρυσοῦ· λαβὼν ὑδράργυρον πῆξον τῷ τῆς μαγνησίας σώματι.

161 ☉ add. in mg. **M** ‖ τῷ om. **MV** et BeRu ‖ **161-2** τετραστοίχω **MVA** : -στίχω **B** ‖
162-3 κρατοῦσα καὶ κρατουμένη om. **BA** ‖ **163** Πιβήχιος **MBV** : Ἐπιβήχιος **A** ‖
συγγένειαν BeRu, coll. Olymp. Alch. (cods. **MV**) : ἀγγελίαν **MBVA** et Olymp. Alch.
(cod. **A**): εὐγένειαν Zur² ‖ **166** βούλομαι **MBV** : -ωμαι **A** ‖ ἀναδραμεῖν **MBV** :
-δρομεῖν **A** ‖ **167** ὑπαρχῆς ἀναλαβεῖν **BA** : ἀπαρχῆς εἰδέναι **MV** ‖ λελοξευμένα ὡς
MV : λελοξευμένως **BA** ‖ **168** pr. et alt. ὑδράργυρος BeRu : ☽ **MBVA** ‖ κινναβάρεως
BeRu : ☉ **MBV** : ☉ **A** ‖ **169** κιννάβαριν Zur¹ : ☉ **MBV** : ☉ **A** : -ις BeRu ‖ **170**
κινναβάρεως BeRu : ☉ **MBV** : ☉ **A** ‖ κιννάβαρις BeRu : ☉ **MBV** : ☉ **A** ‖ **170-1** pr. et
alt. ὑδράργυρος BeRu : ☽ **MBVA** ‖ **172** ὑδράργυρος BeRu : ☽ **MBVA** ‖ **174** ἆρα **M** :
ἄ- **BVA** ‖ **175** οὐράνιοι **MV** : -αι **A** : non legit. **B** ‖ **177** Συνέσιος om. **BA** ‖ **178**
προσδοκώμενον **MBV** : -όμενον **A** ‖ **179** post ἐξερευνῶντες interrogationis signum add.
A ‖ φύσεις σωμάτων **MBVA** : -ιν σώματος prop. BeRu ‖ μαγνησίας BeRu : μῦγ **MBV** :
μᵍΑ ‖ **181** μαγνησίας **BA** : μῦγ **MV** ‖ **182** τουτέστιν om. **BA** ‖ **183** ᵛ𝟃 [i.e. χρυσοῦ
BeRu] **B** : 𝟃 **MVA** ‖ **183-4** ὑδράργυρον BeRu : ☽ **MBVA** ‖ **184** μαγνησίας **BA** : μΤ̅
M : μῦγ **V**

with art takes any form and remains – as I said – fixed and strictly bound together with the *tetrastoichos* [i.e. the alloy formed by the four elements (metals)], which it masters and by which it is mastered. This is why Pebichios also claimed that mercury has a great affinity [for the bodies].

11. Dioscorus. Your explanation is clear, Synesius, and you have instructed me, O philosopher. Therefore I want to return to the discourse of this man [i.e. Democritus] and take up again, from the beginning, exactly the words he said in an obscure way: "Mercury that comes from cinnabar" [Ps.-Dem. Alch. *Cat.* § 1, l. 2]. So every mercury is produced from bodies. He said "cinnabar" to make clear that mercury is produced from cinnabar. However, cinnabar is a yellow [kind of] mercury, while this mercury is white.

Synesius. The mercury is white actually, but yellow potentially.[20]

Dioscorus. Did not the philosopher say these words: "O celestial natures, artificers of natures, which conquer natures with your transformations"? [Ps.-Dem. Alch. *PM* § 15, ll. 150-2].

Synesius. Yes, he did, and that is why he said: "If you do not turn [the natures] inside out, it will be impossible to reach what is expected; those who examine the matter and do not investigate the natures of the bodies of *magnēsia* wear themselves out in vain." Indeed, poets and prose-writers are allowed to fashion the same speeches using different figurative expressions. Therefore he said "the body of *magnēsia*," that is, the mixing of the substances;[21] which is why he said further on in the introduction to [his book on] the making of gold: "Take mercury and make it solid with the body of *magnēsia*" [Ps.-Dem. Alch. *PM* § 5, l. 67].

§ 10 Test. et loci similes 160–3 Zos. Alch. in Olymp. Alch. *CAAG* II 96,9–14 ‖ 163–4 vide SyrL in *CMA* II 18 (textus) et 85 (translatio); SyrC, fol. 58ᵛ4–12 (translatio *CMA* II 245).

§ 11 Test. et loci similes 165–71 vide SyrC, fol. 57ʳ11–57ᵛ5 (translatio in *CMA* II 83–4) ‖ 180–2 Zos. Alch. *CAAG* II 192,19–193,14; Philos. Christ. Alch. *CAAG* II 397,6–399,2.

185   12. Διόσκορος. Ἰδοὺ οὖν προτετίμηται ἡ ὑδράργυρος.

Συνέσιος. Ναί, διὰ ταύτης γὰρ τὸ πᾶν ἀνασπᾶται· καὶ πάλιν προστίθεται (καὶ κατὰ βαθμὸν ἑκάστης οἰκονομίας τετύχηκεν)· χρυσόκολλα ὅ ἐστι βατράχιον· ἐν τοῖς χλωροῖς λίθοις εὑρίσκεται.

190   — Καὶ τίς ἂν εἴη χρυσόκολλα ὅ ἐστι βατράχιον, τίς ἡ σημασία ὅτι καὶ ἐν τοῖς χλωροῖς λίθοις εὑρίσκεται; Ἀναγκαῖον οὖν ἡμῖν ἐστι ζητῆσαι.

— Ὀφείλομεν οὖν εἰδέναι πρῶτον ὅσα ἀπὸ χρωμάτων εἰσὶ χλωρῶν. Φέρε δὴ ὡς ἀπὸ ἀνθρώπου εἴπωμεν· προτετίμηται γὰρ ὁ

195   ἄνθρωπος πάντων τῶν ζῴων τῶν ἐπὶ τῆς γῆς. Λέγομεν οὖν [ὅτι] ὠχριάσαντα τοῦτον χλωρὸν γεγονέναι, καὶ δῆλον ὅτι ὡς ὤχρα τὸ εἶδος μεταβάλλεται, ὅ ἐστι ἐπὶ τὸ χρυσίζον. Μᾶλλον δὲ καὶ αὐτό, τουτέστι τὸ λέπος τοῦ κιτρίου τὸ τῆς ὠχρότητος εἶδος. Τοῦτο δὲ καὶ ὑποκατιὼν εἶπεν· ἀρσενικὸν ξανθόν, ἵνα δείξῃ τὸ

200   τῆς ὠχρότητος εἶδος.

13. Ἵνα δὲ εἰδῇς πῶς μετὰ παρατηρήσεως πολλῆς μερικῶς εἴρηκε τοῦτο, πρόσεχε τὸν νοῦν πῶς λέγει· ὑδράργυρος ἡ ἀπὸ κινναβάρεως, σῶμα μαγνησίας· εἶτα ἐπιφέρει τὴν χρυσόκολλαν, κλαυδιανόν, ἀρσενικὸν ὄνομα. Πάλιν ἐπήγαγεν ἀρσενικόν, ἵνα

205   διέλῃ αὐτὸ ἀπὸ τῶν θηλυκῶν καὶ μετὰ τὸν κλαυδιανὸν ἀρσενικὸν τὸ ξανθόν, τὰ ξανθὰ δύο προσθεὶς ὀνόματα δύο θηλυκά, ἔπειτα δύο ἀρσενικά. Δεῖ οὖν ἡμᾶς ἐξιχνεῦσαι καὶ ἰδεῖν τί ἂν εἴη τοῦτο, ὡς ἐγὼ κεκίνημαι, Διόσκορε· ἐνταῦθα σήπει τὸν χρυσόν, εἶτα ἐπαναλαμβάνει καδμίαν, εἶτα ἀνδροδάμαντα· καὶ ὁ

---

185 ὑδράργυρος BeRu : ☽ **MBVA** ‖ **186** ναί **MV** : καί **BA** ‖ **187** κατὰ βαθμὸν BeRu, ex κατὰβαθμὸν (sic) **A** : κατὰβαθμὸν **MBV** ‖ **188** χρυσόκολλα BeRu : ⚥ **MBVA** ‖ ὅ ἐστι **MBVA** : ἤτοι BeRu ‖ post ἐν add. δὲ **BA** ‖ **188-9** εὑρίσκεται **MBA** : -ισκόμενον **V** ‖ **190** χρυσόκολλα BeRu : ⚥ **MBVA** ‖ ὅ ἐστι **BA** : ἤτοι **MV** ‖ **191-2** ἀναγκαῖον — ζητῆσαι Synesio trib. BeRu ‖ **193** ὀφείλομεν **MBVA** : pr. litt. rubricata **BA** : ὀφέλ- BeRu ‖ **194** post χλωρῶν add. καὶ **V** ‖ δὴ **MV** : οὖν **BA** ‖ ἀπὸ **MBVA** : περὶ prop. Zuber (**Z** 73ᵛ18) ‖ **195** ὅτι secl. BeRu ‖ **196** ὠχριάσαντα **MBV** : ὠχριω- **A** ‖ ὤχρα BeRu : ὦχρα **MV** : ὠχρά **BA** ‖ **197-8** καὶ αὐτὸ δὲ μᾶλλον **BA** ‖ **202** ὑδράργυρος BeRu : ☽ **MBVA** ‖ **203** κινναβάρεως BeRu : ☉ **MBV** : ⊙ **A** ‖ μαγνησίας BeRu : μ̄γ̄ **MBVA** ‖ χρυσόκολλαν BeRu : ⚥ **MBVA** ‖ **205** αὐτὸ om. **BA** ‖ **206** ὀνόματα BeRu : δ̄δ̄ **MBV** : δ̄ō [fort. leg. ὁμοῦ, coll. *CMAG* VIII 14,723; 22,1217] **A** ‖ **207** δεῖ **MBA** : χρὴ **V** ‖ ἐξιχνεῦσαι **MBVA** : ἐπι- Fabr ‖ **208** ἐνταῦθα **MBV** : ἐνταῦτα **A** ‖ σήπει **MV** : σήπτει **BA** ‖ **209** χρυσόν BeRu : ⚘ **MBVA** ‖ καδμίαν **MBV** : -μία **A**

12. Dioscorus. Thus it is that mercury has been put before all other substances.

Synesius. Yes; the *pan* [i.e. the whole, everything] is distilled through mercury; and he adds again (and step by step he touched upon every procedure): "Malachite, that is 'ranunculus' [lit. little frog]: it is found among the green stones" [Ps.-Dem. Alch. *Cat.* § 1, ll. 3f.].[22]

– And what is the malachite that is 'ranunculus'? What is the meaning of "it is found among the green stones"? It is necessary that we investigate.

– Therefore we must know first of all what concerns green colours. Let us start by speaking about human beings, for they are the first to be honoured among all living beings on earth. Now then, the one who turns pale is said to have become green, and it is clear that his complexion turns the colour of yellow ochre, a colour that tends to be similar to gold. Or we can put it better this way, that the peel of a citron looks yellowish in appearance [lit. has the notion/idea of yellowness]. And further on in his work he [i.e. Democritus] said "yellow orpiment" in order to indicate the notion of yellowness.[23]

13. In order to know how perceptively he explained this point in detail, turn your attention to how he says: "Mercury that comes from cinnabar, *magnēsia*'s body"; then he adds "malachite, *klaudianon*," a masculine name. He has introduced a masculine substance again, with the intention of distinguishing it from the feminine ones, and after "*klaudianon*" he [added] "yellow orpiment," and he placed two yellow substances that are referred to by two feminine names closer to [the two yellow substances referred to] by two masculine names [Ps.-Dem. Alch. *Cat.* § 1, ll. 2-4].[24] So we must follow him and know what all this means. I am so excited, Dioscorus: at this point he macerates gold, then he takes up cadmia again, then *androdamas*; *androdamas* and

§ 12. Test. et loci similes 193–4 Olymp. Alch. *CAAG* II 102,10–1.

§ 13. Test. et loci similes 204–7 Zos. Alch. *CAAG* II 216,17.

210 ἀνδροδάμας καὶ ἡ καδμία ξηρά εἰσι καὶ δείκνυσι τὴν ξηρότητα
τῶν σωμάτων. Καὶ ἵνα εὔδηλον αὐτὸ ποιήσῃ, ἐπήνεγκε στυπτηρίαν ἐξιπωθεῖσαν. Βλέπε πόση σοφία τοῦ ἀνδρός, ἵνα καὶ οἱ
ἐχέφρονες νοήσωσι, πῶς αὐτοὺς ἐδίδαξεν εἰπὼν στυπτηρίαν
ἐξιπωθεῖσαν· τάχα δὲ τοῦτο καὶ τοὺς ἀμυήτους ὤφειλε πείθειν.
215 Ἵνα δὲ καὶ βεβαιότερά σοι γένηται, εὐθέως ἐπήγαγε θεῖον
ἄπυρον, ὅ ἐστι θεῖον ἄκαυστον, τὸ πᾶν, τουτέστι τὰ ξηρανθέντα
εἴδη (κάτω, ὅ ἐστι τὰ σώματα ἓν γεγονότα, θεῖον ἄκαυστον
κέκληκεν). Καὶ μετέπειτα ἐπιφέρεται πυρίτης ἀπολελυμένος,
μηδένα τῶν ἄλλων ἀπροσδιορίστως ἐπιβεβαιῶν. Τοῦτο ἀληθὲς
220 ὑπάρχει ὅτι τὰ ἀπομείναντα ξηρά· καὶ ταῦτα ἀποδιαιρῶν ἐπιφέρει σινωπὶν Ποντικήν· μεταβὰς ἀπὸ τῶν ξηρῶν ἐπὶ τὰ ὑγρὰ
σινωπὶν εἴρηκεν, ἀλλὰ [διὰ] τὴν Ποντικήν· εἰ γὰρ μὴ ἦν προσθεὶς
τὸ Ποντικήν, οὐκ ἂν ἐν ἐπιγνώσει ἐγένετο. Ἐπιβεβαιούμενος δὲ
ἐπήνεγκεν ὕδωρ θείου ἄθικτον, τὸ ἀπὸ μόνου θείου θεῖον.
225 14. Διόσκορος. Καλῶς ἐπέλυσας φιλόσοφε, ἀλλὰ πρόσεχε πῶς
εἶπεν· ἐὰν ἀπολελυμένως, τὸ δι᾽ ἀσβέστου.

Συνέσιος. Ὦ Διόσκορε, οὐ προσέχεις τὸν νοῦν· ἡ ἄσβεστος
λευκή ἐστι, καὶ τὸ ἀπὸ ταύτης ὕδωρ τὸ ἀπ᾽ αὐτῆς λευκόν ἐστι
καὶ στυφόν· καὶ τὸ θεῖον θυμιώμενον λευκαίνει. Σαφηνείας οὖν
230 χάριν εὐθέως ἐπήγαγε θείου αἰθάλην. Οὐχὶ δῆλα ἡμῖν ταῦτα
ποιεῖ;

Διόσκορος. Ναί, καλῶς εἴρηκας· καὶ μετὰ τοῦτο σῶρι ξανθὸν
καὶ χάλκανθον ξανθὸν καὶ κιννάβαρις.

Συνέσιος. Τὸ σῶρι καὶ ἡ χάλκανθος ξανθά; Πῶς; Οὐκ ἀγνοεῖς

---

211 ποιήσῃ MBV : -ει A ‖ 211-2 στυπτηρίαν BA : ✳ MV ‖ 212 ἐξιπωθεῖσαν MV :
ἐκσηπτωθεῖσαν ✳ BA ‖ σοφία MV : παρατήρησις B : -ρεισις A ‖ καὶ om. BA ‖ 213-
4 πῶς — ἐξιπωθεῖσαν om. A, sed in mg. add. ‖ 213 στυπτηρίαν BeRu : ✳ MBV et A in
mg. ‖ 214 ἐξιπωθεῖσαν MV : ἐκσεπτωθεῖσαν B et A in mg. ‖ τάχα δὲ om. BA ‖ post
τοῦτο add. γὰρ BA ‖ 215 γένηται MV : -οιτο BA ‖ 216 post ἄπυρον des. V atque rursus
inc. Vᵃ ‖ ξηρανθέντα BVᵃ : -ραθέντα MA ‖ 217 ἓν om. BA ‖ 218 κέκληκεν MVᵃ :
-ται BA ‖ 220 ὅτι MBA : ὁ Vᵃ ‖ ἀποδιαιρῶν MVᵃ : ὑπο- BA ‖ 221 σίνωπιν (sic)
MVᵃ : σίνο- BA ‖ 222 διὰ secl. BeRu ‖ μὴ γὰρ BA ‖ 226 ἀπολελυμένως MBVᵃ : -ος
A ‖ 228 pr. ἀπὸ BA : ἐκ MVᵃ ‖ τὸ ἀπ᾽ αὐτῆς om. BA ‖ 229 στύφον (sic) BVᵃA :
στῦφον M ‖ θυμιώμενον BA : -ούμενον MVᵃ ‖ 231 ποιεῖ prop. BeRu : ποιῶν MB
VᵃA ‖ 233 κιννάβαρις scripsi : ☉ MVᵃB : ☉ A : -ιν BeRu ‖ 234 σῶρι BVᵃA : σωρι
(sic) M ‖ χάλκανθος MVᵃ : ☿ B : -άνθη A ‖ ξανθά BA : ξανθ M : om. Vᵃ ‖ πῶς
Dioscoro tribuere prop. BeRu

cadmia are dry and they indicate the dryness of the bodies. And in order to make it clear, he added "dried out alum" [Ps.-Dem. Alch. *Cat.* § 2, ll. 13-4].[25] Look at how wise this man was who enabled prudent people to understand as well. Look at how he instructed them by saying "dried out alum": perhaps this might have persuaded even those who had not been initiated. In order to make it even more certain, right after these words he inserted "unburnt sulphur," that is, the incombustible sulphur, the *pan* [lit. the whole], that is, the dried up species (further on in the text he gave the name of incombustible sulphur to what comes from the bodies reduced to unity). Afterwards he adds "pyrites" in a general sense [i.e. without further qualification], while he does not add in confirmation [of his argument] any other name of a substance without qualifying it. It is proven that the rest of the substances are dry: and in order to draw a distinction between these substances he adds "Sinope's earth from Pontus." Switching from dry to moist substances he said "Sinope's earth," but [specified] "from Pontus": for if he had not added "from Pontus," he would not have been understandable. And intending to confirm it, he added "untouched water of sulphur, the divine [water] produced only from sulphur."

14. Dioscorus. Your explanation is clear, O philosopher. Turn your attention, however, to how he said: "if [you understand water] in a general sense, it is [the water] produced with quicklime" [Ps.-Dem. Alch. *Cat.* § 1, ll. 6-8].[26]

Synesius. O Dioscorus, you do not pay attention. Quicklime is white and the water that is produced from it is the water made white and astringent by this very substance. Sulphur, furthermore, when it is turned to vapour, has a whitening power. Therefore, in order to be clear he immediately introduced "sulphur vapour." So has he not made these questions clear for us?

Dioscorus. Yes, you spoke well. And thereafter [Democritus says]: "yellow *sōri* and yellow copper flower and cinnabar" [Ps.-Dem. Alch. *Cat.* § 1, ll. 8-9].

Synesius. Are *sōri* and copper flower[27] yellow? How is that possible?

235 ὡς χλωρὰ εἴη. Αἰνιττόμενος οὖν τὴν τοῦ χαλκοῦ ἐξίωσιν ἤτοι
ἐξίσχνωσιν, μᾶλλον δὲ τὴν τοῦ παντός, ὡς ἀπὸ χρωμάτων τοῦτο
εἴρηκεν· καὶ πάλιν ἐπιβεβαιούμενος ἐπὶ τοῦ τέλους ἐπήγαγε·
μετὰ γὰρ τὴν ἀφαίρεσιν τοῦ ἰοῦ, ἥτις καλεῖται ἐξίωσις, τότε
ἐπιβολῆς τῶν ὑγρῶν γενομένης, γίνεται βεβαία ξάνθωσις. Καὶ
240 ὄντως ἡ ἀφθονία τοῦ ἀνδρὸς ἐνταῦθα ἀπεδείχθη.

15. Ὅρα γὰρ πῶς εὐθέως συνῆψε τῷ διορισμῷ χρησάμενος
καὶ εἰπών· τὰ δὲ ⌐ἐν ζωμοῖς εἰσι ταῦτα· κρόκος Κιλίκιος, ἀριστολοχία, κνήκου ἄνθος, ἀναγαλλίδος ἄνθος τῆς τὸ κυάνεον ἄνθος
ἐχούσης. Τούτου πλέον τί εἶχεν εἰπεῖν ἢ καταλέξαι, ἵνα πείσῃ
245 ἡμῶν τὰς καρδίας, εἰ μὴ διὰ τὸ εἰπεῖν ἄνθος ἀναγαλλίδος; Θαυ-
μάσαι γάρ μοι· οὐ μόνον ἀναγαλλίδος, ἀλλὰ καὶ ἄνθος εἶπε· τὸ
γὰρ ἀναγαλλίδος ἐμήνυσεν ἡμῖν τὸ ἀναγαγεῖν τὸ ὕδωρ· διὰ γὰρ
τοῦ ἄνθους τὰς τούτων ψυχὰς ἀναγαγεῖν, τουτέστι τὰ πνεύματα.
Εἰ μὴ γὰρ ταῦτα οὕτως ἔχοι, οὐδέν ἐστι βέβαιον. Καὶ μάτην
250 δυστυχήσαντες οἱ τάλανες εἰς τὸ πέλαγος τοῦτο ὑπορριπιζό-
μενοι πολλοῖς κόποις καὶ μογεροῖς ἐμπεσόντες, ἀνόνητοι καθε-
στῶτες ἔσονται.

16. – Καὶ τί πάλιν ὁ ἄφθονος φιλόσοφός τε καὶ καλὸς διδάσκαλος ἐπήγαγεν ῥᾶ Ποντικόν;

255 – Βλέπε ἀφθονίαν ἀνδρός· ῥᾶ εἶπεν αὐτό, καὶ ἵνα ἡμᾶς πείσῃ,
εὐθέως ἐπήγαγε τὸ Πόντιον. Τίς γὰρ ἀνδρῶν φιλοσόφων οὐκ
οἶδεν ὅτι ὁ Πόντος κατάρρους ἐστὶν ἐκ τῶν ποταμῶν πάντοθεν
περικλυζόμενος;

– Ἀληθῶς, Συνέσιε, ἔφρασας καὶ ηὔφρανάς μου τὴν ψυχὴν
260 σήμερον· οὐκ ἔστι γὰρ μέτρια ταῦτα. Τοῦτο δέ σε παρακαλῶ, ἵνα

---

235 χαλκοῦ BeRu : ♀ MVᵃ : ♀ BA ‖ **236** ἐξίσχνωσιν scripsi : -ίχνευσιν MBA :
-ίεχνευσιν Vᵃ ‖ ὡς om. BA ‖ **239** γενομένης BVᵃA : γι- M ‖ **240** ὄντως MB : ὂν Vᵃ :
ὄντος A ‖ **241** τῷ διορισμῷ BA : -ὸν -ὸν MVᵃ ‖ post διορ. des. Vᵃ atque rursus inc.
V ‖ **242** ✳ ℭι [i.e. σημείωσαι BeRu] in mg. M : ℭ in mg. V ‖ εἰσὶ BA : ἐστὶν MV ‖
supra κρόκος add. ♄ BA ‖ **243** κνήκου BeRu : κνί- MBVA ‖ **243-4** ἐχούσης ἄνθος
V ‖ **244** τούτου MBVA : pr. litt. rubricata A ‖ καταλέξαι MV : κατάλεξιν (sic) B :
καταλέξιν A ‖ **246** post μοι add. ὅτι V s.l. ‖ ἀναγαλλίδος BA : -λιᵟ MV ‖ **247**
ἀναγαλλίδος BA : -λιᵟ M : -λίδα V ‖ alt. γὰρ MBVA : δὲ prop. BeRu ‖ **249** ἔχοι MV :
-ει BA ‖ **251** μογεροῖς BA : μογη- MV ‖ **253** pr. καὶ MBVA : pr. litt. rubricata BA ‖
τε om. BA et BeRu ‖ **256** εὐθέως om. MV ‖ πόντιον MBVA : Ποντικὸν BeRu ‖ ante
πόντιον add. ῥᾶ A s.l. ‖ **259** ηὔφρανας MV : εὐ- BA

You know that they are green! Therefore he has made this claim as if to deal with colours, intending to hint at processes for removing rust from copper (*exiōsis*) – rather than from every metal – or for making it thinner (*exischnōsis*).[28] In order to confirm this point further, he added at the end [of his work]: "in fact, after the rust has been taken away, a process that is called *exiōsis*, then yellowing will be assured, if moist [or liquid] substances are laid on [metallic bodies]." On this point the man has clearly displayed his generosity.

15. For look at how he immediately referred to the distinction [i.e. between dry and moist/liquid substances] he had made, and said: "These are the substances that fall within washes: Cilician saffron, *Aristolochia*, Carthamus flower, flower of the Anagallis that has blue blossoms" [Ps.-Dem. Alch. *Cat.* § 2, ll. 11-3]. What more than "flower of Anagallis" should he have said or listed in order to persuade our hearts? What a wonder! He did not just say "of Anagallis," but also "flower": for the expression "of Anagallis" discloses to us how to distil water; and by saying "flower" he has indeed disclosed how to distil the souls of these substances, that is, their spirits.[29] If this is not the case, nothing is certain, and the wretched people who were so unlucky as to throw themselves in vain into this sea, and to fall into many painful troubles, will be in no condition to gain any advantage from it.

16. – So once again why has this generous philosopher and good master introduced "rhubarb from Pontus"?[30]

– See this man's liberality: he did say "rhubarb," but he immediately introduced "from Pontus," in order to persuade us. Who among the philosophers does not know that Pontus is full of streams since [the country] is washed by rivers on all sides?

– You have spoken truly, Synesius, and today you have delighted my soul: these topics, in fact, are not ordinary. But, please, teach me more: why did he say "yellow

§ 14 TEST. ET LOCI SIMILES 235–40 Agathod. Alch. *CAAG* II 115,7-8 (= *CAAG* II 167,13-5); Zos. Alch. *CAAG* II 161,24-6; Zos. Alch. (?) *CAAG* II 223,23-5 ≈ *De lapide philosophiae*, *CAAG* II 200,24-6; Pelag. Alch. *CAAG* II 254,15-6; Steph. Alch. II 204,29-30 et 205,19-21 Ideler; Philos. Anon. Alch. (Zos. Alch. BeRu) *CAAG* II 127,19-20.

§ 16 TEST. ET LOCI SIMILES 255–8 Steph. Alch. II 234,4-27 Ideler; vide etiam imag. in M, fol. 10ʳ (*CAAG* I 141-2); Philos. Christ. Alch. *CAAG* II 417,8-11.

ἐπιπλεῖόν με διδάξῃς· διὰ τί ἄνω εἶπε χάλκανθον ξανθήν, ὧδε δὲ
ἀπροσδιορίστως μετὰ τῆς κυανοῦ χάλκανθον ἐπήγαγεν;

   – Ἀλλὰ ταῦτα, ὦ Διόσκορε, τὰ ἄνθη μηνύουσιν· χλωρὰ γὰρ
ὑπάρχουσιν. Ἐπειδὴ οὖν τὸ ἀνερχόμενον ὕδωρ δεῖται πήξεως,
265 εὐθέως ἐπήγαγεν· κόμμι ἀκάνθης. Εἶτα ἐπάγει· οὖρον ἄφθορον,
καὶ ὕδωρ ἀσβέστου, καὶ ὕδωρ σποδοκράμβης, καὶ ὕδωρ στυπτηρίας, καὶ ὕδωρ νίτρου, καὶ ὕδωρ ἀρσενικοῦ καὶ θείου. Βλέπε πῶς
πάντα τὰ λυτικὰ καὶ διαφορεῖν δυνάμενα προήνεγκεν, οὕτω
δηλονότι διδάσκων ἡμᾶς τὴν ἀνάλυσιν τῶν σωμάτων.

270 17. – Ναί, καλῶς εἴρηκας. Καὶ πῶς ἐπὶ τέλει εἴρηκε· κυνὸς
γάλα; <Ἢ> ἵνα σοι δείξῃ ὅτι ἀπὸ τοῦ κοινοῦ τὸ πᾶν λαμβάνεται;

   – Ὄντως ἐνόησας, Διόσκορε. Πρόσεχε δὲ πῶς λέγει· αὕτη ἡ
ὕλη τῆς χρυσοποιίας ἐστί. Ποία ὕλη; Τίς οὐκ οἶδεν ὅτι πάντα
275 φευκτά ἐστιν; Οὔτε γὰρ ὄνειον γάλα οὔτε κυνὸς γάλα πυριμαχῆσαι δύναται· τὸ γὰρ ὄνειον γάλα, ἐὰν ἀποθήσῃς ἐν <θερμῷ>
τόπῳ ἱκανὰς ἡμέρας, ἀφαντοῦται.

   – Τί δὲ καὶ τὸ εἰπεῖν· ταῦτά εἰσι τὰ μεταλλοιοῦντα τὴν ὕλην,
ταῦτα καὶ πυρίμαχα ποιεῖ, φευκτῶν αὐτῶν ὄντων. Καὶ τὸ· ἐκτὸς
280 τούτων οὐδέν ἐστιν ἀσφαλές;

   – Ἵνα νομίσωσιν οἱ τάλανες ὅτι ἀληθῆ εἰσι ταῦτα. Ἀλλὰ
πάλιν ἄκουσον αὐτοῦ, τί εἶπεν καὶ ἐπιφέρει· ἐὰν ᾖς νοήμων, καὶ

---

261 ἄνω **MV** : ἄνωθεν **BA** ‖ χάλκανθον **BVA** : -ανθ **M** ‖ ξανθήν **MBV** : -όν **A** ‖ ὧδε
MV : ἐνταῦθα **BA** ‖ δὲ **MBVA** : om. BeRu ‖ **262** κυανοῦ **MBVA** : -νέας prop. BeRu ‖
χάλκανθον Tannery, "Études sur les alchimistes," 287 n. 22 : -κανθ **MV** : -άνθου **BA** ‖
265 ✳ Ἓ [i.e. σημείωσαι BeRu] in mg. **M** : ✳☉ in mg. **V** ‖ κόμμι scripsi : κόμη **MV** et
B a.c. : κόμι **A** et **B** p.c. ‖ ἐπάγει **MV** : ἐπήγαγεν **A** et **B** qui s.l. ✳ add. ‖ **266** pr. ὕδωρ
BVA: ὗ **M** ‖ alt. ὕδωρ **BA** : ὗ **MV** ‖ **266-7** ὕδωρ στυπτηρίας BeRu : ὗ✳ **MV** : ὕδωρ
✳ **BA** ‖ **267** pr. ὕδωρ **BA** : ὗ **MV** ‖ νίτρου BeRu : Ν̄ **MBA** : ℔ **V** ‖ alt. ὕδωρ **BA** : ὗ
MV ‖ ἀρσενίκου BeRu : Ⳕ **MBVA** ‖ tert. καὶ om. **V** ‖ θείου BeRu : ✕ **MBVA** ‖ supra
πῶς add. ταῦτα **A** ‖ **268** πάντα **MBA** : ἄπαντα **V** ‖ τὰ om. **BA** ‖ οὕτω Zur[1] : τοῦτο
MBVA ‖ **270** ναί **MV** : om. **B** : καί **A** ‖ **271** ἢ addidi post BeRu qui prop. ἢ ‖ **273**
ὄντως **MBVA** : pr. litt. rubricata **BA** ‖ ἐνόησας **MV** : ἐννόη- **B** : ἐννόει- **A** : ἐννοεῖς
Fabr ‖ **274** ⳋποίας [i.e. χρυσοποιίας BeRu] **MBVA** ‖ **274-5** ποία — φευκτά ἐστιν
om. **A**, sed in mg. add. ‖ **274** ποία ὕλη Dioscoro trib. BeRu ‖ οἶδεν **MBV** : οἶδα in mg.
A ‖ **275** ἐστὶν **MBV** : εἰσὶ in mg. **A** ‖ **275-6** πυριμαχῆσαι **MBV** : πυριμαχῆ **A** : -εῖν
Fabr ‖ **276** ἀποθήσῃς **M** : -εις **BA** : ἀποθῆς **V** ‖ θερμῷ addidi post Fal : ἐν τῷ πυρὶ
prop. BeRu : post τόπῳ add. τινὶ Garzya ‖ **277** ἀφαντοῦται **MBVA** : ἀφανιοῦται
Fabr ‖ **282** τί εἶπεν Dioscoro tribuere prop. BeRu

copper flower" previously, when at this point, after "azurite," he introduced "copper flower" without qualifying it?

– But these substances, O Dioscorus, indicate the flowers: for they are green. Then he immediately introduced "acacia gum," since the distilled water needs be made solid. Then he adds: "pure urine, and quicklime water, and water of cabbage ashes, and alum water, and soda water, and orpiment water, and sulphur water" [Ps.-Dem. Alch. *Cat.* § 2, ll. 13-6]. Look at how he has brought up all the substances that can dissolve and evaporate bodies: in this way he has clearly taught us how to dissolve them.

17. Yes, you have spoken well. And in which sense did he say at the end: "dog's milk"? [Ps.-Dem. Alch. *Cat.* § 2, ll. 16-7].[31] <Perhaps> in order to show that the *pan* [lit. the whole] comes from that which is common [i.e. from substances of everyday use]?

– You have understood well, Dioscorus. Turn your attention to how he says: "These are the substances for the making of gold" [Ps.-Dem. Alch. *Cat.* § 2, l. 18]. Which substances? Who does not know that all these substances are unstable? In fact neither jenny's milk nor dog's milk can resist fire; in fact, if you put jenny's milk in a <warm> place for enough days, it disappears.

– But why did he say also: "these are the substances that change the matter, these can make [the substances] fire-resisting, although they are unstable"? And also: "Nothing is certain beyond these substances"? [Ps.-Dem. Alch. *Cat.* § 2, l. 19].

– In order that wretched people acknowledge that his words are true. But listen to him again, to what he added by saying: "If you are intelligent and operate according

§ 17 Test. et loci similes 278–80 Zos. Alch. *CAAG* II 168,15–7 et 196,13–4; Philos. Anon. Alch. (Joan.. Alch. BeRu) *CAAG* II 264,5–11 ‖ 283–4 Zos. Alch. *CAAG* II 233,25–6 = I 186–8 Mertens; Steph. Alch. II 235,3–5 Ideler.

ποιήσῃς ὡς γέγραπται – ἀντὶ τοῦ· ἐὰν ᾖς σοφός, καὶ διακρίνῃς τὸν λογισμὸν ὡς δεῖ κεχρῆσθαι – ἔσῃ μακάριος.

285   – Καὶ τί ἀλλαχοῦ εἶπε· τοῖς ἐχέφροσιν ὑμῖν λέγω;

      – Δεῖ οὖν ἡμᾶς γυμνάζειν τὰς φρένας ἡμῶν καὶ μὴ ἀπα-τᾶσθαι, ἵνα καὶ τὴν ἀνίατον νόσον τῆς πενίας ἐκφύγωμεν, καὶ μὴ νικηθῶμεν ὑπ᾽ αὐτῆς, καὶ εἰς ματαίαν πενίαν ἐμπεσόντες δυστυχήσωμεν, ἀνόνητοι καθεστῶτες. Γυμνάζειν τοίνυν τὰς φρέ-

290 νας ὀφείλομεν καὶ ὀξὺν ἔχειν τὸν νοῦν.

      18. – Διὰ τί οὖν ἐπιφέρει τὸ ἐπιβάλλειν;

      – Οὐ διὰ <...> λέγει τὰ προλεγόμενα, ἀλλὰ τὰ ἀπὸ τοῦ νοός. Ἀλλὰ πάλιν λέγει· ποτὲ μὲν χρυσὸν διὰ τὸν χρυσοκόραλλον, ποτὲ δὲ ἄργυρον διὰ τὸν χρυσόν, ποτὲ δὲ χαλκὸν διὰ τὸν χρυσόν,

295 ποτὲ δὲ μόλυβδον ἢ κασσίτερον διὰ τὸ μολυβδόχαλκον. Ἰδοὺ αὐτὸς ὑπὸ τοὺς βαθμοὺς τῆς τέχνης ἀνήγαγεν ἡμᾶς, ἵνα μὴ κενεμβατοῦντες εἰς βόθρον ἐμπέσωμεν τῆς αὐτῆς ἀγνοίας τῶν σημαινομένων παρ᾽ αὐτοῦ. Πολλὴ γὰρ ὑπάρχει τῷ ἀνδρὶ ἡ σοφία· μετὰ γὰρ τὸ εἰπεῖν αὐτόν· αὕτη ἡ ὕλη τῆς χρυσοποιίας εἰρήσθω,

300 ἐπιφέρει λέγων· φέρε δὴ καθεξῆς καὶ τὸν τῆς ἀργυροποιίας λόγον ἀφθόνως ἐξείπωμεν, ἵνα δείξῃ ἡμῖν ὅτι δύο ἐργασίαι εἰσί· ὅτι καὶ ἡ ἀργυροποιία πρὸ πάντων προτετίμηται καὶ προτερεύει, καὶ χωρὶς αὐτῆς οὐδὲν γενήσεται.

283 ᾖς **BA** : ῆ **M** : ῇ **V** ‖ **285** τί om. **B** : add. s.l. **A** ‖ **286-90** Dioscoro trib. BeRu (incuria?) ‖ **286** ἡμᾶς **BA** et Garzya : ὑμᾶς **MV** et BeRu ‖ ἡμῶν **V** et Garzya : ὑμῶν **MBA** et BeRu ‖ **287** νόσον τῆς πενίας **BA** : πενίαν τῆς νόσου **MV** ‖ ἐκφύγωμεν scripsi ex **A** (-ομεν) : -οιμεν **MBV** ‖ **289** δυστυχήσωμεν **MBV** : δυστοι- **A** ‖ γυμνάζειν τοίνυν **V** : τοῦ γυμνάζεσθαι **M** : τὸ -εσθαι **BA** : διὸ -εσθαι Zur[1] ‖ **291** ἐπιβάλλειν **MV** : -βάλειν **BA** ‖ **292** καλὸν (litt. inversae) add. in mg. **M** ‖ post διὰ lacunam ind. **MB** : διὰ τοῦτο **V** : διαλέγει **A** : οὐ διά<φορα> λ. <παρὰ> prop. BeRu, coll. [Plut.] *Prov.* I 12,3 ‖ τοῦ om. **BA** ‖ **293** χρυσὸν BeRu : ☾ **MBVA** ‖ ☾κόραλλον [i.e. χρυσοκόραλλον BeRu] **MBV** : ☾κόραλον **A** ‖ **294** ἄργυρον BeRu : ☾ **MBVA** ‖ pr. et alt. χρυσόν BeRu : ☾ **MBVA** ‖ χαλκὸν BeRu : ♀ **MV** : ♀ **BA** ‖ **295** δὲ om. **MV** ‖ μόλυβδον BeRu : ♄ **MBVA** ‖ κασσίτερον BeRu : ♃ **MV** : ☿ **BA** ‖ τὸ **MV** a.c. : τὸν **BAV** p.c. ‖ μολυβδόχαλκον BeRu : ♄♃ **MV** : ♃☿ **B** : ♄☿ **A** ‖ ἰδοὺ **MBV** : -οῦ **A** ‖ **296** post αὐτὸς add. ἔστιν s.l. **A** ‖ ὑπὸ **MV** : ἐπὶ **BA** ‖ ἵνα Zur[1] : καὶ **MBVA** ‖ **297** βόθρον **MBV** : βάραθρον **A** ‖ **298** σημαινομένων **MBV** : σημε- **A** ‖ αὐτοῦ scripsi, coll. Pizzim 17ᵛ20 (*apud ipsum*) : -ῶν **MBVA** ‖ **299** αὐτόν **MBV** : -ήν **A** ‖ ☾ποιίας [i.e. χρυσοποιίας BeRu] **MBVA** ‖ **300** λέγων **MBV** : -ον **A** ‖ ☾ποιίας [i.e. ἀργυροποιίας BeRu] **MBVA** ‖ **301** ἐξείπωμεν **MBV** : -ομεν **A** ‖ ὅτι δύο non legit. **B** ‖ εἰσί om. BeRu ‖ **302** ☾ποιία [i.e. ἀργυροποιία BeRu] **MBVA** ‖ πρὸ non legit. **B**

to what has been written – equal to 'if you are wise and discern what kind of reasoning we must make use of' – you will be happy" [Ps.-Dem. Alch. *Cat.* § 2, l. 20].

– And why did he say elsewhere: "I am addressing those of us who are prudent"?

– Because we must train our minds and not deceive ourselves, in order to escape the incurable disease of poverty, and to not be overcome by it by being so unlucky as to fall into vain poverty without receiving any profit. So we are required to train our minds and cultivate sharp thoughts.

18. – So why does he add "to throw upon [metallic bodies]"?

– He does not say what has been mentioned above with <... >, but he makes judicious statements. And he says again: "now [throw upon] gold for [making] gold coral, now silver for [making] gold, now copper for [making] gold, now lead or tin for [making] *molybdochalkon* [i.e. lead-copper alloy]" [Ps.-Dem. Alch. *Cat.* § 2, ll. 20-3].[32] You see, he has led us through the various steps of the Art, so that we do not take a false step and fall into the abyss of ignorance over that which he has explained. In fact, great knowledge belongs to this man, for after saying: "These are the substances for the making of gold, let this suffice," he adds the words: "Next let us speak without envy also on the subject of making silver," in order to show us that there are two procedures: and the making of silver is preferred to all [the other procedures] and comes first, since nothing will be accomplished without it.

§ 17 Test. et loci similes **286–90** Olymp. Alch. *CAAG* II 97,5–7 et 103,9–10; Philos. Christ. Alch. *CAAG* II 397,17–8; vide etiam supra, Syn. Alch. § 5, ll. 56–7.

19. Ἄκουσον αὐτοῦ πάλιν ἐνταῦθα λέγοντος· ἡ ὑδράργυρος
305 ἡ ἀπὸ ἀρσενικοῦ ἢ θείου ἢ ψιμυθίου ἢ μαγνησίας ἢ στίμεως
Ἰταλικοῦ. Καὶ ἄνω μὲν οὖν ἐν τῇ χρυσοποιίᾳ· ὑδράργυρος ἡ ἀπὸ
κινναβάρεως· ἐνταῦθα δέ· ὑδράργυρος ἡ ἀπὸ ἀρσενικοῦ ἢ ψιμυθίου καὶ τὰ ἑξῆς.

– Καὶ πῶς ἐνδέχεται ὑδράργυρον ψιμύθιον γενέσθαι;

310 – Ἀλλ' οὐκ ἀπὸ ψιμυθίου ὑδράργυρον εἶπεν ἵνα λάβωμεν, ἀλλὰ τὴν λεύκωσιν τῶν σωμάτων, εἶτ' οὖν ἀνάκαμψιν
αἰνιττόμενος εἴρηκεν· ὧδε γὰρ τὰ λευκὰ πάντα εἶπεν, ἐκεῖ δὲ τὰ
ξανθά, ἵνα νοήσωμεν. Ὅρα πῶς εἶπεν· σῶμα μαγνησίας, χρυσο-
κόραλλον· ἐνταῦθα δὲ σῶμα μαγνησίας μόνον ἢ στίμεως
315 Ἰταλικοῦ. Καὶ ταῦτα μὲν πρὸς βραχύ τι αὔταρκες ὑμῖν εἰρήσθω.
Προγυμνάζεσθαι δὲ τὸν νοῦν χρή, ἵνα διαγιγνώσκωμεν τὰς τῆς
φύσεως ἐνεργείας περὶ τῶν σπουδαζομένων τῇ τοῦ θεοῦ
συνεργείᾳ. Δεῖ οὖν ἡμᾶς γινώσκειν, <ὅτι> ταριχεύεσθαι δεῖ τὰ
εἴδη πρῶτον καὶ ταῖς χω<νεύ>σεσιν ὁμόχροα ἀποτελεῖσθαι εἰς
320 ἓν χρῶμα· καὶ τὰ μὲν δύο ὑδράργυρα ὑδραργυρίζονται καὶ εἰς
σῆψιν ἀποχωρίζονται. Θεοῦ δὲ βοηθοῦντος ἄρξομαι ὑπομνηματίζειν.

---

304 post ἄκουσον add. τοίνυν **V** ‖ αὐτοῦ πάλιν non legit. **B** ‖ λέγοντος om. **B** : supra ἐνταῦθα
add. λέγοντος **A** ‖ ὑδράργυρος BeRu : ☽ **MBVA** ‖ 305 ἀρσενίκου BeRu : ⟁ **MBVA** ‖ θείου
BeRu : ✕ **MBVA** ‖ ψιμυθίου **MV** : non legit. **B** : ψιμμι- **A** : ψιμμυ- BeRu ‖ μαγνησίας **BA**: μ͞γ
MV ‖ στίμεως scripsi post BeRu (στίμμεως) : -ίμη **M** : -ίμης **V** : -ίμμεος **BA** ‖ 306 ἰταλικοῦ **A** :
ἰτα⟩ **M** : -ῆς **V** : non legit. **B** ‖ ἐν τῇ ⸫ποιίᾳ [i.e. χρυσοποιίᾳ BeRu] **MBV** : ἡ τῆς ⸫ποιίας **A** ‖
ὑδράργυρος BeRu : ☽ **MBVA** ‖ 307 κινναβάρεως BeRu : ☉ **MBV** : ⊙ **A** ‖ supra ἐνταῦθα add.
λε͞ atque sub ἐντ. add. λεύκω͞σι **A** ‖ δὲ om. **BA** ‖ ὑδράργυρος BeRu : ☽ **MBVA** ‖ ἀρσενίκου
BeRu : ⟁ **MBVA** ‖ 307-8 ψιμυθίου **MV** : ψιμμι- **BA** : ψιμμυ- BeRu ‖ 309 ὑδράργυρον BeRu :
☽ **MBVA** ‖ ψιμύθιον **MV** : non legit. **B** : ψιμμιθῖ **A** : ψιμμύ- BeRu ‖ 310 ψιμυθίου **MV** : ψιμμι-
BA et BeRu : ψιμμυ- Garzya ‖ ὑδράργυρον BeRu : ☽ **MBVA** ‖ 310-1 λάβωμεν **MBV** : -ομεν
A ‖ 312 ὧδε **MV** : ἐνταῦθα **BA** ‖ pr. τὰ om. **A** ‖ 313-4 σῶμα — ἐνταῦθα δὲ om. **BA**, sed post
μόνον add. χρυσωκόραλον (sic) ἐνταῦθα δὲ σῶμα μ͞γ [i.e. μαγνησίας]· μ͞γ [i.e. μαγνησίαν] μόνον
A in mg. ‖ 313 μαγνησίας BeRu : μ͞γ **MV** : ⸫κόραλλον [i.e. χρυσοκόραλλον BeRu] **MV** :
χρυσωκόραλον add. in mg **A** : *chrysocollam* Pizzim 18ʳ110 ‖ 314 μαγνησίας **BA** : μ͞γ **MV** : post
μαγνησίας add. μαγνησίας BeRu (vide ll. 313-4 **A** in mg.) ‖ στίμεως **MV** : στιμ[..] **B** : στίμμεος
A ‖ 318 συνεργείᾳ **A** : -ία **MBV** ‖ pr. δεῖ **MV** et **B** ut vid. δ[..] : χρεῖ **A** ‖ 318 ἡμᾶς **MVA** : ὑ-
B ‖ ὅτι add. BeRu ‖ ταριχεύεσθαι **MBA** : -χεύειν **V** ‖ alt. δεῖ om. **V** ‖ 319 χωνεύσεσιν BeRu :
χώσεσιν **MBVA** ‖ ὁμόχροα **BA** : -χρόους **MV** ‖ 320 ὑδράργυρα BeRu : ☽☽ **MBV** : ℭℭ [i.e.
ἄργυροι] **A** ‖ ὑδραργυρίζονται **MBA** : -ειν **V** ‖ 321 ἀποχωρίζονται **MBA** : -ειν **V** ‖ 321-2
ἄρξομαι ὑπομνηματίζειν **MV** : τὸ πᾶν τοῦ λόγου τετέλεσται **BA**, sed ἄρξομε (sic) δὲ
ὑπομνηματίζειν add. **A** in mg.

19. Listen to him who says again on this point: "Mercury that comes from orpiment, or from sulphur, or from white lead, or from *magnēsia*, or from Italian stibnite" [Ps.-Dem. Alch. *Cat.* § 3, ll. 26-7]. And above, in the making of gold: "Mercury that comes from cinnabar" [Ps.-Dem. Alch. *Cat.* § 1, l. 1]; here, on the contrary: "Mercury that comes from orpiment, or from white lead" [Ps.-Dem. Alch. *Cat.* § 3, l. 26] and so on.

– But how is it possible that white lead becomes mercury?
– Yet he did not say "Mercury that comes from white lead" in order that we should try to extract it, but by these obscure words he has hinted at both the whitening of [metallic] bodies and their restoration[33] from one to another. Here he mentioned all white substances, there all yellow substances, in order to make us understand. Look at how he spoke [previously]: "*magnēsia*'s body, gold coral;" at this point, on the contrary, he [said] only: "body of *magnēsia* or body of Italian stibnite." And this short explanation should be enough for you.[34] We must train our minds to understand the powers of natures related to that which we zealously pursue with God's aid. Therefore we must know <that> species first must be macerated, melted and made uniform in colour in order to produce a single colour;[35] and the two mercuries are 'mercurized'[36] and separated for their maceration. With God's aid we shall start our commentary.

Περὶ λευκώσεως

1   1. Γινώσκειν ὑμᾶς θέλω ὅτι πάντων ἐστὶν κεφάλαιον ἡ λεύκωσις· μετὰ δὲ τὴν λεύκωσιν, εὐθὺς ξανθοῦται τὸ τέλειον μυστήριον.

2. Ἡ λεύκωσις καῦσίς ἐστιν· ἡ δὲ καῦσις ἀναζωοπύρωσις·
5   αὐτὰ γὰρ ἑαυτὰ καίουσι καὶ ἀναζωοπυροῦσι, καὶ αὐτὰ ἑαυτὰ ὀχεύει, καὶ ἐγκυοποιεῖ καὶ ἀποτίκτει τὸ ζητούμενον ζῷον κατὰ τοὺς φιλοσόφους.

3. Ἐὰν λευκώσῃς, εὐκόλως βάψεις· εἰ δὲ καὶ ἰώσεις ἢ κινναβαρίσεις, μακάριος ἔσῃ, ὦ Διόσκορε· τοῦτο γὰρ ἐστιν τὸ λυτρού-
10   μενον πενίας, τῆς ἀνιάτου νόσου.

M, fol. 118ʳ2-14

B, fols. 90ᵛ18-91ʳ 9

A, fol. 92ʳ16-26

Aᵃ, fol. 14ᵛ20-30

Aᵇ, fol. 250ᵛ13-21

CAAG II 211,3-11

1 ante γινώσκειν add. δεῖ **AAᵃ** et χρεῖ **Aᵇ** ‖ κεφάλαιον ἐστὶν **BAAᵃ** ‖ **2** τελιον μυ φησὶ (sic) add. in mg. **M** ‖ **3** post μυστήριον add. :~ [i.e. signum discriminis] **MBAAᵃ** ‖ **4** pr. ἡ rubricatum **AAᵃ** : om. **B** (rubricator deest) ‖ ἀναζωοπύρωσις scripsi, post ἀναζωπ- BeRu : -ζωπύρησις **M** : -ζωπύρησις **BA** : -ζωοπυρίσεις **AᵃAᵇ** ‖ **5-6** αὐτὰ γὰρ — ὀχεύει **MBAAᵃ** : καὶ αὐτὰ ἑαυτὰ ὀχεύουσι καὶ ἀναζωοπυροῦσιν **Aᵇ** ‖ **5** ἀναζωοπυροῦσι **BAAᵃAᵇ** : ἀναζωπ- **M** ‖ **6** ἐγκυοποιεῖ **M** : ἐγγυο- **BAAᵃAᵇ** ‖ ζῷον **M** : ζωὴν **BAAᵃAᵇ** ‖ **7** post φιλοσόφους add. :~ **MBAAᵃ** ‖ **8** ἐὰν **MBAAᵃAᵇ** : pr. litt. rubricata **AAᵃ** ‖ λευκώσῃς **MBA** : -εις **Aᵃ** : -ις **Aᵇ** ‖ βάψεις εὐκόλως **BAAᵃ** ‖ ἰώσεις **MAᵃ** : -ης **B** : -ις **AAᵇ** ‖ **9** κινναβαρίσεις **MAᵃ** : κιναβαρίσης **BAAᵇ** ‖ διόσκορε **BAAᵃAᵇ** : -ωρε **M** ‖ **9-10** post λυτρούμενον add. ἐκ **Aᵃ** s.l. ‖ **10** post νόσου add. :~ **MBA** : add. :~ τέλος **Aᵃ**

On whitening

1. I want you to know that whitening is the main point of all the procedures; after whitening, the perfect mystery immediately yellows.

2. Whitening consists of a burning process; and burning in a further burning [which regenerates bodies]. In fact, the bodies burn each other and kindle each other again [by regenerating one another], and they couple with one another and bring to birth the living being we are seeking, according to the teaching of the philosophers.

3. If you whiten, you will easily dye; if you accomplish the *iōsis*, that is giving the colour of cinnabar, you will be happy, O Dioscorus. For this is what frees us from poverty, the incurable disease.

[90ᵛ] [1]

1. SyrC, fol. 90ᵛ2-12

 (sic)[4] .
1. *Versio codicum londinensium* (SyrL in *CMA* II 10,4-11)

/ [lege /], vide *CMA* II 10]
2. SyrC, fol. 90ᵛ12-7

/ (sic)
 (sic)[7].
---

[1] SyrL /

[2] SyrC : scripsi

[3] leg. fort. SyrC :

[4] SyrC : fort. leg., coll. *B.B.* I 879,6 :

[5] SyrC : scripsi /

[6] add. SyrC infra lineam

[7] SyrC : fort. leg.

Book by Democritus: *On the Making of Shiny Gold** (= 1SyrC)

First paragraph in Cambridge ms. = *PM* § 5[1]

Take mercury [lit. milk] and make it solid with the body of *magnēsia*, or with Italian stibnite (*stîmi*),[2] or with red[3] sulphur, or with moon foam (*aphroselēnon*), or with lime (*titanos*), or with alum (*styptēria*), or with orpiment (*arsenikon*), or according to your understanding. So get two melting-pots (*koanos*?)[4] ready, and cook [the abovementioned ingredients?] and lay [them] on *Hermēs* [i.e. copper/mercury?]. Measure [i.e. inspect] its rust: if it is red, add silver and it will be gold; but with the gold any [metal?] will be gold coral.[5] Red orpiment and realgar (*sandarachē*) – if you convert them with the body of *magnēsia* and also of malachite (*chrysokolla*) – and red cinnabar (*kinnabaris*) produce the same effect. But mercury [lit. milk] alone makes *Hermēs* [i.e. copper/mercury?] shiny.

First paragraph in London mss. = *PM* § 5

Take mercury [lit. cloud] and make it solid with the body of *magnēsia*, or with Italian stibnite, or with sea sulphur, or with red sulphur, or with moon foam (*aphroselēnon*), or with roasted lime, or with alum (*styptēria*), or with orpiment (*arsenikon*), or according to your opinion. If it turns white, distil it on copper (*Aphroditē*), and you will get 'shadowless' (*askiastos*) white copper (*Aphroditē*). But if it turns red, distil the mercury on lead (= *kronos*?), and you will get gold; and [distil it] on gold, and it will be gold coral (*chrysokorallos*?). Also red orpiment, and processed realgar, and solidified malachite (*chrysokolla*?), and cinnabar (*kinnabaris*) that has been completely transformed [lit. overturned] produce the same effect. But mercury alone makes copper (*aphroditē*) 'shadowless' (*askiastos*). Nature conquers nature.

Second paragraph in Cambridge ms. = *PM* § 6[6]

Pyrite (*pyritēs*) stone: cook it as you are accustomed to doing in order to melt it [lit. so that they (?) melt] by dissolving it by means of resin (*rhētinē*?)[7] and litharge (*lithargyros*), or wax, or also with Italian stibnite (*stîmi*). Sprinkle [the pyrite?] on this [ingredient; i.e. stibnite?], not [on] the one that is usually employed – do not fall into error – but on the one that comes from Samos. You will lay this drug wherever you want [i.e. on whatever metal you want]. Nature masters nature.

---

*   In many cases, the Syriac names of the substances are simple transliterations of the Greek ones. I always indicate the Greek name (usually in Latin transliteration) next to the translation of the corresponding Syriac term. See also the Syriac-Greek index at the end of the volume.

2. *Versio codicum londinensium* (**SyrL** in *CMA* II 10,12-6)

   (sic) .
   (sic)
3. SyrC, fols. 90^{v}17-91^{r}4
≈ SyrL in *CMA* II 10,17-20

   [91^r] [13]
   (sic).[14]
4. SyrC, fol. 91^{r}4-9
omittit SyrL

---

[8] om. SyrL

[9] SyrL

[10] scripsi : SyrC ut plerumque

[11] SyrL

[12] SyrL

[13] [lege , vide *CMA* II 10] SyrL

[14] SyrC SyrL : fort. leg.

[15] SyrL

[16] SyrL

[17] SyrL

[18] tamquam ex glossemate seclusi

Second paragraph in London mss. = *PM* § 6

Process silver pyrite (*argyritēs pyritēs*) that is called also siderites, as you are accustomed to doing, so that it may be spilled out [i.e. be liquefied?]. It is spilled out by means of resin (*rhētinē?*), or by means of white litharge, or by means of Italian stibnite. So scatter it on lead (?), not on the one that is usually employed – do not fall into error – but on the one that comes from Samos. Roast it and when it becomes red lay it on whatever [metallic] body you want: you will dye it. Nature is delighted by nature.

Third paragraph in Cambridge ms. = *PM* § 7[8]
(A similar version is also preserved in the London mss.)

The making of pyrite (*pyritēs*). It must be processed in this way [lit. it takes this treatment]:[9] it is washed with vinegar and salt and with urine, or with vinegar and honey, or with seawater. After washing, mix it with unburnt sulphur (*theion apyron*), or red alum (*styptēria*), or Attic (?) ochre (*ōchra*),[10] or according to your understanding. And lay it on silver and you will find gold. For nature masters nature.

Fourth paragraph in Cambridge ms. = *PM* § 8
(London mss. do not preserve this text)

Make *Klaudianos* gleaming (*marmaros*),[11] and cook it until it turns red; cook it with alum (*styptēria*), or with orpiment (*arsenikon*), or with realgar (*sandarachē*), or with quicklime. If you lay a part of it on silver, you will produce gold; if you [lay it] on gold, it will be gold coral.[12] Nature masters nature.

5. SyrC, fol. 91ʳ9-14

omittit SyrL

6. SyrC, fol. 91ʳ14-20

omittit SyrL

7. SyrC, fol. 91ʳ20-91ᵛ10

≈ SyrL in *CMA* II 10,21-11,4

1 [91ᵛ]
---

[19] SyrC : fort. leg. : ‹ (vide زنجفر) in mg. SyrC

[20] supplevi

[21] in mg. SyrC

[22] SyrC : scripsi

[23] in mg. SyrC

[24] SyrC : fort. vel leg.

[25] Signum alchemicum ꟼ SyrC

[26] SyrL (idem l. 3)

[27] om. SyrL

[28] SyrL

Fifth paragraph in Cambridge ms. = *PM* § 9
(London mss. do not preserve this text)

Again another [recipe]. Make cinnabar (*kinnabaris*)[13] white by means of oil, or honey, <or> alum (*styptēria*), or with water and salt. But it will turn red by means of *misy* (*musidin* = μίσυ) and *sōri*, or copper flower (*chalkanthos*), or with unburnt sulphur (*theion apyron*), or according to your understanding. Then lay it on silver and it will be gold; and lay it on *Hermēs* [i.e. copper/mercury?] and it will take half of the dye.[14]

Sixth paragraph in Cambridge ms. = *PM* § 10
(London mss. do not preserve this text)

Cyprian (?) cadmia (*kadmia*) ...;[15] make it white as you are accustomed to doing. Then, after that, make it red. This is how it turns red: by means of the bile of a bull, or also with terebinth resin, or with radish oil, or with egg yolks, or with any other red substance. This is laid on silver by means of gold, and on gold by means of the ferment of gold (*chrysozōmion*). Nature masters nature.

Seventh paragraph in Cambridge ms. = *PM* § 11
(A similar version is also preserved in the London mss.)

Cook *androdamas* with robust wine, or with seawater, or with urine, or with vinegar and salt, with those [substances] which are found able to quench natures [lit. when natures are quenched]. Then grind it with stibnite, pouring seawater on it, or with urine, or with vinegar and salt; wash in this way until the blackness of the stibnite disappears. Afterwards dry it and cook until it turns red. Boil with

8. SyrC, fols. 91ᵛ10-92ʳ2
≈ SyrL in *CMA* II 11,5-12

 [92ʳ]
9. SyrC, fol. 92ʳ2-7
≈ SyrL in *CMA* II 11,13-5

---

[29] **SyrL**

[30] **SyrL**

[31] **SyrL**

[32] om. **SyrL**

[33] **SyrL**

[34] om. **SyrL**

[35] **SyrC SyrL** : prop. Duval : vide *PM* § 12, l. 120 (ἐν δρόσῳ λέγω καὶ ἡλίῳ)

[36] **SyrC** : om.) **SyrL**

[37] **SyrL**

[38] **SyrL**

[39] **SyrL**

[40] om. **SyrL**

[41] **SyrL**

[42] **SyrL** : difficile lectu **SyrC**

[43] supplevi, coll. **SyrL** : [...] **SyrC**

sulphur water and lay part of it on silver; it will be gold. But if you lay unburnt sulphur (*theion apyron*), it will be ferment of gold (*krysozōmos?*). Nature masters nature.

Eighth paragraph in Cambridge ms. = *PM* § 12
(A similar version is also preserved in the London mss.)

Take the lead that is white (*'aboro ḥeworo* = λευκὸς μόλυβδος),[16] I mean the one that [is composed] of white lead (*psymythion*) and of *helkysma*, and [of ?] the body of *magnēsia*. This is how it turns white: with seawater; and it is ground after being laid out in the sun and in the shade [*sic*; maybe 'under the dew']: it becomes like white lead (*psymythion*). Then cook it and lay the rust of *Hermēs* [i.e. copper/mercury?][17] or burnt copper on it. You must lay also copper and azurite (*kyanos*) on it, until it becomes soft and bright.[18] It will easily become that [metal] which is called by two names, I mean *Hermēs* [i.e. copper/mercury?] and lead. Then test if it is 'shadowless' (*askios*); if not, blame *Hermēs* [i.e. copper/mercury?]; and if it is blameless, you did good work. Then grind it with these substances that are natural to it and cook until it turns red. So lay it on whatever [metal] you want. For nature masters nature.

Ninth paragraph in Cambridge ms. = PM § 13
(A similar version is also preserved in the London mss.)

Mix copper flower (*chalkanthos*) and *sōri* with unburnt sulphur (*theion apyron*). Indeed *sōri* looks like copper flower (*chalkanthē?*) and it is found in

10. SyrC, fol. 92ʳ7-14

≈ SyrL in *CMA* II 11,15-9

11. SyrC, fol. 92ʳ14-92ᵛ7

omittit SyrL

 [92ᵛ]

---

[44] ‌⸀ SyrL

[45] SyrL : difficile lectu SyrC

[46] , supplevi, coll. SyrL : non legit. SyrC

[47] supplevi, coll. SyrL : non legit. SyrC

[48] SyrL

[49] SyrC SyrL : fort. leg., coll. *PM* § 14, l. 144 (οὔρῳ δαμάλεως)

[50] supplevi, coll. **SyrL** : [...] **SyrC** : fort. leg. (vide l. 3)

[51] om. SyrL

[52] supplevi, coll. *PM* § 15, ll. 155-6

[53] scripsi, coll. *PM* § 15, ll. 161-2 (ποῖόν ἐστι θερμόν) : **SyrC**

[54] supplevi, coll. *PM* § 15, l. 163 (ἤ ὑγρὸν ἤ) : non legit. **SyrC**

misy (*b-musidis* = ἐν τῷ μίσυι). Put these substances in a vessel and cook them on a flame over three days, until they turn red. Then lay them on mercury *Hermēs* [i.e. copper/mercury?][19] and it will be gold.

Tenth paragraph in Cambridge ms. = *PM* § 14
(A similar version is also preserved in the London mss.)

Grind Macedonian malachite (*chrysokolla*), the one that looks like copper rust, with the urine of a bull, until it is transformed [lit. overturned]. For nature is hidden inside. After it has been transformed [lit. overturned], cook it with oil several times. Afterwards cook it with alum (*styptēria*), or with *misy* (*musidin* = μίσυ), or with unburnt sulphur (*theion apyron*): in this way it turns red. Then lay it [on a metal] and cook and you will find what you are seeking for.

Eleventh paragraph in Cambridge ms. = *PM* § 15[20]
(London mss. do not preserve this text)

O celestial natures, miracle of nature. O great natures, which set natures in motion and transform them. O natures overcoming natures and different from natures: these are the natures that have a great nature, the natures that are great, when they are transformed by fire. Wise men <know>[21] that they are a wonder: these natures heal any bodily disease when handled with art. But those who work the substances (*hule* = ὕλαι) without experience will fail several times, since they do not know them [i.e. these substances]. They forget that physicians test the drugs [lit. herbs] and then start to prepare them: they [i.e. the physicians] do not set immediately to work, but first of all test which kind [of drug] is hot, which kind ... and has an intermediate effect, or a cold or a <wet [effect], or> of which kind is the disease, and in this way they prepare the drugs [lit. herbs] ... the healing [they are able to heal?] by testing them.

12. SyrC, fols. 92ᵛ7–93ʳ3

ll. 4-12 ≈ SyrL in *CMA* II 1,8-2,2

(Syriac text, lines 1–10, with interlinear footnote markers 55–73)

---

55 ܗܘܝ dubit. supplevi, coll. *PM* § 6, l. 166 (οὗτοι)

56 ܘܢܝܟܕܗ. ܘܚܕܡ difficile lectu **SyrC**

57 ܘܡܣܡ[...] **SyrC**

58 ܘܡ supplevi : non legit. **SyrC**

59 ܡܤܟܠܣ ܚܡܝ fort. leg. : vide *PM* § 16, ll. 168-9 (δοκοῦντες γὰρ ἡμᾶς [...] ἀπαγγέλλειν)

60 ܡܕܡ fort. leg.

61 ܗܘܝ **SyrL** : ܗܘܝ ܕܡ ܘܠܐ – ܚܚܘܡ **SyrC**

62 ܗܐܒ/ om. **SyrL**

63 ܗ/ܣܐ ܡܕܙܝܘ **SyrL**

64 ܗ/ܣܐ ܐܩܐ ܡܢܝܘܐ ܚܣܘܝ **SyrL**

65 ܗ/ܣܐ ܣܟ ܚܕܘܡܡܐ **SyrL**

66 ܗ/ܣܐ ܘܡܡ ܚܐܣܟ ܡܝܝ **SyrC** in mg.

67 ܘܗ om. **SyrL**

68 ܘ addidi, coll. ܡܚܕܝ **SyrL**

69 ܘܚܟܡ om. **SyrL**

70 ܘܚܟܡ om. **SyrL**

71 ܘܚܐܣܡܣ ܘܬܢܟܐ **SyrL** : ܐܠܐ ܠܐ – ܚܣܢܟܐ **SyrC**

72 ܝܘܦܐܠܐ ܘܦܘܙܦܘܘܙܐ ܘܗܝܬܢܐܠܐ ܐܢܚܟܠ supplevi, coll. **SyrL** : difficile lectu **SyrC**

73 ܚܕܘܙܐ ܗܝܬܢܐܠܐ ܦܘܗܘܒܐ supplevi, coll. **SyrL** : difficile lectu **SyrC**

Twefth paragraph in Cambridge ms. = *PM* § 16
(London mss. preserve only ll. 3-11)

<Those people>, on the contrary, although they are foolish and inexperienced, [hasten?] to heal us up to our soul and hasten to take us away from any pain. They want to make mixtures without understanding that they will fail, and they will blush unawares [i.e. without being aware of blushing], since they think to speak to us in riddles and not truthfully, and they do not acquire any knowledge based on experience: [they do not know] which is [the species] that cleanses when applied, and which one softens, and which one makes [dyes?] stable, and which one only acts superficially, and which one penetrates deeply, and which is the one that vanishes from above, and which one vanishes from below, and which is the one that is [fire]-resisting <and> works, and which one makes [substances/dyes?] resist [fire]. Salt cleanses both superficially and inside, and there are those [dyes] that both whiten superficially and whiten the inner parts, and those that both vanish from the surface and vanish from the inner parts. If they [the foolish people] make use of them, they will be completely defeated in their attempts. But now let us come to the natural substances. A drop <of purple> (*pyrphyra*) has the capacity to <destroy many [species]>, and a pinch of sulphur (*theion*) <lays waste to many drugs [lit. herbs]>. ... that one must know ...

13. SyrC, fol. 93ʳ3-12
omittit SyrL

 (sic) . 1
<ܚܘܡܕܐ ܘܗܘܐ> [75] <ܘܡܣܘܙܐ>. [76]
 <<ܡܠܐ>[78] 5
14. SyrC, fol. 93ʳ13-93ᵛ3
omittit SyrL

. [93ᵛ]
15. SyrC, fol. 93ᵛ4-12
≈ SyrL in *CMA* II 11,19-12,2

---

[74] [...] SyrC : Duval vertit "fais boulir": fort. leg., coll. *PM* § 18, l. 186 (λείου)

[75]   supplevi, coll. *ISyrC* § 14, l. 5 : non legit. SyrC

[76]   supplevi, coll. *PM* §17, l. 187 (Δέξαι τὸ πέταλον τὸ μήνης) : non legit. SyrC

[77]   add. SyrC infra lineam

[78]   supplevi : [...] SyrC

[79] [...] supplevi : [...] SyrC

[80]   seclusi

[81]   difficile lectu SyrC

[82]   seclusi

[83]   SyrL : εἰς πυρίτην *PM* § 19, l. 208

Thirteenth paragraph in Cambridge ms. = *PM* § 17
(London mss. do not preserve this text)

Take rhubarb (?)[22] that comes from Pontus [the Black Sea]; [boil/triturate?] it with white wine of vines until it becomes as thick as honey.[23] Then take <silver> leaves and rub part of the wash on them until they turn red. You must ... and put it in a new vessel. ... and set it up so that it will be boiled. You must boil [it ?] ... to the [metallic] body. Place it in that wine and leave it <to soak>. After it cools, you will not lay it out in the sun, but in the shade.[24] And afterwards cook it with (?) and you will find what you are seeking. Nature masters nature.

Fourteenth paragraph in Cambridge ms. = *PM* § 18
(London mss. do not hand preserve this text)

Take saffron that comes from Cilicia and from the flower of safflower (*Carthamomus tinctorius*); throw them into wine and they will become a good wash (*zōmos*). Dip copper leaves into this wash until their color seems beautiful [to you]. If you dip leaves of *Hermēs* [i.e. copper/mercury?],[25] they will become more beautiful. Afterwards take Aristolochia, two minae, and saffron and celandine (*elydrion*), four minae, so that they become as thick as honey. Triturate [them] and rub part of them [on the leaves] and treat them: so you will wonder at the magnificence of God.[26] Cilician saffron has the same power as mercury [lit. milk], just as cassia (*kasia*) has the same power as cinnamon (*kinnamōmon*).

Fifteenth paragraph in Cambridge ms. = *PM* § 19
(A similar version is also preserved in the London mss.)

Recipe. Take our lead that has become harder (*arreuston*) by means of Chian [earth], red soda (*nitron*) and alum. Cook [these ingredients] with [a fire of] chaff and throw them into *swrṭws* (?)[27] and saffron and safflower and *kmwnywn*

16. SyrC, fols. 93ᵛ12-94ʳ3

ll. 1-3 et 7 ≈ SyrL in *CMA* II 12,2-5

 [94ʳ]

---

[84] ، **SyrL**

[85] **SyrL**

[86] **SyrC** scripsi, coll. **SyrL** (

[87] **SyrL**

[88] ، **SyrC** supplevi, coll. **SyrL** : [...]

[89] **SyrL**

[90] **SyrL**

[91] supplevi, coll. **SyrL** : non legit. **SyrC**

(= οἰχομενίου ἄνθος?)²⁸ and celandine (*elydrion*) and saffron-sauce (*krokomagma*) and Aristolochia. Triturate them with sharp vinegar and make them into a wash. Throw lead into them: it will stay [dipped in this solution] for one hour. Take it out from the vessel and you will find what you are seeking. In this wash there will be unburnt sulphur, so that the wash is effective.

Sixteenth paragraph in Cambridge ms. = *PM* § 20²⁹
(London mss. preserve only ll. 1-3 and 7)

Since it overcomes nature, this [nature?] is the greatest that I taught [lit. handed down] to the Egyptians, after their priests came and made me swear to teach them about the power of this book. In fact many are the substances (*hulo* = ὕλη) for the red [i.e. for the making of gold] and for the white [i.e. for the making of silver]. But it will be not a wonder for us that the substance is unique and this is the one that works in place of all the others. Its name indeed is both secret to anyone and evident to anyone. Its value is great, but also very little: you cannot find it [lit. it is not found] in many places, and it is discarded in dung-hills. Now you must look away from the plurality of matters [lit. from the substances; *hule* = ὕλαι] and ... something good. It is <unique,> in fact, [the nature] that masters all the natures.³⁰

End of the first treatise by the wise Democritus.

[94ʳ4]

1. SyrC, fol. 94ʳ5–94ᵛ1

≈ SyrL in *CMA* II 12,5–11

    [94ᵛ]

2. SyrC, fol. 94ᵛ1–7

≈ SyrL in *CMA* II 12,11–5

---

[1] SyrL

[2] SyrL

[3] om. SyrL

[4] SyrL

[5] SyrL

[6] SyrL

[7] om. SyrL

[8] SyrL

[9] SyrL

[10] SyrL

[11] SyrL

[12] (SyrC SyrL) seclusi, coll. *AP* § 1, l. 6 (τὸν μόλυβδον λύσεις)

[13] om. SyrL : fort. leg., post *AP* § 1, ll. 7-8 (Ὁ γὰρ μάγνης ἔχει συγγένειαν πρὸς τὸν σίδηρον)

[14] SyrL

[15] om. SyrL

[16] om. SyrL

[17] om. SyrL

[18] SyrC : SyrL

Second Book by the Philosopher Democritus (= 2SyrC)

First paragraph in Cambridge ms. = AP § 1
(A similar version is also preserved in the London mss.)

Consider the power of the drugs [lit. herbs] that lead [us] to silver; these are as follows [see *PM* § 20, ll. 228-9]:

Mercury [lit. milk] that is found in orpiment (*arsenikon*) or in realgar (*sandar-achē*), or according to your understanding: make it solid as you are accustomed to doing, then lay it on *Hermēs* [i.e. copper/mercury?] or on alum (*styptēria*) and it will turn white.[1] Also the *magnēsia* that you have whitened produces this [effect], as well as transformed [lit. overturned] orpiment (*arsenikon*), and cadmia (*kadmia*), and realgar (*sandarachē*) without fire, and whitened alabaster (*alabas-tros*), and white lead (*psimythion*) roasted together with sulphur (*theion*). Iron is melted with *magnēsia*, when it takes sulphur (*theion*), a half-portion. *Magnēsia* is indeed related to iron.[2] Nature delights in nature.

Second paragraph in Cambridge ms. = AP § 2[3]
(A similar version is also preserved in the London mss.)

Take for yourself the mercury [lit. milk] described above and boil it with oil. After-wards take it and then roast it with three drachmas of alum (*styptēria*). Cook this

3. SyrC, fol. 94ᵛ7-18

≈ SyrL *CMA* II 12,15-21

    <ܕܡ> [27] <ܘܡ> [28] .
---

[19] **SyrL** : **SyrC**

[20] **SyrC** : scripsi, coll. **SyrL**

[21] **SyrL**

[22] om. **SyrL** (idem alt.)

[23] **SyrL** : difficile lectu **SyrC**

[24] **SyrL** : **SyrC**

[25] **SryC** ut plerumque : **SyrL**

[26] om. **SyrL**

[27] supplevi, coll. **SyrL** : non legit. **SyrC**

[28] supplevi, coll. **SyrL** : non legit. **SyrC**

[29] om. **SyrL**

[30] **SyrL** : **SyrC**

[31] om. **SyrL**

[32] **SyrC** : scripsi, coll. **SyrL**

[33] **SyrC SyrL** : fort. leg., coll. *AP* § 3, l. 25 (κατ᾽ ὀλίγον κασσιτέρου)

[34] **SyrL** / : **SyrC**

[35] **SyrL** : **SyrC**

[36] **SyrL**

[37] **SyrC SyrL** : dubit. scripsi

[38] **SyrL**

[39] **SyrL**

[substance?] in an oven and you will find it whitened. Then mingle alum (*styptēria*) with it. For these substances, when they are mutually whitened, dye whatever you want white. Nature masters nature.

Third paragraph in Cambridge ms. = *AP* § 3
(A similar version is also preserved in the London mss.)

Take white *magnēsia*. You whiten it with water and salt, and also with alum (*styptēria*), or with seawater, or with vinegar (?) of citrus (*kitrion*) or with sulphur (*theion*). Since they are white, sulphur fumes whiten all bodies. But foolish persons say that also the fumes of the dyer/of the pigment [scil. whiten them].[4] Mingle an equal part of wine dregs (*spheklēs*) with this [i.e. *magnēsia*], so that it [the *magnēsia*] turns extraordinarily white. They make this [i.e. *magnēsia*] like an amalgam (*malagma*): take copper, four ounces; you must add part of this [metal?] while you are stirring;[5] it must be added until it is burnt. Lay on it also the *magnēsia*: this makes it [i.e. the copper] unbreakable [?lit. that will not be crushed] and removes its rust. Nature masters nature.

4. SyrC, fols. 94ᵛ18-95ʳ9

omittit **SyrL**

 [95ʳ]
5. SyrC, fol. 95ʳ10-95ᵛ3

ll. 9-10 ≈ **SyrL** in *CMA* II 12,21-2

 [95ᵛ] ✢

---

[40] scripsi, coll. *AP* § 5, l. 42 (πυρίτη) : **SyrC**

[41] supplevi, coll. *2SyrC* § 8, l. 4 : non legit. **SyrC**

[42] difficile lectu **SyrC**

[43] difficile lectu **SyrC**

[44] supplevi, coll. *AP* § 5, l. 51 (οὐκέτι ἔσται μόλυβδος) : non legit. **SyrC**

[45] difficile lectu **SyrC**

[46] fort. leg. : Duval (*CMA* II 271) vertit "supérieure à celle
   qui est connu"

[47] seclusi : in rasura **SyrC**

[48] **SyrC**

[49] non legit. **SyrC** : Duval (*CMA* II 271) vertit "Moi, je ferais en sorte que"

Fourth paragraph in Cambridge ms.; l. 1 (ܣܒ ܠܐܘ̈ܢ ܐܘܦܘܢ ܣܘܦܪܙܐ) = AP § 4,l. 31 (λαβὼν θεῖον τὸ λευκόν) ; ll. 1-8 = AP § 5
(London mss. do not preserve this text)

Take white unburnt sulphur (*theion apyron*)[6] and litharge (*lithargyros*) and crush [them] together with cadmia (*kadmia*) or with orpiment (*arsenikon*) or with pyrite (*pyritēs*)[7] or with vinegar and honey. Cook it [this mixture?] on a strong fire; there will be <among these [substances]> also quicklime that has been mixed with vinegar, so that it will turn extraordinarily white. Lay sulphur (*theion*) on them and it will be as white as a clod of clay; but if cooked too much, it turns red; and if it turns red, it is useless and is no longer able to whiten. So whiten it, and then set it on the fire, and make use of it. If it turns white, litharge will no longer be fusible; thus, it will no longer be fully <lead>. In fact, it is easily transformed into [many] species. For nature is satisfied with nature.

Fifth paragraph in Cambridge ms.: not preserved in the Byzantine tradition[8]
(London mss. preserve only ll. 9-10)

For these [substances?] will be sufficient (?) for you, O wise men.

The making of silver. It is indeed one and unique, [the substance] that works in place of all the drugs [lit. herbs]. For this substance, when cooked with many transformations, manifests itself to us in many different forms [lit. shows us many differences/alterations]. White lead (*psimythion*) has a power completely different from the usual; and all these [drugs] have a single nature, even though each of them is cooked according to its own property. If the books do not give proper credit to this nature, more than for all [other drugs], we too will be easily thrown down into the plurality of matter (*hule* = ὕλαι). The philosophers throw us into the plurality of matter, and in this [plurality] many of them have hidden the truth. But I will … so that they try your knowledge.

6. SyrC, fol. 95ᵛ3-13

≈ SyrL in *CMA* II 12,22-13,5

⁶⁰ ⁶¹ ⁶² <ܝܐܠܗܐ>⁶³

5   <ܝܥܝܡ>⁶⁴ .
---

⁵⁰ **SyrC** : **SyrL**

⁵¹ **SyrL** : difficile lectu **SyrC**

⁵² **SyrL**

⁵³ om. **SyrL**

⁵⁴ **SyrL**

⁵⁵ **SyrL**

⁵⁶ **SyrC** : **SyrL**

⁵⁷ **SyrL**

⁵⁸ **SyrL** : difficile lectu **SyrC**

⁵⁹ **SyrL**

⁶⁰ **SyrL** : difficile lectu **SyrL**

⁶¹ **SyrL**

⁶² **SyrL**

⁶³ supplevi, coll. **SyrL** : non legit. **SyrC**

⁶⁴ supplevi, coll. **SyrL** : non legit. **SyrC**

⁶⁵ **SyrC** : **SyrL**

⁶⁶ **SyrL**

⁶⁷ **SyrL**

Let us move on to the waters in which both the white and red drugs [lit. herbs] are boiled down.

Sixth paragraph in Cambridge ms. = *AP* § 6
(A similar version is also preserved in the London mss.)

Take saffron that comes from Cilicia and put it in seawater. Make it a wash (*zōmos*) and dip *Hermēs* [i.e. copper/mercury?] in it.[9] It will turn white. Then take also the following drugs [lit. herbs] such as realgar (*sandarachē*), orpiment (*arsenikon*) and unburnt sulphur (*theion apyron*), or according to your understanding. Crush until it becomes like wax and make use of it. After rubbing it on the [metallic] leaves, put them in pure water and they will turn white.[10] Make use of them as a craftsman. The appearance [of metals?] is made white by means of Cilician saffron with sea-water; wine [i.e. saffron with wine] makes it red. Nature delights in nature.

7. SyrC, fols. 95ᵛ13-96ʳ3

omittit **SyrL**

 [96ʳ] ⁷⁰ 5
8. SyrC, fol. 96ʳ3-9

omittit **SyrL**

9. SyrC, fol. 96ʳ9-16

omittit **SyrL**

---

⁶⁸ **SyrC** : fort. leg., coll. *AP* § 7, l. 66 (μετὰ... Κιμωλίας)

⁶⁹ **SyrC** : scripsi

⁷⁰ difficile lectu **SyrC** : Duval (*CMA* II 272) vertit "Ceci"

⁷¹ **SyrC** : fort. leg., coll. *AP* § 7, ll. 71-2 (χωρὶς πυρός... πυρίμαχα)

⁷² difficile lectu **SyrC**

⁷³ circumscripsit **SyrC** (fort. expungere voluit)

⁷⁴ difficile lectu **SyrC** : Duval (*CMA* II 272) vertit "Tu broieras sur le feu"

⁷⁵ ؛ **SyrC**

⁷⁶ fort. leg.

Seventh paragraph in Cambridge ms. = *AP* § 7
(London mss. do not preserve this text)

Here for you is white litharge (*lithargyros*); whiten it as follows: it is added to shrubbery leaves, flour, honey and realgar (*sandarachē*). Crush them and they will become thick. Rub them on the surface [of the metallic leaf?] and leave a half [of the drug] to one side.[11] Boil it [the metallic leaf?] as you are accustomed to doing and, after boiling, put it into one of the natures [i.e. the part of the dyeing substances left aside?]. Empty out this [solution] into water that contains wood ashes. What has been well mixed ..., this is done (?) by means of this effective and inexpensive water. For nature overcomes nature.

Eighth paragraph in Cambridge ms. = *AP* § 8
(London mss. do not preserve this text)

Take mercury [lit. milk] and crush alum and a bit of *misy* (*musidin* = μίσυ) with it; crush it with vinegar. You must also add white cadmia (*kadmia*), or *magnēsia*, or quicklime, so that it becomes a white body. Crush them over a fire and roast over live coals;[12] among these [ingredients] there must also be unburnt sulphur (*theion apyron*). For nature masters nature.

Ninth paragraph in Cambridge ms. = *AP* § 9
(London mss. do not preserve this text)

Here for you is orpiment (*arsenikon*), one ounce; soda (*nitron*), four drachmas; bark of *Persea* or peel of tender leaves, two ounces; cow's milk, one ounce; and the same quantity of alum (*styptēria*). Crush them with vinegar or urine or quicklime, until they become as a wash (*zōmos*). Make this wash boil and dip all [the metals] that have rust, and their rust will be removed. Nature masters nature.

10. SyrC, fol. 96ʳ16-96ᵛ2

ll.1-3 ≈ **SyrL** in *CMA* II 13,6-8

 [96ᵛ] .[82]

_____

[77] ، SyrC a.c.

[78] SyrL

[79] SyrL

[80] ، om. SyrL

[81] SyrL

[82] SyrL

[83] om. SyrL

Tenth paragraph in Cambridge ms. = *AP* § 10
(ll. 1-3 are preserved in the London mss.)

Now we receive all the methods for preparing (*skeuasia*) gold and silver; here you do not lack anything except the vapour (*aithalē*) of mercury [lit. milk] and how water is distilled [lit. rises up], since I will hand down these things to you, without jealousy, in another book.

End of the second book by the philosopher Democritus.

[96ᵛ3] ²¹

1. SyrC, fol. 96ᵛ4-6
≈ SyrL in *CMA* II 13,9-12

    ⁶<ܠܠ> . ⁵

2. SyrC, fols. 96ᵛ6-97ʳ12
≈ SyrL in *CMA* II 13,12-23

    [97ʳ] ¹⁶. ¹⁵

---

¹ et om. SyrL

² SyrC : SyrL

³ SyrL

⁴ om. SyrL (idem alt. et ll. 2-3)

⁵ SyrL : difficile lectu SyrC

⁶ / supplevi, coll. SyrL : non legit. SyrC

⁷ SyrL

⁸ : SyrC SyrL

⁹ supplevi : [...] SyrC

¹⁰ ll. 2-3 () om. SyrL

¹¹ om. SyrL

¹² SryC : SyrL, coll. scripsi

¹³ SyrL

¹⁴ : SyrC SyrL

¹⁵ om. SyrL

¹⁶ : SyrC SyrL

Again by Democritus: I Greet you Wise Men[1] (= 3SyrC)

First paragraph in Cambridge ms.: not preserved in the Byzantine tradition
(A similar version is also preserved in the London mss.)

While we are approaching other useful things, he [Democritus?] brings forth different kinds of dyes.[2] He brings forth something wonderful, since it works without fire. This is not its only [property], but it alone performs all the dyeing processes and accomplishes [the making of] all the stones. Here it is for you: it is presented without jealousy.

Second paragraph in Cambridge ms. = Phil. Anonym. Alch. *CAAG* II 122,4-17
(A similar version is also handed down in the London mss.)

Here for you is the stone that is not a stone; it has no value, but I claim that it is precious; it is not known and manifest to everyone; it has many names and its name is only one. This is not a stone, although it is a stone; although it is precious, there is no place where it is sold; although its name is only one, it is called by many names. This is not [said] in vain, but according to its nature; this is how someone could say: "take [the substances] that vanish and are white; and I have also the white *Hermēs* [i.e. copper/mercury?]." He does not lie when he makes such a statement concerning mercury [lit. milk], since it is made to vanish by the fire; this is about the vapour (*aithalē*) that is called cinnabar (*kinnabaris*). This is the only

 [23](sic)

3. SyrC, fol. 97ʳ12-97ᵛ2

≈ SyrL in *CMA* II 13,23-14,4

5   . [33]   [97ᵛ]   [34]

---

[17] **SyrL** : **SyrC**

[18] om. **SyrL** (idem l. 10)

[19] **SyrL**

[20] **SyrL** (fort. mel.) : **SyrC** (ut semper)

[21] **SyrL**

[22] **SyrL**

[23] **SyrL** (fort. mel.)

[24] **SyrL**

[25] **SyrL**

[26] **SyrL**

[27] **SyrC** : **SyrL**, coll., scripsi [27]

[28] **SyrL**

[29] **SyrL** (fort. mel.) : **SyrC**

[30] / : scripsi / **SyrC SyrL**

[31] **SyrC** : **SyrL**, coll., scripsi [31]

[32] **SyrL** (fort. mel.)

[33] **SyrC** : **SyrL** : fort. leg., coll. *CAAG* II 357,11-2 (ἐπίχριε ὅσον βούλει λίθον, λειώσας αὐτόν, καὶ ἔσται μαργαρίτης)

[34] **SyrL**

[35] **SyrL** : **SyrC**, coll., scripsi [35]

one that whitens *Hermēs* [i.e. copper/mercury?] in this way. Boil this stone that has many names until it turns white. Throw it into jenny's milk or goat's milk;[3] in this milk there is cadmia (*kadmia*). Add marble quicklime or [quicklime of] wine dregs (*spheklē*); take care that there is not too much quicklime. Put cabbage ashes [lit. burnt cabbage] in another vessel according to your understanding. Filter the water of both [the vessels] and water the stone; set it on a gentle fire for one day. Strain away the water from the stone and you will find it (?) black. Take this [material?] and work whatever you want; when changed it changes the colours.

Third paragraph in Cambridge ms.: not preserved in the Byzantine tradition (except ll. 3-4 = *CAAG* II 347,10-2)
(A similar version is also preserved in the London mss.)

Here is for you *komaris* from Scythia, which is a region.[4] But that one which comes from Scythia is strong and deadly for men and kills easily. That is why they keep its power secret. Throw it into quicklime by mixing with wine dregs (*spheklē*) and pound [these ingredients] since they are moist by nature [lit. in their natural moisture]; when it [i.e. the *komaris*] gets soft and watery, rub it on whatever stone you want. Crush this [material?] and it will be similar to marble [*marmaritis*? Perhaps "to pearl"].[5] *Komaris*, after being diluted, gives its beauty to stones.

4. SyrC, fol. 97ᵛ2-14
omittit SyrL

 (sic)³⁶
 (sic)³⁷. ³⁸(sic) (sic). (sic).
 (sic). ³⁹
5. SyrC, fols. 97ᵛ14-98ʳ1
omittit SyrL

 (sic)⁴⁰ .
. [98ʳ]
6. SyrC, fol. 98ʳ1-7
omittit SyrL

---

³⁶ SyrC (ut semper) : Duval *CMA* II 273 n. 4 (fort. mel.)

³⁷ fort. leg.

³⁸ fort. leg.

³⁹ scripsi : SyrC

⁴⁰ SyrC (idem l. 3) : fort. leg.

⁴¹ SyrC

⁴² fort. leg.

Fourth paragraph in Cambridge ms.: not preserved in the Byzantine tradition
(London mss. do not preserve this text)

Here is for you the snake (? *yqdkwn*) which is found in the rocks of the Nile river.[6] Remove his bones and crush them in hare's blood, or in hog's blood, or in vinegar of citrus, or according to your understanding. Make it like chickpeas. Put it in a vessel of moon foam (*aphroselēnos?*), or in [a vessel] of Cimolian earth (*kimōlia?*), or in *'stwrtwqys* (?),[7] or according to your understanding. It will become beautiful. Take care that it is not in large pieces. Take the small pieces away, their little black stones. Crush them only in vinegar, but mix with it a wash (*zōmos*) made of *solanum nigrum* and mud (*pēlos?*) of *qlpyswn* (?),[8] so that we will make them more and more sodden [i.e. we will let them drink]. After we have made them sodden [lit. they have drunk], place them into moulds. Throw into them juice (*chylos*) of comfrey (*symphyton*), or of wild cabbage, or cumin.

Fifth paragraph in Cambridge ms.: not preserved in the Byzantine tradition (even though it presents some similarity with *PM* § 1)
(London mss. do not preserve this text)

How to make purple.[9]

Take [purple] and triturate it well. Throw it into water and make it into a paste (*mazin* = μᾶζα). Throw it into water and throw seaweed (*phykos*) in with it: one part for two parts of water. Plunge wool into it and it will turn purple. You will also have quicklime water: after you have plunged in [the wool], wash it in this water.

Sixth paragraph in Cambridge ms.: not preserved in the Byzantine tradition
(London mss. do not preserve this text)

Again another [recipe].

Take seaweed (*phykos*) and throw it into quicklime, or vinegar, or alum (*styptēria*) for three days.[10] Filter this water [i.e. solution] and boil it; mix *'srtws* [perhaps a kind of clay?][11] with this water. As soon as it boils, plunge in wool or white woolen cloth (*hgywn* = ἔριον?).[12] Take them out and wash with seawater. You will find a beautiful purple [dye].

7. SyrC, fol. 98ʳ7-13

omittit **SyrL**

_____

⁴³ **SyrC** a.c.

Seventh paragraph in Cambridge ms. : not preserved in the Byzantine tradition
(London mss. do not preserve this text)

Again another [recipe for] cold purple [dyeing].

It is the purple made by means of that water which, after boiling, makes its flowers [i.e. the flowers of the purple dye] soft. Wool is plunged in: it takes alum and it is left [to soak] for one day. Afterwards wash it. You must throw it into flour (*rgmn* = ἐρεγμός?) and wash it again.[13] This is the purple [dyeing].

End of the book by the philosopher Democritus. Whoever works, shall accomplish [his work]; whoever seeks, shall find.

Democriti naturalia et arcana
(Z, fols. 62ᵛ1–67ᵛ21)

[1] Proiiciens in λῑ ᾱ [*i.e libram unam?*] purpurae [*blank*]*¹ libram*² scoriam*³ ♂ [*i.e. ferri?*] in cuius ρᵛ ζζ [? εἰς οὔρου ⅄ (*i.e.* δραχμάς) ζ **M**] impone in rogum ut appraehendat fervores (ebullitiones). Deinde accipiens ad igne decoctum proiice in catillum adiiciens ostreum, et infundens decoctum ostreo, sinas unum diem ac noctem macerari: deinde accipiens muscorum marinorum libras δ̄ [*i.e. quatuor*] proiice aquam ut sit supra muscos quatuor digitorum, et habe donec addensetur, et defaecans per colum fove, et componens lanam affunde: laxiora*⁴ vero componantur, ut antevertat [assequatur *above the line*] iusculum usque ad fundum, et sinas per duos dies ac noctes stare: deinde post illa accipiens sicces in umbrâ, iusculum vero non effundito: deinde proiiciens in idem iusculum duas libras muscorum, mitte in iusculum aquam, ut fiat proportio prima, et similiter habe, donec crassescat: deinde defaecans proiice lanam seu primum, et fac per diem unum et unam noctem: deinde accipiens lava in urinâ et sicces in umbrâ: deinde accipiens [*blank*]*⁵ et accipiens lapathi libras quatuor, excoque cum lanâ,*⁶ ut lapathum solvatur, et defaecans [63ʳ] aquam iniice [*blank*: τὸν λακχάν **M**] et coque, donec densetur, et ubi defaecaveris haec lava urinâ, deinde rursus aquâ et postea rursum [*blank*: τὸν λακχάν **M**] iniice T̂ ♋ ε̃ [? τὴν ἐρέαν **M**]: deinde siccans similiter [*blank*]*⁷ unguibus marinis perfusam in urina duos dies: [2] quae vero veniunt ad praeparationem purpurae, sunt haec: fucus marinus, anchusa, [*blank.*]*⁸ praecipitium,*⁹ rubia Italica, Phylanthium occidentale, vermis purpureus e lanâ confectus, rhodium Italicum. Hi flores magni aestimati sunt a veteribus, et sunt fugaces, non pretiosi. Est autem Galatiae vermis, et flos Achaiae, quem nominant Lacham, et Syriae flos, quem vocant Rhizium (ℵ parvam radiculam), et ostrinum, et ostrinum Libycum, et Aegyptiaca concha marina quae vocatur pinna, et isatis herba, et*¹⁰ superioris et Syriae quod

Zuber's comments in the margin:
¹ * διαβολοῦ
² * λίτραν non λύτραν
³ * σκωρίαν non σκορέαν
⁴ * lege χαυνότερα: alterum enim non est Graecum
⁵ * λακχάν
⁶ * ἐρίου non ἔρου
⁷ *The asterisk does not refer to any comment in the margin* (ἐν σκιᾷ **M**)
⁸ * λαδικήνη
⁹ * κρημνός extat in Graeco, vix tamen in sensu habebit locum
¹⁰ * defectus esse creditur

ostrum vocant marinum. Haec sunt immobilia, neque apud nos valde pretiosa, exceptâ isatide. [3] Haec igitur a praedicto Praeceptore cum didiceris, et quum materiae differentiam cognoveris [*blank*]*¹¹ ut applicem (accommodem) naturas. Quamvis enim noster Praeceptor humanis concesserit, nobis nondum usque adeo perfectis, tamen aliquid superest ad cognitionem materiae circa [63ᵛ] eam occupatis. Ex inferis, inquit, conabar ipsum reducere. Quamprimum autem in eum impetum feci, illico revocavi dicens: praebes mihi praemia (dona) pro quibus tibi sum operatus: quo dicto, tacui.¹² Postquam vero saepe evocabam, interrogabam, quomodo applicabo naturas? Dixit mihi, difficile est dictu: non permittente ipsi daemone, tantum autem dixit: "In templo libri sunt." Reversus ad templum indagabam, sicubi poteram, ut libros mihi compararem. Neque enim circumiens hoc dixerat; sine testamenti enim factione, ut quidam dicunt, mortuus est, veneno usus ad dissolutionem*¹³ animae e corpore: ut autem filius dicit, praeter expectationem convivio exceptus. Ante mortem vero animo erat confirmato, filio suo tantum hosce libros manifestare, si primam transgressus esset adolescentiam.¹⁴ Horum vero nihil quicquam nullus nostrum sciebat. Investigantes igitur postquam nihil inveniebamus, gravem sustinuimus laborem, donec consubstantiarentur et una intromitterentur substantiae et naturae. Postquam vero perfecimus compositiones materiae progressu temporis, circa nundinas, in Templo omnes sumus convitati. Quum igitur essemus in delubro, forte fortuna columna erat, quam [64ʳ] disrumpit, quam nos vidimus intus nihil habere: neque quisquam dicebat reconditos ibi instar gazae esse libros, et pergens in medium duxit, interpellantes autem admirabamur, quod nihil relinqueremus, praeterque verbum hoc valde utile nihil ibi inveniremus: natura delectatur naturâ, et natura naturam vincit, et natura naturam regit. Valde admirabamur, quod brevi sermone omnem scripturam collegisset.

[4] Venio et ego in Aegyptum afferens naturalia, ut multum curiositatem et non confusam materiam despiciatis.

[5] Accipiens ☽ [*i.e. mercurium*] fige Magnesiae corpore, aut Italici stibii corpore, aut sulphure crudo, (mortuo), aut spumâ maris, aut calce assâ, aut alumine a pomo, aut arsenico, aut ut ipse putas, et iniice terram albam ♀˙ [*i.e. aeri*] et habebis ♀˙ [*i.e. aes*] sine umbrâ; flavam autem iniice lunam et habebis ˋθ [*i.e solem/aurum*] auro, et erit aureum corallium corporatum.

Hoc ipsum facit etiam arsenicum flavum, et sandaracha praeparata, et cinnabaris valde eversa, aes autem sine umbrâ solum argentum vivum facit. Natura naturam vincit.

[6] Pyriten, Argyriten, quem et Syderiten nominant, praepara pro more, ut fluere possit: fluet autem per [*blank*: νίθεως M] aut album lithargyrum, aut Italicum stibium, et sparge plumbo: non simpliciter dico, ne erres, sed praecise, lithargyro

¹¹ * ησκούμην non est Graecum
¹² *Zuber writes in the margin*: Ostani manes evocati a Democrito
¹³ * emigrationem melius est
¹⁴ *Zuber writes in the margin*: quod non disces a Deo vel a vivis hominibus, non disces a mortuis vel daemonibus

nostro nigro, aut ut cogitas, et assa [64ᵛ] et iniice materiae flavum factum et tinget. Natura enim naturâ delectatur.

[7] Pyriten praepara, donec fiat incombustibilis, abiiciendo nigredinem. Praepara autem aceto salso aut urinâ incorruptâ, aut mari, aut oxymelite, aut ut ipse putas, donec fiat quasi incombustibile auri ramentum, et si factus fuerit taliter, admisce ipsi sulphur crudum, aut alumen flavum, aut Atticam ochram, aut quod ipse putas, et iniice argento, per ϴ [*i.e. solem/aurum*] aut ᵂϴ [*i.e. soli/auro*] propter aureum ostreum. Natura enim naturam regit.

[8] Claudianum fac marmor,*¹⁵ et praepara pro more, donec flavum fiat, flavefacito igitur non lapidem dico, sed lapidis utile, f<l>avefacies autem cum alumine expressô, sulphure, arsenico, sandarachâ, aut calce, aut quod ipse putas, et si inieceris argentum [《 *above the line*], facis °ϴ [*i.e. solem/aurum*], si vero ᵛϴ [*i.e. solem/aurum*] facis aureum ostreum. Natura enim naturam vincens regit.

[9] Cinnabari album fac per oleum aut acetum, aut mel, aut muriam, aut alumen, deinde flavum per misy aut sori, aut chalcanthum, aut mortuum sulphur, aut quod ipse putas, et iniice 《ᵂ [*i.e. lunae/argento*] et ϴ [*i.e. sol/aurum*] erit, si aurum tinxeris, si aes, electrum. Natura delectatur naturâ.

[10] Cypriam vero Cadmeam*¹⁶ expulsam dico, dealba pro more, fac deinde flavam [65ʳ]. Flavefacies autem felle vitulino, aut terabinthinâ, aut cocino aut raphanino, aut iis quae possunt ipsam flavefacere, et iniice 《ᵂ̇ [*i.e. lunae/argento*] aurum enim erit propter aurum et auream aquam. Natura enim naturam vincit.

[11] Androdamanta praepara vino austero, aut mari, aut urinâ, aut aceto salso, quae possunt exstinguere naturam eius. Laeviga cum stibio Chalcedonensi: praepara rursus aquâ marinâ vel salsilagine, vel acidâ muriâ lava, donec stybii nigredo fugiat, torre aut assa dum flavefeceris, et coque aquâ divinâ mundâ, iniice vero argento [《 *above the line*], et quando sulphur crudum adieceris, fac aureum iusculum. Natura enim naturam regit. Hic est lapis qui dicitur Chrysites.

[12] Accipiens terram albam (dicens) a cerussâ*¹⁷ et [*blank*: ἑλκύσματος M] aut stibio Italico et Magnesiâ, aut albo lithargyro dealbato. Dealbato autem ipsam mari, aut salsilagine*¹⁸ aut aquâ aereâ, in rore dico et sole, ut ipsa laevigata fiat alba instar cerussae. In fornacem itaque conflatoriam coniice, et iniice ei ♀´ [*i.e. aeris*] florem et aeruginem rasam praeparatam dico, aut ♀` [*i.e. aes*] ustum valde corruptum, aut Chalcitidem, et [*blank*]*¹⁹ iniice, donec fiat influxibilis et imperforatus. Facile autem fiet. Hoc est illud plumbeum aes. Probes igitur si factum est sine umbrâ et si non fuerit factum, ne repraehendas aes, sed magis teipsum, quod non recte prae [65ᵛ]paraveris. Fac igitur sine umbrâ et laeviga, et iniice quae possunt flavefacere, et assa, donec flavum fiat, et iniice omnibus corporibus. ♀ [*i.e. aes*] enim omne tingit. Natura enim naturam vincit.

---

¹⁵ * μάρκαρον non est Graecum: ideoque aut μάρμαρον legendum aut μάργαρον ut significet unionem, ein peerle |*sic*|
¹⁶ * Graecum vocabulum hoc in loco mihi non placet
¹⁷ * ψιμμύθου non ψυμιθίου
⁸ * τεθρεωμένη falsum est
¹⁹ * κυανὸν

[13] Sulphure crudo collaeviga Sori et Chalcanthum. Sori autem est [*blank*]*²⁰ scabiosus, qui invenitur in misy. Hoc est viride et Chalcanthum vocant. Assato igitur illud mediis luminibus dies γ̄ [*i.e. tres*] donec fiat flavum pharmacum. Iniice ♀ʷ [*i.e. aeri*] aut ☾ʷ [*i.e. lunae/argento*] e nobis genito, et erit °θ [*i.e. sol/aurum*]. Hoc repone factum folium in acetum Chalchanti et misy et alumen, et sal Cappadocicum, et triticum, aut et quod et ipse putas, ad dies γ̄ [*i.e. tres*] aut ε̄ [*i.e. quinque*] aut ϛ̄ [*i.e. sex*] donec fiat aerugo*²¹ et tingas. Aurum enim facit chalcanthum, ἰὸν aurum. Natura naturâ delectatur.

[14] Chrysocollam Macedonum similem ἰῷ ᵛ♀ [*i.e. aeris*] praepara laevigans urinâ buculae donec evertatur. Natura enim intus occultatur. Si igitur fuerit eversa, tinge in oleum cicinum crebro incendens et tingens: deinde da assandam cum alumine, ante laevigans misy aut sulphure mortuo fac flavam, et intinge omne corpus auri. [15] O naturae naturarum architectatrices! O naturae omnino magnae mutationibus vincentes naturas!

O Natura supra naturam oblectans naturas! + Haec itaque sunt magnam habentia [66ʳ] Naturam. His Naturis non sunt aliae maiores in tincturis, non aequales, non inferiores.

Haec resoluta omnia operatur. Vos igitur tanquam meos collegas,*²² scio non diffisuros esse, sed potius admiraturos. Novistis enim materiae vim: iuniores vero valde periclitaturos et diffisuros scripturae, quia ipsi materiam ignorant, neque sciunt, quod medicorum pueri, quando*²³ salutare medicamentum velint conficere, non impetû quodam ancipiti hoc facere aggrediantur, se prius probantes quale sit calidum,²⁴ quale cum hoc congrediens medium perficiat temperamentum frigidum aut humidum, aut ubi sit passio, in congruum mediae mixturae, et sic offerant medicamentum ad ipsam sanitatem probe iudicatum. [16] Hi vero insubido et irrationali impetû animae remedium, et omnis laboris redemptionem comparare volentes, non sentiunt se laedi. Visi enim sibiipsis, nos non mysticum, sed fabulosum profiteri sermonem, nullam in formam inquirunt, exempli gratiâ, an hoc sit detersorium, an vero superiniiciendum, an hoc sit tinctorium, an applicandum, et quodam timore facit apparentiam, et si secundum apparentiam erit fugax, etiam e profundo fugiet, et si hoc igni resistit, applicatum quoque igni resistere facit. Veluti si sal ea quae sunt supra ♀ᵛ [*i.e. aes*] et quae intus*²⁵ detergit, et si dealbat [66ᵛ] quae sunt foris post abstersionem, quaeque sunt intus dealbat, et si quae sunt extra ♀θ [*i.e. aes aureum*] dealbat et detergit. ☾° [*i.e. mercurius*] etiam quae sunt intra dealbat, et si fugit extrinsecus, et ab internis fugiet. Si in his exercitati erunt adolescentes, non adversâ utebantur fortunâ, iudicio ad res gerendas impulsi.

---

²⁰ * ὡσκυανὸς
²¹ * ἰὸς vocabulum hoc in chymiâ certum requirit terminum, et inde natum verbale ἴωσις. Quod nisi mihi ex professo chymicus dixerit, qui communem hunc tenetur scire terminum, frustra ego laboravero N.B.
²² * vocabulum συμπροφῆται sic reddidi libenter: Germanice Mitpropheten
²³ * lege ὁπηνίκα coniunctim
²⁴ *Zuber writes in the margin*: Considerate agendum, et res prius diligenter cum omni distinctione vestiganda
²⁵ * ἐξάπαντος

Nesciunt enim naturarum contrarias passiones, quomodo una species decem subvertat. Olei enim gutta novit (potest) multam purpuram exstinguere, et exiguum sulphur multas comburere species. Haec itaque de medicamentis aridis, et quomodo Scripturae sit attendendum, dicta sunto.

Agite sane et iuscula deinceps dicamus. [17] Accipiens rhaponticum laeviga in vino Aminaeo austero et fac crassitiem ceroti: accipe folium Lunae, ut facias `ʘ [*i.e. solem/aurum*] operans [*blank*]*²⁶ et huius iterum gracilius auri pharmaci, et pone in novum vasculum camo constrictum undique, et sensim succende donec perventum sit ad medium: deinde pone folium in pharmaci reliquias, remitte vino ordinato, donec tipi appareat iusculum succosum,*²⁷ in hoc actutum repone folium, nequa refrigeratum sinas combibere. Deinde accipiens in fornacem coniice conflatoriam, et invenies Aurum.

Si vero rha sit tempore vetustum, admisit ipsi elydrii aequale, prius sale macerans pro [67ʳ] more. Elydrium enim habet cognationem cum rha. Natura delectatur naturâ.

[18] Accipe crocum Cilicium, remitte sanguinem flore croci ordinato succo vineae, fac iusculum pro more, tinge argentum e foliis, donec placuerit color.

Si vero folium erit aereum, melius. Praepurga autem plumbum pro more. Deinde proiiciens Aristolochiae herbae ʮ̃ β̃ [*i.e. duo partes*] et croci et elydrii duplum, fac crassitiem ceroti, et ungens folium operare primâ educatione, et miraberis. Crocum enim Cilicium eandem habet efficaciam cum argento vivo, quemadmodum casia cum cinnamomo. Natura vincit naturam.

[19] N.B.²⁸ Accipe nostrum plumbum influxibile per terram, latibulum, transitum, et alumen, coniice in fornacem conflatoriam acera et elutria (decapula) in pyriten et croci cicini et Orchomenii florem, et elydrium, et crocini unguenti retrimentum, et Aristo-lochiam, laeviga aceto mordacissimo, et iusculum fac pro more, et [*blank*]*²⁹ plumbum sine combibere, et invenies Aurum `ʘ [*i.e. solem/ aurum*]. Habeat autem compositio parum quoque sulphuris crudi. Natura naturam regit.

[20] Haec est [*blank*]*³⁰ quam sacerdotibus in Aegypto demonstravit, usque ad naturalia haec est auri conficiendi materia.³¹ Ne vero [67ᵛ] miremini, si una species tantum perficit Mysterium. Annon videtis quam multa medicamenta vix aliquo tempore sectionem e ferro prohibeant? Stercus autem humanum non facit hoc tempore. Et multa allata (oblata) medicamenta per cauteris saepe nihil perficient, sola vero calx praeparata sanat affectum: et oculorum morbus saepe variis oblatis molestia est, et novit laedere. Spina*³² vero alba planta ad omnem talem

---

²⁶ *The asterisk does not refer to any comment in the margin* (ὀνυχοπαˣ **M**)

²⁷ * χυλώδης, non χηλώδης

²⁸ N.B. In Graeco quoque textû sequentia verba obelisco sunt notata. Nemo igitur miretur, si ego a sensû longinscule deflexi

²⁹ *The asterisk does not refer to any comment in the margin* (τῇ ῥᾷ **M**)

³⁰ * παμμένους

³¹ *Zuber writes in the margin*: Materia una vel lapis unus

³² * forte melius si vocabulum Graecum reliquero ῥάμνος ᴂ spinosus frutex

faciens passionem non frustratur. Oportet igitur despicere vanitatem et intempesti-
vam illam materiam, et solis uti naturalibus. Nunc vero etiam ex hoc iudicate sine
praedictis naturis, quis aliquando sit operatus? Si vero sine his nihil licet facere,
cur amamus multâ praeditam materiâ imaginationem? Quid nobis multarum spe-
cierum in unum et idem concursus conducit, quum una sit natura quae omnia
vincat?*[33] Sciamus nempe etiam formarum compositionem in Argenti confectione.

[33] * νικώσης lege pro κινώσης

De obscurâ Confectione
(Z, fols. 67ᵛ22–69ᵛ21)

[1] Argentum vivum ab Arsenico aut sandarachâ, aut quod ipse censes, fige pro more, et iniice ♀ω [*i.e. aeri*] ferro sulphurato, et candescet.¹ Hoc ipsum facit etiam Magnesia dealbata, et Ar[68ʳ]senicum eversum, et Cadmea assa et sandaracha cruda, et pyrites dealbatus, et cerussa cum sulphure assata: ferrum autem solves Magnesiam iniiciens aut sulphuris ⅂ [*i.e. dimidium*] aut magnetis parum. Magnes enim habet cognationem cum ♂ [*i.e. ferro*]. Natura delectatur naturâ.

[2] Accipiens praescriptam nubem [☽ *above the line*] coque oleo cicino aut raphanino, prius abstergens parum alumen. Deinde accipiens stannum*² purga sulphure pro more, aut pyrete [*pyrite?* θ *above the line*], aut quod ipse censes, et transvasa in nubem et fac mixturam, da assandam luminibus*³ eximiis, et invenies cerussae simile. Pharmacum hoc dealbat omne corpus, admisce vero ipsi in iniectonibus terram, latibulum, aut asteriten, aut spumam Lunae, aut quod ipse putas.

Spuma enim Lunae ☽ [*i.e. mercurio/argentum vivum?*] mixtum omne corpus dealbat. Natura naturam desiderat [*sic*]⁴ vincit.

[3] Magnesiam albam, dealbabis autem ipsam muriâ et alumine ✳ [*i.e. alumine fissili*] in aquâ marinâ, aut succo citrino (dico) aut sulphuris favillâ. Fumus enim sulphuris quum sit albus, omnia dealbat. Nonnulli autem dicunt fumum [*blank*]*⁵ dealbare ipsam, admisce ipsi post dealbationem, et faeculae*⁶ tantandem, ut valde fiat alba, et recipiens ♀ᵛ [*i.e. aeris*], orychalchi subalbidi dico, ⌐σ δ [*i.e. quatuor uncias*] in fornacem coniice conflatoriam, iniiciens infra parum stanni praepurgati ⌐σ ᾱ [*i.e. unam unciam*] subinde manibus movens, donec substantiae intra sese nubant, erit ruptum. Iniice igitur medicamenti albi ⅔ [*i.e. dimidium*] et [68ᵛ] erit primum. Magnesia enim dealbata non sinit rumpi corpora, neque umbram plumbi inferri. Natura naturam regit.

[4] Accipiens sulphur album, dealbes illud autem urinâ laevigans in sole aut alumine et muriâ salsâ [*blank*]*⁷ candidissimum, laeviga cum sandarachâ, aut

---

Zuber's comments in the margin:

¹ *Zuber writes in the margin:* Praeparatio Argenti secundum veritatem

² * κασσίτερον vel καττίτερον Attice legendum, non κασίτηρον

³ * εἰλικτοῖς forte ἐκλεκτοῖς

⁴ *Zuber probably intended to delete* desiderat *by underlining the verb.*

⁵ * κωβαθίων

⁶ * φέκλης lege, non σφέκλης

⁷ * ἀνθιγηπάνυ, iurarim vel daemones exhorrescere audito hoc vocabulo

urinâ buculae per sex dies, donec fiat medicamentum marmori simile, et si fuerit factum, magnum est Mysterium. Plumbum enim dealbat, emollit ♂ [*i.e. ferrum*], imperforatum facit stannum, Saturnum influxibilem, infragiles facit substantias, inevitabiles tincturas. Nam sulphur cum sulphure mixtum, Divinas facit substantias, multam inter se habentia*[8] affinitatem. Delectantur enim naturae naturis.

[5] Si vero dealbatum Lithargyrum ☉ [*i.e. laeviga*] cum sulphure aut Cadmeâ, aut Arsenico, aut Pyrite, aut oxymelite,*[9] ut non amplius fluat: assa igitur ipsum splendidioribus luminibus muniens vas. Habeat autem compositio calcis assae madefactae aceto dies ⊤ [*i.e. tres*] ut fiat detersior. Iniice igitur ipsi album factum magis quam cerussam. Fit autem frequenter et album, si lumina augeantur. Sed si fiat flavum, non commodat [*blank*]*[10] dealbare enim vis corpora. Combure igitur ipsum modo quodam et porportione [*proportione*], et iniice omni corpori opus habenti dealbatione. Lithargyrum enim si fiat influxibile, non amplius erit plumbum. [69ʳ] Fit autem citra negotium: celeriter enim in multa transfertur Saturni natura. Naturae enim naturas vincunt.

[6] Accipiens crocum Cilicium ☉ [*i.e. laeviga*] et maris aut muriae, et fac iusculum, in quod ignis demerge □ ᵅ□ ♀ᵘ υ ☿ [*i.e. folia aeris, plumbi, ferri*] donec tibi placeat, fiunt autem alba: deinde accipe medicamenti ⾕ [*i.e. dimidium*] et unâ laeviga sandarachâ, aut arsenico albo, aut sulphure crudo, aut quod ipse censes, et fac ceroti crassitiem aurum □ [*i.e. folium*], et pone in novum vasculum obstructum undique pro more pone in [*blank*: πρισματοκαύστην **M**] per totam diem: deinde efferens (eruens) pone in mundum iusculum, et erit candidum candidissimum plumbum. Operare reliquum ut Artifex. Cilicium enim Crocum mari quidem dealbat, vino autem flavefacit. Natura delectatur naturâ.

[7] Recipe album lithargyrum et ipsum laeviga cum foliis daphnes, Cimoliae, mellis et sandarachae albae ♋ [*i.e. fac*] strigmentum indigens medicamento ⾕ [*i.e. dimidium*] et succende pro more, demerge in medicamenti reliquias, resolvens aquâ cineris populi albae lignorum.

Quae enim substantiâ carent mixtiones eleganter (probe) operantur sine igne ♋ [*i.e. fac*] ipsa iusculis igne ferventia. Natura enim naturam vincit.

[8] Recipiens praescriptam nubem contere ipsâ [69ᵛ] alumen et misy, acetoque undique purgans iniice ipsi exiguam albam Cadmeam, aut Magnesiam, aut calcem, ut fiat corpus a corpore [*blank*: ☉ (*i.e. τρῖψον*) **M**] cum melle, albissimum fac iusculum, in quod ignem non demergens, quemvis, sinas infra et fiet. Habeat autem compositio etiam exiguum sulphur crudum, ut intus penetret medicamentum. Natura naturam regit.

[9] Accipe arsenici ℔ ᾱ [*i.e. unam unciam*] et ℔ ⾕ [*i.e. nitri unciae dimidium*] et corticis foliorum Perseae tenerorum ℔ β̄ [*i.e. duas uncias*] et salis ⾕ [*i.e. dimidium*] et Sycamini (mori) succum ℔ ᾱ [*i.e. unam unciam*] aequaliter fissi, laeviga simul in

---

⁸   * neutrum est positum in Graeco, malim esse foemininum
⁹   * desideratur Graecum verbum
¹⁰ * σκυκῶν non est Graecum verbum

aceto aut urinâ, aut calce liquidâ, donec fiat iusculum, in hoc umbrosa igne demerge
□ ᵅ□ [*i.e. folia*] et obumbrabis. Natura naturam regit.

[10] Referte omnia auro et argento utilia. Nihil excedit, nihil deficit, praeter nubis et aquae evectionem. Sed haec volens tacui quia insunt copiose etiam in aliis meis scripturis. Valete in hâc Scripturâ.

Synesii philosophi ad Dioscorum in librum Democriti tanquam in scholiis (Z, fols. 70ʳ1–76ʳ19)

Dioscoro Sacerdoti magni Sarapidis Aelexandrini [*sic*], Deo consentiente, Synesius Philosophus salutem. [1] Epistolam meam a te repraehensam esse, de Divini Democriti Libro, non tuli aegrius, sed multo studio et labore meipsum torquens excurri ad te, in quo propositum est nobis dicere, quisnam sit ille vir Philosophus Democritus, qui venit ab Abderitanis, Physicus, et qui omnia naturalia inquisivit, et ea quae secundum naturam sunt, conscripsit.

Est autem Abdera urbs Thraciae. Factus est autem vir ille doctissimus, qui venit peregrinatus, et in Aegypto sacris initiatus est a magno Ostane in Templo Memphitico unâ cum omnibus Sacerdotibus. Ab ipso accipiens occasiones conscripsit libros δ̄ (quatuor) tinctorios de ☉ (Sole) et ☽ (Luna) et lapidibus et purpurâ. Dico sane occasiones nactus conscripsit de magno Ostano. Ille enim erat primus qui scripsit quod Natura delectetur Naturâ, et Natura Naturam regat, et Natura Naturam vincat, etc.

[2] Sed nobis necessarium est Philosophi investigare et discere quae mens sit, qualis ordo consequentiae in ipso. Quod igitur duos fecerit Catalogos, [70ᵛ] nobis factum est planum, albi et flavi, et primum quidem solida delegit (conscripsit), deinde iuscula, h. e. [*hoc est*] humida, nullo licet horum assumpto in arte. Ipse enim testatur dicens de magno Ostane, quod hic vir non usus fuerit Aegyptiorum immissionibus, neque assationibus, sed extrinsecus iunxerit substantias, et fulvum intromiserit medicamentum.[1] Dixit autem quod Persae in more haberent hoc facere, quod autem dicit, hoc est, quod nisi attenuaveris substantias et resolveris et aqueam in naturam converteris, nihil sis effecturus. [3] Veniamus igitur ad viri huius dictionem, et audiamus ipsum dicentem: dicitur autem et rhaponticum: ecce tanta viri observatio, ab herbis obscure incepit, ut indicet florem. Herbae enim sunt floriferae. Dixit insuper rhaponticum, quod pontus a fluminibus perfundatur, et omnes fluvii in ipsum defluant. Manifestum igitur nobis faciens indicat

Zuber's comments in the margin:
[1] *Zuber writes in the margin*: Intellego particularia per imperfectarum praeparantia: sed lapide extrinsecus superiecto fecit aurum

conversionem in aqueam naturam et offuscationem et attenuationem corporum, vel substantiarum.

[4] Dioscorus dixit, et quomodo dixit quod iuramenta nobis iniunxerit ne cui aperte contraderemus? Bene dixit, nulli, h. e. [*hoc est*] nulli profanorum.[2]

Nulli enim (sc. vocula) non de omni praedicatur. Ipse enim de initiatis*[3] et exercitatis et cordatis*[4] dixit. [5] Vide enim quid in immissione confectionis Auri dixerit: ☽ (Mercurio) a ☉ (Sole) ♍ chrysocolla. Dioscorus, et horum est indigentia (utilitas). Synesius, nonne Dioscore? Dioscorus, sed opus est aliquo, audivisti, et rursus audi: [71ʳ] resolutio est corporum ut resolvas ipsa et aquas eorum vice facias, ut fluant, et caligentur, et attenuentur. Hoc autem vocatur ♨✕ [*i.e. aqua sulphuris*] et ☽ [*i.e. mercurius*] et ♍ [*i.e. chrysocolla*] et ✕ [*i.e. sulphur*] mortuum et quaecumque alia sunt nomina. Dealbatio*[5] enim ustio est, et flavefactio ignis exsuscitatio. Ipsa enim seipsa urunt,*[6] et ipsa iterum seipsa excitant. Philosophus autem multis ipsa nominavit nominibus, aliquando enim singulariter, aliquando vero pluraliter, ut exerceat nos et videat an simus intelligentes. Dixit enim sic paulatim discendens: si fueris attentus et feceris, sicut scriptum est, eris beatus. Vinces etenim viâ compendiariâ paupertatem incurabilem morbum. Reiiciens itaque et avellens nos a vano errore, fac, ut liberemur a multae materiae phantasiâ.[7]

Advertit autem in libri iniectione quid dixerit. Eram*[8] sane et ego in Aegypto afferens naturalia, ut multam illam materiam despiceretis. Naturalia autem dixit solida corpora: nisi enim illa resolvantur et rursus figantur, nihil ad finem ex hâc re perduceris. [6] Et ut intelligamus quod e solidis sumantur aquae, h. e. [*hoc est*] flos, vide quomodo dixerit, et quae in iusculis, crocum cilicium, Aristolochia, etc. Flores nominando, planum nobis fecit quod e solidis aquae sumantur, et ut nobis persuadeat, haec se ita habere, postquam dixit urinam incorruptam (virgineam), dixit etiam aquam calcis, [71ᵛ] et aquam cineris crambini et ♨ [*i.e. aquam*] faeculae et ♨✳ [*i.e. aquam aluminis*] et in fine dixit, canis lac. Et manifestum nobis est quod ex communi relatum: attulit enim quae corpora solvunt, ♨ [*i.e. aquam*] nitri et ♨ [*i.e. aquam*] faeculae. Et vide quomodo dixerit. Haec est Materia confectionis Auri: haec sunt quae alterant: extra haec enim nihil est tutum. Si igitur fueris attentus, et feceris sicut scriptum est, eris Beatus. [7] Dioscorus. Et quomodo possum intelligere, Philosophe, Methodum? Abs te volo discere. Si enim sequar Dicta, nihil ab illis lucri fecero. Audi Dioscore, et vide, quid dicat. Everte ipsorum Naturam. Natura enim intus delitescit. O Synesi, qualem dicit eversionem? Corporum dicit eversionem. Et quomodo ipsam evertam? Aut quomodo educam Naturam foras? Acue tuam mentem, Dioscore, et attende quomodo dicat. Si igitur

---

2 *Zuber writes in the margin*: Ars secrete tenenda
3 * lege μεμυημένων, non μεμνημένων
4 * τὸν νοῦν, non νοῦν
5 * sic Hermogenes in Turba vel potius Lucas: comburere est dealbare, et rubrum facere, est vivificare
6 * καίουσι non καίουσα
7 *Zuber writes in the margin*: Materia una
8 * ἤμην |*sic*| potius quam ἤνον. In Flamello pag. 168 sic citatur hoc Democriti. Ego autem venio in Aegypto naturalia ferens ut materiam superfluam contemnatis

praeparaveris ut oportet, educis naturam foras, *terra,⁹ latibulum, asterites, cadmea alba, etc.

Ecce quanta viri animadversio? Quomodo omnia alba obscure innuit, ut ostendat Dealbationem.

Quod dicit igitur, Dioscore, tale est. Proiicere corpora cum ☽ [*i.e. mercurio*], et lima ad exilitatem, et resume ☽ [*i.e. mercurium*] alteram. Omnia enim ☽ [*i.e. mercurius*] in seipsam trahit, et sinas coqui tres aut quatuor dies, et proiice ipsam in doliolum ad calidos cineres non habentes ignem vehementem, [72ʳ] sed ad mansuetiorem cineris calorem, velut est cerotatis.*¹⁰ Hâc igitur ignis eruptione applicat doliolo vitreum instrumentum habens mamillam superiora advertens: ponantur in capita, et quae inde resurgit Aqua, recipe, habe, et in putredinem verte. Hoc dicitur ⚖✕ [*i.e. aqua sulphuris*]. Haec est Eversio, hâc eductione extrahis Naturam intus absconditam. Haec vocatur Solutio Corporum. Quod quum putruerit, vocatur acetum, et Vinum Aminaeum, et similia. [8] Et ut admireris Viri huius sapientiam, vide quomodo duos fecerit Catalogos confectionis Auri, et confectionis ☽ [*i.e. argenti*], et rursus duo iuscula, unum quidem in flavo (rubeo), et unum in albo, hoc est, ☉ [*i.e. sole/auro*] et ☾ [*i.e. luna/argento*], et vocavit ☉ [*i.e. solis/auri*] catalogum ☉ [*i.e. solis/auri*] confectionem, ☾ [*i.e. lunae/argenti*] vero confectionem Argenti. Admodum bene dixisti, Philosophe Synesi: et Artis est certique modi dealbare aut flavefacere [rubefacere *above the line*]. Synesius. Magis dealbare. Dioscorus. Quare autem flavefactionem [rubefactionem *above the line*] dixit primum? Quia pluris aestimatum est Aurum quam Argentum. Et sic oportet facere Synesi. [*blank*]*¹¹ Dioscore. Sed ut exerceremus Animum nostrum et Mentes, sic ordinata sunt. Audi ipsum dicentem: tanquam intelligentibus*¹² vobiscum conversor, exercens vestrum Animum. Quod si volueris accurate scire, ausculta duos Catalogos [72ᵛ], quod prae omnibus ☽ [*i.e. mercurius*] ordinata fuerit et in flavo, h. e. [*hoc est*] Auro, et in albo, h. e. [*hoc est*] Argento, et in Auro quidem dixit ☽ [*i.e. mercurius*] a ☉ [*i.e. cinnabari*], in albo vero dixit*¹³ ☽ [*i.e. mercurius*] a ☊ [*i.e. arrhenici*] erant [☊ *above the line*], et deinceps. [9] Dioscorus dixit, praestans igitur est ☽ [*i.e. mercurius*]. Synesius: nae, praestans est quum sit una. Dioscorus : et si una est, quomodo erit praestans? Synesius: certe praestans est, et maximam habet potentiam.

Non audivisti Mercurium dicentem ceram albam et ceram flavam. Dioscorus: profecto audivi: quod autem cupio discere, Synesi, hoc me laboris genus doce.¹⁴ Omnino haec omnium formas recipit? Intellexisti Dioscore. Quemadmodum enim cera qualem assumit colorem recipit, sic etiam ☽ [*i.e. mercurius*] Philosophe, haec omnia dealbat, omnium animas trahit, et coquit et accersit (assumit). Partibus igitur praedita instrumentalibus, et in seipsâ habens humores, omnino etiam

---

⁹   * videmus esse hic defectus
¹⁰ * forte cera liquefacta
¹¹ *The asterisk does not refer to any comment in the margin* (οὐ M)
¹² * νοήμοσιν ὑμῖν, lege
¹³ * ☽ a ☊ puto intelligi, aquam a ☾
¹⁴ *Zuber writes in the margin*: vide hoc dictum Democriti in Flamello pag. 168

putridinem sustinens, mutat omnes colores, et fit substantialis (permanet) ipsis sub-stantiâ carentibus (non permanentibus). Magis vero ipsâ substantiam non habente (ipsâ non permanente), tunc etiam fanatica fit (continetur) praeparationibus per corpora et materias ipsorum (subiecta ipsi addita). [10] Dioscorus: et qualia sunt ista corpora et materiae ipsorum? Synesius: quatuor corporum compages et eorum cognata. Dioscorus: et qualia sunt eorum cognata? Synesius: audivisti, quod materiae ipsorum animae sint illorum? [73^r] Synesius: sane. Quemadmodum enim Faber, si acceperit lignum, aut currum aut aliquid aliud fabricat, sic etiam Ars illa, ô Philosophe, postquam secuit ista. Audi, Dioscore, qui lapides scalpit, radit lapidem aut secat, ut idoneus fiat ad suum usum.[15] Sic etiam Faber lignum findit et radit (dolat), ut inde fiat sella aut currus, et Artifex nihil demeretur (de se exhibet) nisi solam formam. Nihil enim aliud est praeter lignum. Similiter plumbum fit statua aut aliud vas Artificis, solam ipsi formam conciliantis. Ita etiam ☽ (mercurio) a nobis artificiose disposita, ut dictum est, omnem formam ipsi (sibi) assumit et compedibus adstricta corpore quatuor serierum, firma et sine persecutione manet, vincens et victa. Propterea etiam Pibechius multum mandati (affinitatis) habere dicebat. [11] Dioscorus: scite solvisti Philosophe, et me docuisti. Volo igitur ad viri huius properare dictionem, et statim cognoscere quae ad ipso oblique tradita tanquam dicta. ☽ [*i.e. mercurius*] a ☉ [*i.e. cinnabari*]: omnis igitur ☽ [*i.e. mercurius*] a corporibus fit. Hic autem dicebat [☉ *above the line*], scire se, quod esset a ☉ [*i.e. cinnabari*]. Nimirum ☉ [*i.e. cinnabaris*] ☽ [*i.e. mercurius*] flava est, haec vero alba ☽ [*i.e. mercurius*] potentiâ vero flavescit. Dioscorus: nonne igitur dixit Philosophus? O Naturae coelestes Naturarum Architectatrices? Mutationibus vincentes Naturas? Synesius: profecto. Propterea dixit: nisi fuerit eversa, fieri non potest, ut quod [73^v] expectatur fiat, et frustra laborant qui mate-riam investigant, et corporum naturas μ͞γ (Magnesiam)[16] non inquirunt. Licet enim Poetis et Scriptoribus easdem dictiones aliter atque aliter formare. Corpus igitur μ͞γ (Magnesiam) dixit, h. e. [*hoc est*] mixturam substantiarum: et propterea paulatim descendens (procedens) inquit: in immissione confectionis ♉ [*i.e. solis/auri*] recipiens ☽ [*i.e. mercurius*] fige Magnesiae (magnesiam) cum corpore. [12] Dioscorus: ecce igitur pluris aestimatum est ☽ (humidum seu aqua). Synesius: etiam. Propter hoc enim omne divellitur, et rursus apponitur, et descensum omnis praeparationis fabre-fecit. ♃ (chrysocolla) quod est ranunculus qui in lapidibus viridibus inveniretur? Necessarium igitur nobis est quaerere, oportet igitur scire primum quae a viridibus sint coloribus. Eia igitur tanquam *ab[17] homine dicamus. Pluris enim aestimatus est homo, quam omnia animalia terrae. Dicimus igitur pallescentem hunc viridem fuisse factum, nimirum quod pallor formam immutat, quod pariter in deaurando fit, quinimo multo hoc magis. h. e. [*hoc est*] est cortex citrii, palloris forma, hoc etiam paulatim descendens dixit Arsenicum flavum, ut ostenderet palloris

---

[15] *Zuber writes in the margin*: Omnis Ars introducit formam suae intentionis. Ita Alchymista dat Mercurio formam, fixam et figentem et tinctam et tingentem

[16] *Zuber writes in the margin*: Magnesia quid

[17] * malim περὶ quam ἀπό

formam. [13] Ut autem videas, cum quam multâ observatione hoc particulatim dixerit, attende quomodo dicat. ☽ [*i.e. mercurius*] a ☉ [*i.e. cinnabari*] corpus Magnesiae: deinde infert ♉ [*i.e. chrysocollam*] Claudianum Arsenicum nomine: rursus adduxit Arsenicum, ut videat hoc a Theriacis.*¹⁸ Et post [74ʳ] Claudianum Arsenicum flavum, duo flava opponens o͞o [*i.e. nomina*], duo foeminea, duo deinde masculea. Oportet igitur nos investigare et cognoscere, quidnam hoc sit, ut ego motus sum, Dioscore, hic putrefacit `β [*i.e. solem/aurum*], deinde resumit Cadmeam, deinde androdamantem: et androdamas et Cadmea sunt sicca: et ostendit siccitatem corporum, et ut hoc faceret planam intulit ✳ [*i.e. alumen*] expressam (expurgatam). Vide quanta viri sit Sapientia, ut etiam cordati intelligant quomodo ipsos docuerit, dicens ✳ [*i.e. alumen*] expressam.

Fortassis et hoc profanis*¹⁹ etiam persuadere debuit: quo autem firmior tibi fieret, statim adduxit sulphur crudum, quod est omnino incombustum sulphur. h. e. [*hoc est*] siccatas species infra, quae sunt corpora unum facta. Sulphur incombustum vocavit, et deinde infertur pyrites dimissus, nullum aliorum inde confirmans (corroborans). Hoc verum est, quod quae sicca manent et haec dividens infert [*blank*]*²⁰ ponticam transiens a siccis ad humida. Sinopin dixit, sed propter ponticam. Nisi enim adiectum esset Ponticam, non cognosceretur. Corroboratus autem intulit aquam sulphuris mundam, a sulphure solo divinam. [14] Dioscorus: bene solvisti, Philosophe, sed adverte quomodo dixerit, si absolute per calcem. Synesius: ô Dioscore, non advertis Animum. Calx est alba, et aqua ex eâ alba est et spissans, et sulphur exhalans dealbat. Declarationis igitur gratiâ statim adduxit sulphuris favillam. Nonne nobis haec manifesta facit? [74ᵛ] Dioscorus: profecto, bene dixisti.*²¹ Et post hoc sori flavum et Chalcanthum flavum et ☉ [*i.e. cinnabaris*]. Synesius: sori et flavum Chalcanthum quomodo scis esse viridia? Innuens obscure plumbi dealbationem aut investigationem, magis vero universi a coloribus hoc dixit.*²² Et iterum confirmatus ad finem adduxit. Post albationem enim Veneni*²³ quae vocatur [*blank*]*²⁴ quando immissio sit humidorum, stabilis quoque gignitur citrinatio. Et vere viri huius animus expers invidiae hîc est ostensus.

[15] Etenim vide quam celeriter connexuerit determinationem utendo dicendoque. Quae vero in iusculis, sunt haec: crocum Cilicium, Aristolochia, cicini flos, Anagallidos flos coeruleum habentis florem. Quid amplius potuisset dicere aut recensere, ut persuaderet nostris cordibus, nisi dixisset florem Anagallidos?

Admiratus non tantum mihi Anagallidem, sed etiam florem dixit. Significavit enim nobis Anagallidem reducere aquam: per florem enim Animas eorum reducere h. e. [*hoc est*] spiritus. Nisi enim haec ita se habeant, nihil est firmum, et frustra infeliciterque agunt miseri isti, in mare cratibus vimineis natantes, multis et anxiosis

---

¹⁸ * θηριακῶν non θηρυκῶν. Sic ego legendum censeo
¹⁹ * non initiatis
²⁰ * εἰσίνωπιν, non est Graecum: puto σίνωπιν legendum
²¹ * non est integer sensus
²² * sensus iste omnes meos sensus dementat
²³ * redit vocabulum τοῦ ἰοῦ, cuius significatio est varia. Chymicum esse necesse est, qui apte explicet
²⁴ *The asterisk does not refer to any comment in the margin* (ἐξίωσις M)

laboribus involuti, inconsiderata*²⁵ proponendo, erunt. [16] Et quid rursus minime invidus Philosophus et Praeceptor introduxit rhaponticum? Vide liberalitatem viri. ῥᾶ ipsum dixit, et quo nobis persuaderet, introduxit ponticum. Ecquis enim virorum Philosophorum nescit, [75ʳ] quod pontus sit defluxio, undique e fluminibus circumluta? Vere, Synesi, assequutus es, et exhalarâsti hodie meum Animum. Non enim modica sunt haec. Hoc vero te adhortor, ut in pluribus me edoceas. Quare superius dixit Chalcanthum flavum? Hîc vero indeterminate cum Chalcantho caeruleo adduxit. Sed haec Dioscore, indicant flores. Virides enim sunt. Postquam igitur adscendens aqua indiget fixione, statim adduxit. Coma (cacumen) spinae,*²⁶ deinde inducit urinam incorruptam et ⏚ [*i.e. aquam*] calcis et ⏚ 𝈁 [*i.e. aquam arrhenici*] et ✕ [*i.e. (aquam) sulphuris*]. Vide quomodo crambes cinereae et ⏚ ✳ [*i.e. aqua aluminis*] et ⏚ Ⲛ [*i.e. aqua nitri*] et ⏚ 𝈁 [*i.e. aqua arrhenici*] et ✕ [*i.e. (aqua) sulphuris*]. Vide quomodo omnia solventia et ad disiiciendum idonea protulerit? Nimirum docens nos resolutionem corporum. [17] Et probe dixisti. Et quomodo in fine dixit, canis lac? Ut tibi ostendat, quod a communi universorum accipiatur. Vere intellexisti, Dioscore, ausculta vero quomodo dicat. Haec materia est confectionis Auri: neque enim asininum lac, neque caninum lac potest igni ferventi resistere. Asininum enim lac quando seposueris in loco per sufficientes dies evanescit. Quid autem hoc est dicere, haec sunt quae alterant materiam? Haec etiam igni faciunt resistere, ipsis rebus exsistentibus fugientibus, et id quod extra haec est, non est sine periculo: ut intelligant miseri, quod haec [75ᵛ] sint vera. Verum enim vero rursus eum audi quid dicat: et infert, si fueris attentus, et feceris, sicut scriptum est: pro quo, si fueris sapiens, et diiudicaveris ratiocinium, ut oportet uti, eris Beatus.²⁷ Et quid alibi dixit? Cordatis vobis dico. Oportet igitur mentes nostras exercere, et non decipi, ut incurabilem paupertatis morbum effugiamus, et ab ipsâ non vincamur, et in difficilem elapsi egestatem infeliciter agamus stulti reputati: animos excitare debemus, et acutum habere intellectum. [18] Cur autem infert iniicere,*²⁸ non dicit propter prolegomena, sed ea quae a mente. Sed iterum dicit: aliquando ☉ [*i.e. solem/aurum*] propter ☉ corallium [*i.e. auri corallium*], aliquando autem propter ☉ [*i.e. solem/aurum*], aliquando plumbum propter Aurum, aliquando verum ♄ [*i.e. plumbum*] aut ♀ [*i.e. plumbum album*] propter μ̃♀ [*i.e. molybdochalcum*]. Ecce ipse sub gradûs Artis nos reduxit, et ne pedem ponentes in foveam incidamus eorum inscitiae. Multa enim est in hoc viro sapientia, postquam enim dixit ipse, haec est Materia confectionis Auri, esto sane dictum. Infert dicens: eia deinceps quoque rationem conficiendi ☾ [*i.e. lunae/argenti*] liberaliter tradamus, ut ostendat nobis, duas esse Operationes, quod etiam confectio ☾ [*i.e. lunae/argenti*] pluris sit aestimata, et priores partes obtineat, et absque eâ nihil fiat. [19] Audi eum rursus hîc dicentem ☽ [*i.e. mercurium*] a 𝈁 [*i.e. arrhenici*] aut ✕ [*i.e. sulphure*] aut cerussâ aut Magnesiâ aut stibio Italico, et supra quidem in confectione Auri,

---

²⁵ * ἀνόητα lege, non ἀνόνητα
²⁶ * perplexus sensus et defectus apparet. Prorsus enim non cohaerent
²⁷ *Zuber writes in the margin*: Lectio et meditatio et exploratio per laborem
²⁸ * defectus est in Graeco

☽ [*i.e. mercurium*] a ☉ [*i.e. cinnabari*], [76ʳ] hîc autem ☽ [*i.e. mercurium*] a ⟁ [*i.e. arrhenici*] aut cerussâ, etc. Et quomodo contingit ☽ [*i.e. mercurium*] fieri cerussam? Sed non *a²⁹ cerussâ ☽ [*i.e. mercurium*] dicit, ut acciperemus, *sed³⁰ dealbationem corporum. Deinde igitur reflexionem obscure innuens dixit: hîc enim omnia alba dixit, illic vero flava, ut intelligamus, vide quomodo dixerit, corpus Magnesiae solum aut stibii Italici. Et haec quidem brevibus dixisse sufficiat. Oportet autem Mentem praeexercere, ut dignoscamus Naturae operationes et de iis quae studio assequi volumus, Divinâ gratiâ cooperante.

Oportet igitur nos cognoscere, sepelire nos oportet formas primum, et aggeribus similes colore perficere (redigere) in unum colorem: et haec duo quidem ☽ ☽ [*i.e. mercurios*] in Argentum vivum vertunt et in putredinem separantur. Juvante autem Deo, incipiam commentari.

De dealbatione
(Z, fol. 123ʳ 3–13)

[1] Volo vos scire, quod Caput sit omnium Dealbatio. Post dealbationem vero statim flavefit [*flavescit*] perfectum Mysterium. [2] Dealbatio est ustio: ustio autem resuscitatio. Ipsa enim seipsa urunt et revivificant, et ipsa secumipsis coeunt et uterum gestant, et parit quaesitum animal secundum Philosophos. [3] Si dealbaveris, facile tinges: si vero etiam laevigaveris aut cinnabari tinxeris, eris beatus, ô Dioscore. Hoc enim est quod liberat a paupertate morbo incurabili.

---

²⁹ * malim περὶ pro ἀπὸ
³⁰ * defectum animadverto

Notes on the Greek texts

Ps.-Democritus' *On Natural and Secret Questions (PM)*

1] The first two paragraphs concern purple dyeing. The text of their incipit is problematic, as may also be inferred from Zuber's translation (see *supra*, p. 188). It seems possible nevertheless to recognize in ll. 1–3 the description of a dyeing technique involving three different ingredients (which were mixed and boiled together), although we cannot distinguish the different phases of the procedure, namely "washing" (πλύσις), "staining" (στῦψις), and "dyeing" proper [βαφή; see Robert J. Forbes, *Studies in Ancient Technology*, 2nd ed., 9 vols. (Leiden: Brill, 1965–1972), vol. 4, 133–4]:

(a) The term 'purple' (πορφύρα) must be understood as a general term referring to any possible substitute for the true purple (i.e. the purple extracted from *murex* or *buccinum* sea-snails). More than 12,000 shellfish were needed in order to prepare 1.4 grams (about one-twentieth of an ounce) of the dyestuff, which was consequently very expensive (see Halleux, *Papyrus de Leyde*, 225). Our text, in contrast, prescribes using quite a substantial amount of 'purple' (1 *libra*; see Halleux, *Papyrus de Leyde*, 20–1), which is unlikely to refer to the true purple. Likewise, *PLeid.X.* and *PHolm.* usually indicate similar quantities of purple substitutes (*PHolm.* 99; 101).

(b) Iron and its by-products (iron slag, iron acetates), and alum, sometimes mixed with acid adjuvants (e.g. vinegar), were often used as mordants and staining agents (see *PLeid.X* 96, 99; *PHolm.* 98, 103; Halleux, *Papyrus de Leyde*, 44–5).

(c) Sheep's urine was usually employed both for washing wool and for diluting dyeing ingredients, especially woad and alkanet (although camel's urine was also highly valued; see *PHolm.* 93).

In addition, the text does not specify dipping wool into the decoction. At l. 3, the author seems to start describing a second procedure, based on seaweed macerated in water. Unless we take the term 'water' (ὕδωρ) to refer to the preceding solution, this procedure seems to be separate from the first one and based on two different dyestuffs. Despite the adverb εἶτα (l. 3, 'then'), which seems to link the incipit to the following part, we cannot rule out the possibility that the first three lines represent what is still extant of a different recipe.

2] *Seaweed* (*bryon thalassion*). According to Theophrastus (*HP* IV 6,2), the term *bryon* (βρύον) refers to an alga with "a leaf green in colour, but broad and not unlike lettuce leaves" [transl. by Arthur Hort, *Theophrastus, Enquiry into Plants and Minors Works on Odours and Weather Signs*, 2 vols. (London and New York: Loeb, 1916), vol. 1, 333]. Scholars identify it either with the common sea lettuce (*Ulva lactuca* L.; see *CGL* V 592,3: *broia ulva marina*) or with any kind of alga belonging to the *Enteromorphae* genus (Amigues, *Théophraste, Recherches sur les plantes*, vol. 2, 242 n. 12). However, neither of these seaweeds has dyeing properties and they are usually mentioned by ancient sources because of their astringent properties (Diosc. IV 98; Plin. *NH* XXXII 36). Consequently, since ps.-Democritus was clearly referring to a dyestuff expected to penetrate wool, scholars have proposed two different solutions: (a) a dyeing lichen, namely the 'orchella weed' or orchil [*Roccella tinctoria* DC; see *CAAG* III 43 n. 1; the *Roccella* genre includes several lichens with dyeing properties, see Dominique Cardon, *Natural Dyes. Sources, Tradition, Technology and Science* (London: Archetype, 2007), 495–505], which usually occurs on wet rocks that are exposed to rain or are close to the sea (thus *thalassion* in the Greek); or (b) a dyeing seaweed, such as *Rytiphlaea tinctoria* (Clemente) C. Agardh or *Plocamium coccineum* Lyngbye [René Pfister, "Teinture et alchimie dans l'Orient hellénistique," *Seminarium Kondakovianum*, 7 (1935): 11]. The same two hypotheses have also been proposed for the ingredient called *phykos* (φῦκος) or *phykion* (φυκίον; see *infra*, n. 7), which is connected to *bryon thalassion* by later sources (Hsch. φ 961 H-C; Phot. β 289,1 Th.). In particular, two *scholia* in Theocritus (VII 58) are significant in this respect: (1) "*Phykos* is the so-called *bryon* [lit. 'alga']"; and (2) "*Phykos* is a kind of grass cast away by the sea. Some people claim that *phykion* is a plant growing up in the depths of the sea … Others say that *phykia* are the purple algae on the beaches." The references to colour in these passages seem to support the second abovementioned hypothesis, since algae such as *Plocamium coccineum* or *Rytiphlaea tinctoria* are usually purple/red (and are often deposited on the shore by the current), while the *Roccella* lichens are usually grey. A technical detail could further support this identification: ps.-Democritus specifies macerating the dyestuff in water in accordance with the usual treatment required for processing seaweeds; lichens, by contrast, were macerated in a solution rich in ammonia (e.g. old urine; see Pfister, "Teinture et alchimie," 7 and 11).

3] *Lakcha* (λακχά). Most lexicons [*ThGL* V 59 *s.vv.* λάκχα and λάκκα; LSJ⁹ 1025 *s.v.* λακχά; *DELG* 615 *s.v.* λακχά; Jacques André, *Les noms de plantes dans la Rome antique* (Paris: Les Belles Lettres, 1985), 135 *s.v. lacca*] agree in identifying this dyestuff with alkanet (*Alkanna tinctoria* Tausch), a plant of the *Borraginaceae* family, whose roots produce a fine colouring material. Berthelot (*CAAG* III 43 n. 2) and Pfister ("Teinture et alchimie," 22 n. 92) followed the same interpretation as well as several ancient Latin glosses that read *lacca anchusa* (*CGL* III 547,48; 584,13; 592,35; 613,66). Thus the term would be equivalent to the Greek *anchousa* (ἄγχουσα), which is often attested by *PLeid.X.* and *PHolm.* (see Halleux, *Papyrus de Leyde*, 205). The few technical details given by ps.-Democritus' passage, which prescribes macerating the ingredient in urine, could support this identification. In fact, the Leiden papyrus mentions urine among the liquids used for macerating alkanet (*PLeid.X.* 90), and Pfister ("Teinture et alchimie," 13) noted that alkanet could be macerated in oil, vinegar, or urine in order to obtain different colours in the dyeing process (red or blue-green).

However, the information given by some later sources does not accord with this identification. Aetius (II 68 Oliv. in *CMG* VIII/1, p. 175) listed *lachcha* (λαχχά *sic*) and alkanet (ἄγχουσα) as two different ingredients involved in making a cosmetic (namely *rhoidarion*). In addition, the later alchemist Christianus (*CAAG* II 418,21–2) pointed to an Eastern origin for the ingredient, by saying that Indian dyers are familiar with *lacha* (λαχά). We cannot rule out the possibility that this term refers to the dyestuff usually called lac-dye, that is, the pigment extracted from the scale insect *Kerria lacca* Kerr that inhabits trees in Pakistan, India, Nepal, etc. It is likely that this substance was already known in the fifth/fourth century BC, since Aelianus (*NA* IV 46) quotes a passage by Ctesias (*FGrH* 688 F 45q) which mentions the large beetles that inhabit amber trees and are used by Indians for dyeing fabrics. In addition, the same dyestuff seems to be mentioned by the *Peryplus maris Erythraei* (6,22) under the name of *lachchos chromatinos* (λάχχος χρωμάτινος), one of the Indian luxury goods imported at the Adylis harbour. Unfortunately, archaeological finds seem to not attest cases of Egyptian textiles dyed with lac-dye before the Arabic period [see Pfister, "Teinture et alchimie," 36ff.; André Verhecken, "Relation between Age and Dyes of 1st Millennium AD Textiles Found in Egypt," in *Methods of Dating Ancient Textiles of the 1ˢᵗ Millennium AD from Egypt and Neighbouring Countries*, ed. Antoine De Moor and Cäcilia Flück (Tielt: Lannoo, 2007), 208–9]. However, textiles found in Palmyra have provided examples dyed with ingredients extracted from scale insects, among which the *Kerria lacca* has been identified along with the so-called Polish and Armenian cochineals (*Porphyrophora Polonica* L. and *Porphyrophora hamelii* Brandt); see René Pfister, *Textiles de Palmyre*, 3 vols. (Paris: Les éditions d'Art et d'Histoire, 1940), vol. 1, 20–7; Harald Böhmer and Recep Karadag, "Farbanalytische Untersuchungen," in *Die Textilien aus Palmyra*, ed. Andrea Schmidt-Colinet, Annamarie Stauffer and Khālid Al-As'ad (Mayence: Philipp von Zabern, 2000), 84–5.

4] *Monk's rhubarb* (*lapathon*). The term usually refers to different plants of the *Polygonaceae* family (species *Rumex*), such as *Rumex patientia* L. ('herb patience' or 'monk's rhubarb'), *Rumex aquaticus* L. and *Rumex crispus* L. (see André, *Les noms de plantes*, 137 *s.v. lapathum*). Although ancient sources mention the ingredient only as a laxative, Cardon (*Natural Dyes*, 93–4) specifies that several plants of this species are rich in tannins and were perhaps used for dyeing fabrics.

5] *Seashells* (*onyx thalassios*). The identification of this ingredient is not clear (LSJ⁹ 1234, *s.v.* ὄνυξ). Dioscorides (II 8) mentioned different kinds of *onykes* ('fingernails' or 'claws'), namely the shells (πῶμα κογχύλιου) coming from India, Eritrea and Babylonia. The term seems to have referred to an aromatic substance which was often used in fumigation and is already mentioned in the Bible (LXX, *Exodus* 30:34; see also Plin. *NH* XXXII 134; Gal. XII 421,17 K.; Paul. Aeg. III 2,2 Herb. in *CMG* IX/1, p. 132). Commentators on the Biblical passage (e.g. Anonymous in *PL* 117,154C; Petrus Comestor, *Historia Ecclesiastica*, in *PL* 188,1188a) stressed the literal meaning of the Greek word *onyx*, 'fingernail,' which is supposed to have referred to a fish whose bones smelled pleasant when triturated. On the other hand, according to the Aramaic translation of *Exodus* 30:34 [see Roger Le Déaut and Jacques Robert, *Targum du Pentateuque. 2. Exode et Lévitique* (Paris: Éd. du Cerf, 1979), 244 n. 12], the substance was identified with the root of the *Sausurrea lappa* Clarke (κόστος in Greek), an Indian plant used for making essential oils (see Theophr. *HP* IX 7,3 and *Od.* 32; Diosc. I 16). However, Aetius (I 135 Oliv. in *CMG* VIII/1, 68f.) seems to have drawn a distinction between this root and the aromatic shells (ὄνυχες ἀρωματικοί), since he mentions both ingredients in making the same cosmetic.

6] The paragraph draws up a catalogue of dyestuffs, comparable to *PHolm.* 125, which lists similar substances, such as woad, madder and *phykos* (see *infra*, n. 7). A difficult nomenclature seems to have been adopted, perhaps borrowed from ancient craftsmanship. The author often specifies the geographical origin of the ingredients and makes a distinction between (a) dyestuffs that were highly valued by the ancients, but did not guarantee a lasting colour, and (b) substances that were not considered valuable in his time, but that were effective:

Dyestuffs (lit. 'flowers') esteemed by ancients, but not effective	Dyestuffs (lit. 'flowers') not esteemed at the time of ps.-Democritus, but effective
1a. φῦκος called ψευδοκογχύλιον (dyeing alga)	1b. ὁ τῆς Γαλατίας σκώληξ (dyeing cochineal, maybe lac-dye)
2a. κόκκος (dyeing cochineal, maybe lac-dye)	2b. τὸ τῆς Ἀχαίας ἄνθος called λακχάν (dyeing cochineal or alkanet)
3a. ἄνθος θαλάσσιον (perhaps dyeing alga)	3b. τὸ τῆς Συρίας ἄνθος called ρίζιον (madder)
4a. ἄγχουσα Λαοδικηνή (alkanet from Laodicea)	4b. τὸ κογχύλιον (dyeing sea-snail)
5a. κρημνός (not identified)	5b. τὸ κοχλιοκογχύλιον τὸ Λυβικόν (dyeing sea-snail)
6a. ἐρυθρόδανον τὸ Ἰταλικόν (Italian madder)	6b. ὁ Αἰγύπτιος κόγχος called πίννα (dyeing sea-snail)
7a. φυλάνθιον τὸ δυτικόν (not identified)	7b. ἰσάτις (woad)
8a. σκώληξ ὁ πορφύριος (dyeing cochineal, maybe lac-dye)	8b. τὸ τῆς ἀνωτέρας Συρίας called κόγχος (dyeing sea-snail)
9a. ρόδιον τὸ ἰταλικόν (Italian pomegranate or rose extract)	

The criteria for such a classification are not evident, since the same ingredients seem to have been listed in both classes (4a ≈ 2b; 2a ≈ 8a ≈1b; 6a ≈ 3b; see Pfister,

"Teinture et alchimie," 22), although with a different name or a different geographical origin. The inefficacy of alkanet (4a) may be understood if we recall the trouble ancients had in fixing the dye on fabrics, since it requires a specific treatment with iron salts (see Cardon, *Natural Dyes*, 61–2). Furthermore, ps.-Democritus did not trust several dyeing algae (1a, 3a) and cochineals (2a, 8a), even if some of them also fell into the second class (1b; 2b?). The ancients certainly knew many cochineals with dyeing properties (Cardon, *Natural Dyes*, 607–19 and 635–66), but we are not able to understand the distinctions made in the list. Similar doubts also surround the classification of the different kinds of madder: against other ancient sources (Diosc. III 143; Plin. *NH* XIX 47) that usually praised Italian madder (6a), ps.-Democritus seems to have preferred the Eastern kinds (3b). Finally, it is more understandable that both woad and dyeing sea-snails have been listed among the effective ingredients: the first one, in fact, was very popular among Egyptian dyers (Halleux, *Papyrus de Leyde*, 138–40), and the second is probably to be identified with the sea-snails (such as *murex* and *buccinum*) that produce true purple.

7] *Seaweed (phykos)*. This ingredient has been identified by scholars either with the orchil lichen (*CAAG* III 43 n. 1; Lagercrantz, *Papyrus Graecus Holmiensis*, 184; Cardon, *Natural Dyes*, 500–1) or with a dyeing alga, namely the *Rytiphlaea tinctoria* (Clemente) C.Agardh (Pfister, "Teinture et alchimie," 7–12; Halleux, *Papyrus de Leyde*, 233; Amigues, *Théophraste, Recherches sur les plantes*, 241 n. 9). Several ancient sources refer the term *phykos* to different kinds of algae (Ar. *HA* 568a 5; Gal. XII 152,13–5 K.). Theophrastus (*HP* IV 6,4–5) specified that the sea *phykos* (πόντιον φῦκος) – which was usually collected by divers in the open sea, although in Crete it was abundant on the rocks close to the shore – was used for dyeing ribbons, wool, and cloth. If the mention of the open sea seems to point to a dyeing alga, the Cretan variety could be understood as lichen, since lichens often grow on wet rocks. Moreover, a certain ambivalence regarding the term is evident in later sources; both Dioscorides (IV 99) and Pliny (*NH* XXVI 103) listed three kinds of *phykos*, the third of which is said by Pliny to be used in Crete for dyeing cloth ("tertium, crispis foliis, quo in Creta vestes tingunt"). Dioscorides – who claimed that a small dyeing root was also called by the same name of *phykos* – described the third type of *phykos* more extensively: "another that is curly (οὖλον; *varia lectio* λευκόν, 'white') growing in Crete, near the shore; it is very colorful (εὐανθές), and not prone to decay" [transl. by Lily Y. Beck, *Pedanius Dioscorides of Anazarbus, De Materia Medica* (Hildesheim: Olms-Weidmann, 2005), 290]. The variant curly/white (οὖλον/λευκόν) deserves special attention, especially if we consider that lichens are usually white, while dyeing algae are red. The reading 'curly' (οὖλον) was preferred by the sixteenth-century physician Andrés Laguna, *Pedacio Dioscorides Anazarbeo, acerca de la materia medicinal, y de los venenos mortiferos, traduzido de lengua griega en la vulgar castellana* (Salamanca: Mathias Gast, 1563), who translated the adjective with "muy cabelluda" (440; see also *crispis foliis* in Pliny). If this word choice (which Wellmann also accepted) is correct, we could identify this *phykos* with one kind of dyeing alga, such as the *Plocamium coccineum* or the *Rytiph-laea tinctoria* (see Pfister, "Teinture et alchimie," 11), whose intense red colour Dioscorides stressed with the adj. 'colourful' (εὐανθές).

8] *Fake little shell* (*pseudokonchylion*). The term is a *hapax* composed of the adjective 'false' (ψευδής) and the name 'small shell' (κογχύλιον, diminutive of κογχύλη, 'mussel, shell,' a term which is sometimes used as a synonym for 'purple'; see *Kyranid.* IV 53). *Konchylion* is a general term for any kind of mollusc or its shell (LSJ⁹ 966). In Aristotelian taxonomy, sea snails that produce purple (namely πορφύραι and κήρυκες) fall into the *ostracoderms* (lit. 'shell-skinned') genre (*HA* V 15), although they could also be referred to by the general name of 'small shells' (*konchylia*). For instance, in *HA* IX 25, 622b 16–17 when speaking of spontaneous generation, Aristotle made a comparison with the other *konchylia* (ὥσπερ τἆλλα κογχύλια), by referring to *HA* V 15, 546b-548a, where different kinds of molluscs, purples included, were taken into consideration. A similar association between *konchylia* and purples is often found in ancient sources, such as medical literature (Hp. *Vict.* II 48,15–16 Joly in *CMG* I/2,4, p. 170 = VI 550,4–5 Littré), technical texts (Epich. fr. 42 K.; see also Clem. Al. *Paed.* II 10bis, 115 and *Physiol.*, *redactio prima*, 44b, 13), and lexicography (Hsch. κ 3194 L.). Moreover, the adj. *konchylios* (κογχύλιος, 'purple') may denote specific shades of colour, such as the 'purple colour' (χρῶμα κογχύλιον) mentioned by *PLeid.X.* 94. Similarly, the Latin *conchylium* may refer both to purple molluscs (Cic. *Ver.* II 4,59; Vitr. VII 13,1) and to specific dyes or methods of dyeing: in particular, Pliny (e.g. *NH* IX 130f.; XXII 3) often used the term *konchylia* to refer to those shades (*NH* XXI 46) which were produced by using not the true purple, but some substitute, such as *heliotropium* [identified by some scholars with sunflower; see Jacques André, *Pline l'Ancien, Histoire naturelle, livre XXI* (Paris: Les Belles Lettres, 1969), 110 n. 1; Halleux, *Papyrus de Leyde*, 213)], mallow and sweet violet (*Mathiola incana* L.). In *NH* XXVI 103, Pliny also listed *fucus* among these purple substitutes ("focus ... qui conchyliis substernitur"), by associating *phykos* and *konchylia* in a way similar to ps.-Democritus' usage in our passage.

9] *Kermes.* The term *kokkos* (κόκκος; mss. have κόκκον; correction already proposed by Lagercrantz, *Papyrus Graecus Holmiensis*, 114 n. 1) may refer to various kinds of cochineals that produce a dyestuff. It has usually been identified with the *Kermes vermilio* Planchon, a parasite on the *Quercus coccifera* L. or *Quercus ilex* L., which is quite common in Europe and in Eastern Mediterranean countries (Greece, Crete, Turkey; see Cardon, *Natural Dyes*, 611, and Halleux, *Papyrus de Leyde*, 217). Ancient sources enumerate different kinds of *kokkoi*: Diosc. IV 48, for instance, listed the Galatian, Armenian, Cylician varieties (see also Plin. *NH* XVI 32; XXII 3). Besides, we must note that several insects that fall today into the *Coccoidea* family have dyeing properties (Cardon, *Natural Dyes*, 607–66):

(a) *Kermes vermilio* Planchon, sometimes considered by ancient sources as a fruit of the oak tree (Theophr. *HP* III 7,3; Plin. *NH* XVI 32).

(b) Different cochineals of the *Porphyrophora* genre, such as the *Porphyrophora polonica* L. (or Polish cochineal), the *P. crithmi* Goux (which grows on the *Crithmun maritimum* L.) and the *P. armeniaca* Burmeister (Armenian cochineal).

(c) Indian cochineals, such as *Kerria lacca* Kerr and *Kerria chinensis* Mahdihassan.

10] *Sea flower (anthos thalassion)*. Plato (*Rep.* 429d) employed the term 'flower' (ἄνθος) to emphasize the brightness of purple dyes. The same term was used by Aristotle (*HA* V 15, 547a 15) with reference to the part of the molluscs – namely the hypobranchial gland – from which the mucous dyeing substance was secreted. *PHolm.* 106,8 also mentions the 'dyers' flower' (τὸ ἄνθος τὸ ἀπὸ τῶν βαφέων), which is probably to be identified with the dyestuff used by expert craftsmen (see Halleux, *Papyrus de Leyde*, 200). Nevertheless, the identification of this ingredient is not clear. According to *PHolm.* 106 it has the same properties as kermes and *krimnos* (see *infra*, n. 12); however, the specification "from the sea" seems to exclude such identifications, since neither of these two comes from the sea (the same problem arises if we identify the 'sea flower' with *Alkanna tinctoria* L., as proposed by Berthelot, *CAAG* III 44 n. 3). Perhaps the expression could refer to some kind of dyeing alga (see Pfister, "Teinture et alchimie," 21–2).

11] *Alkanet from Laodicea*. The term *anchousa* (ἄγχουσα) refers to alkanet (*Alkanna tinctoria* L.), a plant of the borage family whose roots produce a red dyestuff (see, e.g., Pfister, "Teinture et alchimie," 13–4; Halleux, *Papyrus de Leyde*, 205). I have slightly modified the readings Λαδικήνη (**MV**) and Λαδικίνη (**BA**) into the standard form Λαοδικηνή, attested by *Perypl. M. Erytr.* 49,2 and Jo Mal. *Chron.* 217,1 (Lagercrantz, *Papyrus Graecus Holmiensis*, 114 n. 1, and Halleux, *Papyrus de Leyde*, 108 = *PLeid.X.* 97,2 preferred the spelling Λαοδικινή). Laodicea on the Lycus was an important commercial city in Phrygia widely known for the intensive trade in wool. Strabo (XII 8,16) praised the black colour of the wool produced in the city (see also Plin. *NH* VIII 190). This blackness was often explained by ancient sources as a consequence of the colour of the Phrygian rivers flowing in the area around Laodicea (e.g. Vitr. VIII 3,14; Strab. XIII 4,14, although in reference to Hierapolis).

12] *Krēmnos* (κρημνός). The usual meaning of 'overhanging bank' (LSJ⁹ 994) does not fit into the context; the term seems related instead to the *krimnos* (κριμνός) mentioned by the Leiden and Stockholm papyri (Halleux, *Papyrus de Leyde*, 219). Various hypotheses have been proposed: (a) a dyeing cochineal, such as the so-called 'lac-dye' or 'Körerlack' (= *Kerria lacca* Kerr; see Diels, *Antike Technik*, 146) or the *Porphyrophora hirsutissima* Hall (Cardon, *Natural Dyes*, 655–6), a parasite of the blady or cogon grass (*Imperata cylindrica* Beauv.) common in Egypt; (b) hulled barley (Lagercrantz, *Papyrus Graecus Holmiensis*, 183), if we consider the usual meaning of *krimnon* (κρῖμνον), 'coarse barley meal' (see also Diosc. II 90; Gal. XII 45,5 K.; Aet. I 228,1 Oliv. in *CMG* VIII/1, p. 96,14; however, barley is used in *PHolm.* 148,3 e 153,4 as a mordant); (c) some flowering plants – such as *Lawsonia alba* Lam. or *L. inermis* L – from which henna is extracted (Pfister, "Teinture et alchimie," 14–5). This last plant is usually called *kypros* (κύπρος) by ancient authors (Diosc. I 95; Plin. *NH* XII 109; Gal. XII 54 K.) who often mention its white flowers; a similar association between white flowers and *krimnoi* is made by Hsch. κ 4116 L. (see Halleux, *Papyrus de Leyde*, 219): "*krimnoi*: some white plants" (κριμνούς· λευκάς τινας βοτάνας).

13] *Italian madder.* The term *erythrodanon* (ἐρυθρόδανον or ἐρευθέδανον, see Diosc. III 143) usually refers to the roots of *Rubia tinctorum* L., from which the so-called dyer's madder was extracted. The roots of this plant, after being carefully cleaned and triturated in water or white vinegar, release their dyeing principle (alizarin; see Pfister, "Teinture et alchimie," 19–20, Forbes, *Studies*, vol. 4, 106–7; Halleux, *Papyrus de Leyde*, 213). This plant was also called by the common name of 'root' (ῥίζα; see *PHolm.* 159,7f.; Hsch. ε 6078 L.; Phot. ε 18,1–2 Th.). The Italian species (particularly the one from Ravenna) were well known by ancient authors (Diosc. III 143; Plin. *NH* XIX 47).

14] *Phylanthion* (φυλάνθιον). Without providing any explanation, Berthelot (*Origines de l'alchimie*, 361 n. 4) and Pfister ("Teinture et alchimie," 21–2) proposed identifying this ingredient with some kind of *phykoi*. The term – standardized as *phyllanthion* (φυλλάνθιον) by Berthelot-Ruelle – is a *hapax*, usually interpreted as synonymous with the adj. *phyllanthes* (φυλλανθές, LSJ⁹ 1961). However, this last term is problematic as well, since it is attested only by Plin. *NH* XXI 99, where it is conjectural (mss. have various forms in –os, such as *phyllanthos* Ve, *phyllantos* Rep). It was first introduced by Ermolao Barbaro [*Castigationes plinianae*, (Rome, 1492), vol. 1, Book XXI, ch. 16] and then accepted by many later editors. Émile Littré, *Pline l'Ancien, Histoire naturelle*, 2 vols. (Paris: Dubochet, 1848–1850), vol. 2, 53, proposed identifying this *phyllanthes* with the *Centaurea nigra* L. (a perennial herb with purple flowers), but more recently André (*Les noms de plantes*, 133 s.v. *phyllanthes*) has been very cautious about the real existence of a plant with a similar name. He suggests, in fact, that Pliny's passage was based on an incorrect reading of Theophr. *HP* VII 8,3 (regarding the textual problems of which, see the recent edition by Amigues: *Théophraste, Recherches sur les plantes*, vol. 4, 26). Here the author seems to distinguish between two kinds of a plant called *anthemon* (ἄνθεμον); the first one of which, named *anthemon aphyllanthes* (ἄνθεμον ἀφύλλανθες), has been identified either with a kind of chamomile (*Matricaria chamomilla* L.) or with the *Anthemis rigida* (Sibth. & Sm.) Boiss. & Heldr. (= *A. cretica* Nyman; see Amigues, *Théophraste, Recherches sur les plantes*, vol. 4, 133 n. 8).

15] *Italian pomegranate (rhodion to Italikon)*. In Byzantine Greek the term *rhodion* could refer to 'pomegranate' [Charles du Fresne Du Cange, *Glossarium ad scriptores mediae et infimae graecitatis* (Leiden, 1688), 1304], usually called ῥόα (*rhoa*) in the alchemical papyri (Halleux, *Papyrus de Leyde*, 227). We cannot exclude the possibility that a similar form was already used in the first centuries AD. However, the adj. *rhodios* (ῥόδιος) could also refer to a perfume "made of roses" (see e.g. Hsch. ρ 391 H.), usually called *rhodinon* (ῥόδινον) and identified by Vittorino Gazza, "Prescrizioni mediche nei papiri dell'Egitto greco-romano, parte II," *Aegyptus*, 36 (1956): 95, with the extract of red rose petals (*Rosa gallica* L.; see, e.g., Diosc. I 43; Gal. XII 114 K.). No dyeing properties of this ingredient are attested by other recipes (only *PHolm.* 112 uses the adj. *rhodobaphēs*/ῥοδοβαφής with the meaning of 'rose-coloured'). However, Cardon (*Natural Dyes*, 251) notices that the *Alcea rosea* L. [= *Althaea rosea* (L.) Cav.], a plant of the *Malvaceae* family, was used in Central Asia for dyeing silk and leather.

16] The term 'worm' probably refers to a kind of dyeing cochineal and may be considered as synonymous with *kokkos* (see *supra*, n. 9), which is usually called *vermiculus* by medieval writers. Pliny claims that *coccus ilicis* quickly changes into a worm (*NH* XXIV 8: "est autem genus … celerrime in vermiculum id se mutans, quod ideo scolecium vocant"). On the interpretation of the gloss ἐκ τοῦ ἐρώ᾽ (sic) γενόμενος (handed down by **MV**), see Martelli, *Pseudo-Democrito*, 274–5.

17] On the identification of *lakcha*, see *supra*, n. 3 (on its geographical provenance, see also *CAAG* II 5,7).

18] *Little root (rhizion)*. The term probably refers to the root of *Rubia tinctorum* L., usually called *erythrodanon* (ἐρυθρόδανον; see *supra*, n. 13). Although Pliny (*NH* XIX 47) and Dioscorides (III 143) valued the Italian type, the toponym 'Syrian' may be explained by considering the Near Eastern origin of the plant, a native of Syria, Palestine, Egypt, etc. (see Forbes, *Studies*, vol. 4, 106–7). In any case, various roots could produce dyestuffs; for instance, Theophrastus (*HP* VII 9,3) lists both dyer's madder and alkanet (ἄγχουσα = *Alkanna tincoria* L.) among the *erythrai rhizai* (ἐρυθραὶ ῥιζαί), 'red roots.'

19] The terms *konchylion* (κογχύλιον, see *supra*, n. 8), *kochliokonchylion* (κοχλιοκογχύλιον, *hapax*, lit. 'spiral *konchylion*'), *konchos* (κόγχος, lit. 'shell') and *pinna* [πίννα, a bivalve mollusc; see Ar. *HA* V 15, 547b 15–16 and Eugène de Saint-Denis, *Le vocabulaire des animaux marins en latin classique* (Paris: Klincksieck, 1947), 87 *s.v. pinna*] probably refer to four species of molluscs (Pfister, "Teinture et alchimie," 22) with dyeing properties. According to modern taxonomy, all the purples fall into the large *Muricidae* family; 1,300 different molluscs are known today, and it is not clear how many of them have dyeing properties. Even if scholars usually agree that the ancients knew only a few species (Cardon,

Natural Dyes, 565–87), their exact identification remains very difficult because of both the obscurity of the ancient sources and the complexity of modern classifications. See Odone Longo, "La zoologia delle porpore nell'antichità greco-romana," in *La porpora. Realtà ed immaginario di un colore simbolico. Atti del Convegno di studio, Venezia, 24 e 25 ottobre 1996*, ed. Oddone Longo (Venice: Istituto Veneto di scienze lettere ed arti, 1998), 79–90.

20] *Woad plant (isatis)*. To be identified with the *Isatis tinctoria* L. (Pfister, "Teinture et alchimie," 18–9; Forbes, *Studies*, vol. 4, 109–10; Cardon, *Natural Dyes*, 367–78), a flowering plant from whose leaves the so-called dyer's woad was extracted. The complete method for processing the plant is described by *PHolm.* 109–111 (see also Diosc. II 184 and Plin. *NH* XX 59).

21] The author here gives a detailed first-person narrative of the initiation he received from his master (never mentioned by name), presumably Ostanes. Two main questions arise from this paragraph: (a) the identification of the two characters in the story, and (b) the authenticity of the account.

(a) Although Hershbell ("Democritus," 11–2) and Vereno (*Studien zum ältesten alchemistischen Schrifttum*, 91–4) proposed recognizing the story of Ostanes' initiation into alchemy in this paragraph (see *supra*, p. 18), most commentators (e.g. Berthelot, *Origines de l'alchimie*, 151–2; Festugière, *Révélation d'Hermès*, vol. 1, 228–9) agree on indentifying the Persian magus with Democritus' master. Zuber (see *infra*, p. 189) comments in the margin of his Latin translation: "Ostani manes evocati a Democrito" ("Ostanes' soul invoked by Democritus"). Moreover, several later alchemists stressed that Democritus had learned the saying about natures from the Persian magus, and emphasized the story's Egyptian milieu (according to Syn. Alch. § 2 and Syncell. p. 297 Mosshammer, the initiation took place in the temple at Memphis). The author himself perhaps reworked narrative patterns which were typical of Egyptian traditions (see Vereno, *Studien zum ältesten alchemistischen Schrifttum*, 94f.). Quack ("Les Mages Égyptianisés?," 280) draws attention to an unedited Coptic papyrus, Yale University, *P.CtYBR* inv. 422 (B), which concerns the discovery of an (astrological?) book by Imhotep, son of Ptah. Such an Egyptian background still seems recognizable in later traditions, which present Ostanes as an Alexandrian alchemist rather than a Persian magician (see, in particular, al-Nadīm's *Kitāb al-Fihrist* IX 5 in Fück, "The Arabic Literature on Alchemy," 91; Bladel, *Arabic Hermes*, 54–7, on the relationship between Ostanes and Hermes).

(b) The corrupt and dubious reading of some passages (see, in particular, Martelli, *Pseudo-Democrito*, 283–5 and 289ff.) and the lack of any reference to the main points of the story (the invocation of Ostanes' soul and the breaking of the column) in later alchemical texts allow us to question whether this paragraph belonged to the original four books by ps.-Democritus. Festugière (*Révélation d'Hermès*, vol. 1, 229–31) recognised in this passage various "thèmes hellénistiques" that were quite common in late Hellenistic or early Roman magical and astrological literature. Thus, no anachronistic elements preclude this passage from dating back to the same period; that is, to the time when the *Four Books* were very probably composed. As far as I know, the most ancient explicit reference to this paragraph is attested in an Arabic treatise, the so-called *Book of Crates* [Marcellin Berthelot and Octave Victore Houdas, *La chimie au Moyen-Âge*, vol. 3: *L'alchimie arabe* (Paris: Imprimerie Nationale, 1893), 57 (hereafter *CMA* III) = Bidez-Cumont, *Mages*, vol. 2, 320–1], which explicitly mentions the inquiries that Democritus had to carry out after the death of his master. Unfortunately, the origin and date of *The Book of Crates* are uncertain (perhaps ninth/tenth century). Berthelot-Houdas (*CMA* III 9–12) considered it the translation of a lost Greek alchemical treatise, while Julius Ruska [*Tabula Smaragdina. Ein Beitrag zur Geschichte der hermetischen Literatur* (Heidelberg: Carl Winter's Universitätsbuchhandlung, 1926), 52] supposed a Coptic original behind the Arabic text. In addition, the *Corpus alchemicum* preserves two interesting 'quotations' of *PM* § 3: (a) a very short version encapsulated by the catalogues handed down by the so-called *Chemistry of Moses* (see Appendix, table 4), whose date of composition is unfortunately still debated (third/fourth century in Berthelot's opinion; sixth/seventh century according to Letrouit, "Chronologie," 85–7); (b) the incipit of the paragraph, which is probably cited in a passage by the alchemist Olympiodorus, reading: Ἡμεῖς μὲν γὰρ πάντων τούτων καταφρονήσαντες, κατὰ τὸν Δημόκριτον· ἴσμεν γὰρ τῆς ὕλης τὴν διαφορὰν καὶ ἐπὶ τὰ χρησιμώτατα χωροῦμεν' (M, fol. 177ʳ11–3; V, fol. 30ᵛ5–8; A, fol. 213ʳ16–8; B om. = *CAAG* II 100,6–8). Though Olympiodorus did not quote the text of *PM* word for word, his explicit mention of Democritus allows us to consider this passage as dependent upon *PM* § 3, ll. 35f. (καὶ τῆς ὕλης τὴν διαφορὰν ἐγνωκώς, ἠσκούμην ὅπως ἁρμόσω τὰς φύσεις). In conclusion, although we do not have clear evidence for tracing this paragraph back to the *Four Books*, we cannot exclude the possibility that the *Books* included an account of Democritus' initiation, perhaps in the book on purple as the Byzantine tradition seems to suggest. Later interpolations by anonymous readers perhaps changed or expanded this original narrative 'nucleus.'

22] *aphroselēnos* (ἀφροσέληνος, 'moon foam'). According to Dioscorides (V 141 = Orib. XIII ch. λ 19,1 Raeder in *CMG* VI/1,2, p. 171,25f.), the term refers to *selēnitēs* (σεληνίτης), a stone related to moon's phases, which should be collected when the moon is waxing (see also Plin. *NH* XXXVII 181). Berthelot (*CAAG* I 267) explains that "ce mot désigne notre sulfate de chaux [$CaSO_4$; see also Mertens, *Zosime de Panopolis*, 152] et notre mica, ainsi que divers silicates, lamelleux et brillant." In the *Corpus alchemicum* the term *aphroselēnos* is interpreted in various ways: (a) as referring to a silver ore, as in *CAAG* II 5,26: *argyrolithos* [ἀργυρόλιθος; lit. 'silver stone'] is the *aphroselēnos*"; (b) it is associated with *koupholithos* (κουφόλιθος), lit. 'light stone' (*CAAG* II 5,15), a mineral to be identified with talc or modern selenite (see Halleux, *Papyrus de Leyde*, 219); (c) Zosimus (*CAAG* II

357,2; see also *CAAG* II 13,10) considers the term as equivalent to *pheklē* (φέκλη or σφέκλη; see LSJ⁹ 1921, from the Latin *faecula*), 'wine dregs' or tartar (Halleux, *Papyrus de Leyde*, 231); (d) the philosopher Anonymous (*CAAG* II 123,10–124,4) etymologizes the term *aphroselēnos*, which is said to stem from *Aphroditēs* (i.e. copper) and *selēnē* (associated with both silver and quicksilver). According to the indirect tradition, ps.-Democritus seems to have dealt with this substance (see *CAAG* II 353,4f. and 12–15). In particular, a citation is preserved by the philosopher Anonymous, where ps.-Democritus gives the name of *aphroselē-nos* to a mysterious stone (*CAAG* II 122,5–7): "Take the stone that is not a stone, the one that has no value and is precious, that has many forms and no form, that is not known and manifest to everyone, that has many names and only one name [I read μονώνυμον with M, fol. 84ʳ15], I mean the *aphroselēnos*." However, the closing specification of the name of this anomalous stone is missing in the Syriac tradition that hands down the translation of this passage (*3SyrC* § 2; see *infra*, n. 3).

23] Mercury has to be fixed (or solidified) by means of various solid substances (σώματα; lit. 'bodies'). The verb *pēgnymi* (πήγνυμι; see also *DELG* 894–5) usually refers to the solidification of various liquids: the gods were considered able to solidify/freeze rivers in Aesch. *Pers.* 495–7, but otherwise the middle-passive form could refer to frozen water (Hdt. IV 28,4; Thuc. III 23,5), coagulated blood (Aesch. *Cho.* 67), curdled milk (Aristot. *PA* IV 1, 676a 14), or crystallized salts [Hdt. IV 53,3; (Aristot.) *Pr.* III 16, 873a 29]. Aristotle (*Mete.* IV 8, 385a 2–18) listed the *pēkton-apēkton* pair (πηκτόν-ἄπηκτον; 'capable/incapable of solidification') among the eighteen pairs of passive qualities that distinguish homoeomerous bodies. Watery substances (such as ice and metals) were supposed to be solidified by cold and liquefied by heat; earthy substances, in contrast, were supposed to be solidified by heat (which eliminated their moisture) and liquefied by cold (that is, they are water-soluble). However, according to Aristotle's view, some substances could not be solidified and hardened at all (*Mete.* IV 8, 385b 1–5): "Incapable of solidification are bodies which contain no watery moisture [...] and also bodies in which, though they contain water, air predominates, like oil, quicksilver," etc. [transl. by Henry D. P. Lee, *Aristotle, Meteorologica* (Cambridge, Mass. and London: Loeb, 1952), 343]. Empirical observation is probably at the basis of this statement, since mercury has a very low freezing-point (–39 degrees C.). However, the difficulty of reaching such a temperature seems to have been bypassed by ps.-Democritus, who prescribed solidifying mercury by mixing it with solid bodies, which were probably thought able to transfer their hardness to the liquid metal. Later alchemists (Zos. Alch. *CAAG* II 192,19–193,14; Syn. § 11, ll. 181–2; Christ. *CAAG* II 397–9) at least seem to have interpreted ps.-Democritus' expression "*magnēsia*'s body" (τῆς μαγνησίας σῶμα) in this way, for this term would have referred to any kind of solid substance that could interact with mercury and counteract its volatile and 'fleeing' nature. For instance, a particular technique for fixing mercury was performed by means of a specialized device called *phanos* (φανός), as Zosimus explains (II 6–10 Mertens):

"In the *phanos* and similar instruments, which are equipped with a snake-shaped device, it is possible to solidify mercury (πήσσειν τὴν ὑδράργυρον) and make it yellow again by means of the exhalation of sulphur, as is explained by the ancient writings, provided that the *phanos* does not contain lead." Despite the exegetical problems arising from the identification of the snake-shaped device (ἐγκάθισμα ὡσεὶ δρακοντῶδες; see Mertens, *Zosime de Panopolis*, CLIII-CLXI), the process sounds quite clear: mercury and sulphur are sublimed together, causing them to react to produce a solid compound, presumably mercuric sulphide, whose colour can vary according to its manner of preparation and particle size (Mertens, *Zosime de Panopolis*, 122 n. 12). However, among the solid ingredients listed by ps.-Democritus, only sulphur, orpiment, and stibnite fit such a procedure. The other ingredients (such quicklime and alum) are non-volatile: they were probably triturated and mixed with mercury without being sublimed. A similar technique seems to be attested by *CAAG* II 37,17–38,6 (*On the Making of Cinnabar*), where sulphur is ground, mixed with mercury and heated in order to produce a body the colour of iron (see also *CMA* II 31 rec. 1).

24] Mercury is processed with various ingredients in order to produce two compounds that are called white and yellow mercury. The white compound could whiten copper and make it like silver, while the yellow one could both yellow silver (making it like gold) and change the colour of gold (making it red; see *infra*, n. 25). These operations were regulated by a criterion based on colour [Arthur J. Hopkins, "Transmutation by Colour. A Study of Earliest Alchemy," in *Studien zur Geschichte der Chemie: Festgabe Edmund O. Von Lippmann zum siebzigsten Geburtstage dargebracht aus Nah und Fern und im Auftrage der deutschen Gesellschaft für Geschichte der Medizin und der Naturwissenschaften*, ed. Julius Ruska (Berlin: Julius Springer, 1927), 9–14], which is clearly formulated in a passage (*CAAG* II 159,3–12) perhaps ascribable to Zosimus (Mertens, *Zosime de Panopolis*, LVIII; *contra* Letrouit, "Chronologie," 36). The author of that passage, who seems to quote a lost part of ps.-Democritus' *Four Books*, lists several ingredients used by the ancient alchemist (mercury, *magnēsia*, antimony, litharge), then claims that: "these kinds of substances are common for [the making of] gold and silver: in fact, after being whitened they whiten and after being yellowed they yellow" (ll. 11–2). A similar 'formula' is often repeated by both later Greek and medieval alchemists. For instance, a recipe in the *Mappae clavicula* – a medieval recipe book perhaps stemming from an earlier Greek work [see Robert Halleux and Paul Meyvaert, "Les origines de la Mappae clavicula," *AHMA*, 62 (1987): 7–25] – seems to describe the same operation as ps.-Democritus: "The recipe for the most gold. Take copper that has been hammered out when hot, and grind filings of it in water with 2 parts of crude orpiment so that it becomes as viscous as glue, and roast it in a small pot for 6 hours and it will turn black. Take it out and wash it off, then add an equal portion of salt and grind them together. Then roast it in the pot and watch what happens: if it is to be white, mix in silver, if yellow, mix in gold in equal portions, and it will cause wonder" [rec. 5 in Cyril S. Smith and John C. Hawthorne, "Mappae Clavicula: A Little Key to the World of Medieval Techniques," *Transactions of the American Philosophical Society*, n. s. 64 (1974): 30]. In this case as well, the base metal was chosen in accordance with the colour of the dyeing compound. As noted by Berthelot, the procedure was based on the idea that a unique 'alchemical compound' could accomplish different chromatic changes depending upon its own colour: see Marcellin Berthelot, "Sur les alliages d'or et d'argent et sur les recettes des orfèvres au temps de l'Empire Romain et du Moyen Âge," *Annales de chimie et de physique*, 22 (1891): 152–3.

25] *Solid gold coral.* The term *chrysokorallos* (χρυσοκόραλλος) is attested only in alchemical literature. The second element of this word recalls the term *korallion* (κοράλλιον, 'red coral,' more common than κόραλλος; see LSJ⁹ Suppl. 182), which could hint at a specific reddish colour assumed by the gold after being processed. The various ingredients listed by ps.-Democritus (first of all orpiment, realgar, and cinnabar) could react with the small amounts of silver or copper usually mixed with gold and produce a superficial, coloured patina. Similar processes are also attested by *PLeid.X.* 14 and 24 for the purpose of purifying gold [*CAAG* I 264; Robert Halleux, "Méthodes d'essai et d'affinage des alliages aurifères dans l'Antiquité et au Moyen Âge," in *L'or monnayé.* Vol. 1: *Purification et altérations de Rome à Byzance*, ed. Cécile Morrisson (Paris: éd. du CNRS, 1985), 45–8; *Papyrus de Leyde*, 170 n. 9 and 172 n. 4) and *PLeid.V.* 193–201 (= *PGM* XII) in order to produce 'gold rust' (Halleux, *Papyrus de Leyde*, 165–6; see *PM* § 12 n. 46).

26] Mercury alone would be sufficient to whiten copper, as attested by Pliny (*NH* XXXIII 20; 32; 42), who described a gilding process in which a gold amalgam is applied to copper leaves. Ottavio Vittori, "Pliny the Elder on Gilding: A New Interpretation of His Comments," *Gold Bulletin*, 12 (1979): 36, explained the technique as follows: "Mercury is rubbed on the surface of the copper substrate when it is cold. Some copper dissolves in the mercury and forms a very thin layer of copper-mercury amalgam." The same technique is also described in *PLeid.X.* 41: "*Coating of Copper.* If you desire that the copper shall have the appearance of silver; after having purified the copper with care, place it in mercury and white lead; mercury alone suffices for coating it" [rec. 42 in Earle Radcliffe Caley, "The Leiden Papyrus X: An English Translation with Brief Notes," *Journal of Chemical Education*, 3 (1926): 1156; see also the *Corpus Syriacum* in *CMA* II 28 rec. 8].

27] A clear understanding of this process is hampered by the uncertain identification of the ingredient called *pyritēs* (πυρίτης); see Halleux, *Papyrus de Leyde*, 226; John F. Healy, *Pliny the Elder on Science and Technology* (Oxford: Oxford University Press, 1999), 213. This substance, almost absent from the Leiden and Stockholm papyri, is quite common in the *Corpus alchemicum*, where it seems to play an important role. The *Lexicon on the Making of Gold* reads (*CAAG* II 11,11): "*Pyritēs* is the mystery of any metallic mineral." Scholars have proposed two different interpretations of the recipe. Berthelot (*CAAG* III 47 n. 1) identified *pyritēs* with a silver ore which would have been processed with litharge, lead and antimony in order to make a white alloy. This alloy was then treated and turned yellow by means of an additional ingredient, which is not mentioned in the recipe. This interpretation, however, is based on an incorrect translation of the last sentence, which Berthelot interpreted in *CAAG* III 47: "Faites chauffer, puis mettez dans la matière du jaune factice et teignez." Halleux ("Méthodes d'essai," 54), on the other hand, supposed that ps.-Democritus was describing a process for extracting gold from auriferous pyrites. The first step would then have been to make a gold-copper-lead alloy, from which the gold was afterwards isolated by cupellation.

Yet alchemical texts often associate *pyritēs* with copper ores, especially in the *Lexicon on the Making of Gold*, where the mineral is related to 'copper flower' (ἄνθος χαλκοῦ; *CAAG* II 5,8) and to two copper and iron sulphides, namely *androdamas* (ἀνδροδάμας; *CAAG* II 5,12) and *sōri* (σῶρι; *CAAG* II 12,15). Furthermore, the alchemist Pelagius (*CAAG* II 255,12–3) explicitly states that ps.-Democritus employed the term *pyritēs* to refer to the burning nature of copper (see also Diosc. V 125 and Plin. *NH* XXXVI 125). Following these identifications, we can guess that ps.-Democritus is here describing how to process copper ores with lead and antimony (both can lower copper's melting point) in order to produce copper-lead or copper-lead-iron alloys, which are supposed to be yellow according to Taylor's interpretation [F. Sherwood Taylor, "A Survey of Greek Alchemy," *Journal of Hellenic Studies*, 50 (1930): 128]. Similarly, the tenth recipe of the *Mappae clavicula* (Smith and Hawthorne, "Mappae Clavicula," 31) describes how to process *pyrites* and lead in order to produce a lead-copper-iron alloy (perhaps partly oxidized), which is supposed to be yellow according to the Latin text.

28] On the term *nitheōs* (νίθεως) see *1SyrC* § 2 n. 7.

29] The term *lithargyros* (λιθάργυρος, 'litharge') usually refers to lead monoxide (PbO; see Gazza, "Prescrizioni mediche," 102; Halleux, *Papyrus de Leyde*, 220) that "presents two allotropic forms, different in colour, namely yellow and red" (Healy, *Pliny the Elder*, 321). Consequently, the adj. 'white' (λευκή) does not seem appropriate for qualifying this substance. Ancient sources, on the other hand, often referred to different methods for purifying and whitening it. According to Pliny (*NH* XXXIII 107–110) and Dioscorides (V 87), litharge was either processed with salt and sodium carbonate (νίτρον); or washed with vinegar, to produce white lead acetates. For instance, a recipe edited in *CAAG* II 248,13–6 seems to describe how to process litharge with vinegar in order to produce a by-product as white as *psimythion* (ψιμύθιον; see also *CAAG* II 10,17: "white litharge is *psimythion*"). On producing white lead chlorides, see Kenneth C. Bailey, *The Elder Pliny's*

Chapters on Chemical Subjects, 2 vols. (London: Edward Arnold & Co, 1929–32), vol. 1, 216.

30] After the adj. *Koptikos* (Κοπτικός; on its unusual meaning of 'from Coptos,' not registered in the lexica, see Mertens, *Zosime de Panopolis*, 150 n. 6) we must read the implied term *stîmi* (στῖμι, 'stibnite'), as already suggested by Mertens with reference to Zos. Alch. IV 69 (see also *CAAG* II 18, 8–9 and 159,7; *PGM* IV 1071; V 66 and VII 336; Aet. VII 41,39 and 100, 48 Oliv. in *CMG* VIII/2, pp. 294, 8–9 and 345, 24). This helps us better understand ps.-Democritus' specification about the kind of lead, μέλανι (*scil.* μολύβδῳ) τῷ ἡμῶν ('our black [lead]') involved in the process. In fact, various passages of the *Corpus alchemicum* (Zos. Alch. *CAAG* II 154,1–2; Olymp. Alch. *CAAG* II 94,2; *CAAG* II 360,29) deal with the lead produced from stibnite and litharge (presumably a lead-antimony alloy). For instance, the *Lexicon on the Making of Gold* (*CAAG* II 11,9) reads: "Our lead is the one which derives from the two antimonies and from litharge" (on the two kinds of *stîmi*, see Plin. *NH* XXXIII 101).

31] Analogous procedures for roasting *pyritēs* are attested by other sources. In particular, Diosc. V 125 reads: "Copper pyrite is a type of stone from which copper is mined. You must choose that which is copper-colored and which emits sparks readily. You must burn it this way: after coating it with honey place it on gently burning coals and fan it continuously until it becomes orange-tawny in color. But some people, after moistening it well with honey, bury it among many very hot coals" (transl. by Beck, *Pedanius Dioscorides*, 391). Similar techniques were also explained by Plin. *NH* XXXIV 135 and Diosc. V 103, especially with regard to twice-burned pyrite (usually called *diphryges*, διφρυγές), which was expected to become as red as cinnabar. Bailey (*Pliny's Chapters*, vol. 2, 183), commenting on Pliny's passage, claimed that "the roasting of a pyritic mineral no doubt gave a mixture of oxides of copper and iron, owing its red colour to the latter." Similarly, a recipe from *The Chemistry of Moses* prescribed a "treatment of *pyritēs*. Boil it for one day after having triturated it in sea water; let it dry and make use of it" (*CAAG* II 302,7–8). In light of these passages – which convinced me to adopt the **BA** reading καὶ ὄπτησον (omitted by **MV**) at l. 85 – it is possible to read the recipe as the explanation of a roasting process for iron and copper sulphides (perhaps pre-treated with liquid substances). The mention of their black colour can be understood by taking into account the dark shades of either the ores themselves or the by-products that were probably formed during the process. At the end of the recipe, the orange colour of the iron and copper oxides obtained by roasting the minerals was finally made more intense by mixing them with various yellow substances (sulphur, yellow alum, Attic ochre).

32] The same criterion of colour at the core of the first recipe (see *supra*, n. 24) is here applied again: the yellow compound was expected to turn silver yellow and to change the colour of gold, transforming it into *chrysokonchylion* (χρυσοκογχύλιον). This term, attested only by the *Corpus alchemicum* (*CAAG* II 16,2 and 9; 282,13), probably refers to a specific material prepared by processing gold with substances that changed its mechanical and chromatic properties. Since the dyeing compound was rich of iron, we might suppose that the metal combined with the gold, hardened it and perhaps gave it a reddish hue (see *supra*, n. 25). According to Halleux' interpretation (*Papyrus de Leyde*, 170 n. 2), a similar process – aimed at producing a gold-iron alloy – is also attested by *PLeid.X.* 16 (= *PLeid.X.* 86).

33] *Klaudianon* (Κλαυδιανόν). This term is attested in both masculine and neuter forms by the *Corpus alchemicum*. The masculine form in two cases seems to be the name of an alchemist, who is mentioned in the list of *auctoritates* handed down by A, fol. 195ᵛ (*CAAG* II 26,1), and a passage by Stephanus (II 208,7 Ideler). However, scholars generally agree in interpreting *Klaudianon* (*-os*) as the name of a metallic alloy similar to the *aes Sallustianum, aes Livianum* and *aes Marianum* mentioned by Pliny (*NH* XXXIV 3–4). Berthelot proposed either a lead-tin or a copper-lead-zinc alloy, perhaps with a small amount of tin (*Origines de l'alchimie*, 223, and *CAAG* I 244); Taylor a copper-lead alloy ("A Survey of Greek Alchemy," 123); and Hershbell "an alloy resembling gold, and containing copper, lead, tin or brass" ("Democritus," 11). It is nevertheless impossible to choose between these interpretations, since the *Corpus alchemicum* never mentions the components of the alloy. Olympiodorus alone claims (*CAAG* II 73,18) that Zosimus used the expression 'klaudianon leaves' (πέταλα κλαυδιανά) to refer to 'gold leaves' (πέταλα χρυσοῦ; see Stéphanidès, "Petites contributions,"203f.). On the other hand, ps.-Democritus specifies that *klaudianon* is a stone (λίθος), a term that seems more appropriate if it refers to an ore rather than to a metallic alloy. This ore might have taken its name from the area in which it was mined. In this regard, recent archeological investigations have explored an area situated in the north-east of Egypt, where the so-called 'granito del foro' was mined. The area, called *mons Claudianus* (from the name of the emperor Claudius) was starting to be exploited by the Romans during the first century AD: see, e.g., Patrizio Pensabene, "Le cave del Mons Claudianus. Conduzione statale, appalti, e distribuzione," *Journal of Roman Archaeology*, 12 (1999): 721–36; Valerie A. Maxfield, "Stone Quarrying in the Eastern Desert with Particular Reference to Mons Claudianus and Mons Porphyrites," in *Economies beyond Agriculture in the Classical Word*, ed. David J. Mattingly and John Salmon (London: Routledge, 2001), 143–170. Several documents (*ostraka*) were found in the urban area around the mines, where the term *klaudianon* is used to indicate both the mines and the village (*O. Claud.* 7,3; 8,4; 11,4; 159,4; 177,6; 241,8; 371,1; 373,4; 374,2; 375,3; 376,4, etc.). Unfortunately, there is no evidence that metallic ores were mined in the same area.

34] *Make cinnabar white.* Berthelot (*CAAG* I 244; II 48 n. 1) supposed that the term *kinnabaris* (κιννάβαρις) was not used here with its usual meaning of 'cinnabar'

(HgS), but was intended to refer to a red lead oxide (Pb_3O_4, called *minium* by Pliny; see Diosc. V 94). This oxide, when processed with vinegar, could produce white lead acetates. However, it is also possible to suppose a similar reaction for cinnabar. On the one hand, it is well known that ancients used to treat cinnabar with vinegar in order to obtain mercury (usually described as 'white'; see *infra*, *Cat.* § 1 n. 1). On the other, ps.-Democritus also prescribed the use of salts (ἅλμη) in the process, perhaps in order to produce white mercury chlorides (a procedure also attested by Syriac alchemical recipes; see *CMA* II 47 rec. 8). At the end of the recipe, this 'white cinnabar' is mixed with yellow ingredients such as sulphur, iron sulphides and *chalkanthon* (χάλκανθον). This last term perhaps refers to a reddish copper oxide, which is often called 'copper's flower' (ἄνθος χαλκοῦ) in alchemical literature (e.g. *PLeid.X.* 73 and *CAAG* II 5,8–9).

35] *If you dip gold [into the solution?].* The second part of the recipe is unclear (for the philological questions, see Martelli, *Pseudo-Democrito*, 312). The dyeing drug was used for processing silver in order to make it gold. However, ps.-Democritus seems to prescribe adding gold to this drug. We must remark that similar procedures are also attested by other recipes, where scholars have noticed the need to add gold to the dyeing solution in order to turn silver yellow. For instance, *PLeid.X.* 53 reads: "*Coloration in Gold.* How one should prepare gilded silver. Mix some cinnabar with alum, pour some white vinegar upon this, and having brought it all to the consistency of wax, press out several times and let it stand overnight" (rec. 55 in Caley, "Leiden Papyrus X," 1157; see also Smith and Hawthorne, "Mappae Clavicula," 39 rec. 86–D). Though gold is not mentioned, Halleux (*Papyrus de Leyde*, 177) thought it must surely be involved in the making of the dyeing solution. This missing ingredient is attested in a recipe of the *Corpus Syriacum* (*CMA* II 206 rec. 11), where a dyeing drug is prepared by mixing cinnabar, salt, gold filings, alum and vinegar.

36] The term *kadmia* (καδμία) may refer to both zinc ores and a by-product of the combustion of these ores in vaulted ovens (Bailey, *Pliny's Chapters*, vol. 2, 66). Since ancient sources specify that *kadmia* rises up upon being separated from its ores (Plin. *NH* XXXIV 101, "fit autem [cadmia] egesta flammis"; Diosc. V 74,3 τὰ ἀνα-φερόμενα σώματα ἀπὸ τοῦ χαλκοῦ), I have interpreted the participle *exōsmenēn* (ἐξωσμένην, from ἐξωθέω, 'thrust out, force out,' LJS⁹ 600) as referring to the process that forced the substance out of the ores (see Martelli, *Pseudo-Democrito*, 313–5). When roasted, these ores produce volatile zinc oxide (Halleux, *Papyrus de Leyde*, 215; Healy, *Pliny the Elder*, 204) which collects in cooler parts of the furnace, and whose colour depends on its purity. Dioscorides (V 74,6–7) and Pliny (*NH* XXXIV 103–4) remind us that this by-product was often roasted over burning coals and cooled with vinegar in order to increase its whiteness. The Cyprian type was often praised by ancient sources (Diosc. V 74; Plin. *NH* XXXIV 103; Gal. XIV 7 K.).

37] The whitened *kadmia* was then dissolved in yellow liquid substances, which are also frequently mentioned in other alchemical texts: (a) the bile of a calf was employed both as a fixing substance (especially with gum arabic; see Halleux, *Papyrus de Leyde*, 178 n. 2) and for producing yellow dyes (*PLeid.X.* 61; while the bile alone was also used for dyeing bronzes called χολοβάφινα; see Halleux, *Papyrus de Leyde*, 41 n. 8); (b) the greenish resin produced by the *Pystacia tere-benthinus* L. was used in dyeing baths for fabrics and stones (Halleux, *Papyrus de Leyde*, 190 n. 4), and for producing paints for metals (*Corpus Syriacum* in *CMA* II 81 rec. 19); (c) similar uses are attested both for castor oil, a dense brown-yellow oil (see, for instance, *PHolm.* 35; 43; 63) and for its substitute, called radish juice (*PM* § 10, l. 104; *CAAG* II 54,19; 55,8). This last ingredient is said to be greenish-yellow by Zosimus (IX 38 Mertens).

38] *Androdamas* (ἀνδροδάμας). This ingredient (never attested by the Leiden and Stockholm papyri) was mentioned by Sotacus (end of the fourth century BC) in his book on precious stones (Plin. *NH* XXXVI 146) and was described by Pliny (*NH* XXXVII 144) as an ore as shining as silver and composed of small cubes ("androdamas argenti nitorem habet ut adamas, quadratis semper tessellis similis"). On the basis of this description, scholars [e.g., David E. Eichholz, *Theo-phrastus, De lapidibus, Edited with Introduction, Translation and Commentary* (Oxford: Clarendon Press, 1965), 281] have proposed identifying the mineral with an iron pyrite, which often has cubic crystals (see also *CAAG* III 48 n. 6). Such a hypothesis seems to be supported by an entry in the *Lexicon on the Making of Gold*, which identifies *androdamas* with pyrite and orpiment (*CAAG* II 5,12). According to ps.-Democritus' recipe, this mineral was triturated with various liquids, mixed with stibnite, and roasted. These procedures could perhaps produce some yellowish iron oxides. These by-products were then processed with a highly reactive liquid that the ancient alchemists called 'divine' or 'sulphur water' (whose preparation is described in *PLeidX.* 87); see also Viano, "Acqua divina"; Martelli, "Divine Water"; Principe, *The Secrets of Alchemy*, 10–1.

39] Comparison with the Syriac translation reveals an interesting detail: where the Greek text reads "take white earth" (λαβὼν γῆν λευκήν), the Syriac version preserves (*1SyrC* § 8) "take white lead" (ܐܚܕ ܐܒܪܐ ܚܘܪܐ = λαβὼν μόλυβδον λευκόν). Curiously, the Greek recipe has retained masculine forms to refer back to the feminine 'earth' (see critical apparatus at l. 117 and 122). This ungrammatical situation could reflect an older – or at least a different – version of the recipe which began by mentioning the masculine term 'lead' (μόλυβδος) rather than the feminine 'white earth' (γῆ λευκή).

40] *Psimythion* (ψιμύθιον). Scholars do not agree over the identification of this substance, which might refer either to a lead carbonate (Gazza, "Prescrizioni mediche," 105; Halleux, *Papyrus de Leyde*, 235) or a lead acetate (Healy, *Pliny the Elder*, 262). Several ancient sources prescribe processing lead leaves with vinegar (e.g. Theophr.

Lap. 36; Diosc. V 103; Plin. *NH* XXXIV 175f.), and Pliny (*NH* XXXIII 109) recommended adding *nitron* (sodium carbonate) to litharge. Since both the compounds are white solid substances, I have kept the traditional translation of 'white lead' (LSJ⁹ 2024).

41] The term *helkysma* (ἔλκυσμα, lit. 'that which is drawn'; LSJ⁹ 534) is likely to have referred to a by-product of silver extraction. Pliny (*NH* XXXIII 105) and Dioscorides (V 86,1) claimed that it has the same properties as *molybdaena* (μολυβδαίνη), which has been identified with both silver-lead ores and litharge (obtained by roasting silver-lead ores; see Bailey, *Pliny's Chapters*, vol. 2, 203). The relation between *helkysma* and (silver-)lead metallurgy is supported by an entry in the *Lexicon on the Making of Gold* (*CAAG* II 7,2): "*helkysma* is the roasted lead."

42] See *supra*, n. 29

43] It is possible to recognize three different steps in this procedure (*CAAG* I 71). A copper-lead-antimony alloy is first produced. This alloy is then made yellow by adding several yellow ingredients, and finally the yellow alloy – which was supposed to have a yellowing property (see *supra*, n. 24) – is used to colour any kind of base metal.

44] We should note that the same association between *sōri* (σῶρι), copper flower (χάλκανθον) and *misy* (μίσυ) is stressed by Plin. *NH* XXXIV 117–122. Bailey, in commenting upon this last passage, writes that "an approximate identification of these related minerals is easy, an exact identification difficult. The chief sulphide compounds of copper and iron occurring naturally are copper glance or chalcocite (Cu_2S), copper pyrites or chalcopyrite ($CuFeS_2$), iron pyrites or pyrite (FeS_2), and marcasite (FeS_2). Marcasite and iron pyrites readily, and copper pyrites more slowly, weather on exposure to air and moisture, the most important alteration products being the sulphates of copper and iron, that is to say blue vitriol ($CuSO_45H_2O$) and green vitriol ($FeSO_47H_2O$)" (*Pliny's Chapters*, vol. 2, 175 n. 117). Similar colours are also mentioned by ps.-Democritus, who spoke of azurite and green copper flower. The roasting of similar substances (sometimes carried out in vessels; see *1SyrC* § 9, which mentions an unspecified ܡܐܢܐ, *mo'no*, 'vessel') could possibly produce small amounts of sulphuric acid. Iron and copper sulphides, in fact, could oxidize and release sulphur dioxide upon prolonged heating, which, reacting with water and atmospheric oxygen, would produce sulphuric acid. If the metals to be treated contained an amount of gold, this corrosive would attack the non-precious metals in the alloy and remove them from the surface; thus, the alloy could become yellow because of the enrichment of gold at the surface (a process called 'depletion gilding'): see Halleux, "Méthodes d'essai," 55–7; David M. Jacobson, "Corinthian Bronze and the Gold of Alchemists," *Gold Bulletin*, 33 (2000): 61–4.

45] The first part of the recipe explains how to produce a yellow compound with which to treat copper and a specific kind of silver. Taylor interpreted the process as a method for "tinting the metal with solutions which form a thin superficial layer of sulphides" ("A Survey of Greek Alchemy," 130).

46] The second part of the recipe explains how to make 'gold rust.' This section is not included into the Syriac translation and was considered by Berthelot to be a later interpolation. However, a similar procedure is already attested by *PLeid.V.* (J 384) = *PGM* XII 193–9: "*How to Rust Gold* (Ἴωσις χρυσοῦ). Take thickened pungent vinegar and also have ready 8 drachmas of ordinary salt, 2 drachmas of rock alum that has clear cleavage, 4 drachmas of litharge, and triturate them [together] with vinegar for 3 days, and strain off [and] use. Then add one drachma blue vitriol [cupric sulphate] to the vinegar, ½ obol in weight of sōri, 8 obols of chalkitis, ½ obol in weight of misy, a carat (viz., $1/1728$ lb = siliqua) of ordinary salt, 2 [carats] of Cappadocian [salt]. Make a leaf [of metal] of 2 fourths by weight, dip [it] 3 times into fire until the leaf [breaks up] into fragments. Then take up the pieces [and] assume them as rusted gold (ἔχε ὡς ἐξίωσιν τοῦ χρυσοῦ)" (transl. by Betz, *Greek Magical Papyri*, 160–1, slightly modified). According to Halleux' interpretation (Halleux, *Papyrus de Leyde*, 165–6; "Méthodes d'essai," 60–2), the sulphur contained in the alum, vitriol, and various copper ores (*sōri*, *misy*, *chalkitis*) together with the chlorine contained in the salts could react with the baser metals in the gold (small amounts of copper and silver) and produce a superficial layer of 'rust.'

47] Gold is treated with a gluey compound of *chrysokolla* (χρυσόκολλα, 'malachite'), perhaps in order to make a silver-copper alloy, often used by the ancients for soldering the precious metal. The preparation of similar alloys (Cu/Ag/Au) is attested by both the Leiden papyrus (rec. 30 and 32) and the *Mappae clavicula* (Smith and Hawthorne, "Mappae Clavicula," 59). In addition, a glutinous copper compound (*santerna*) used for soldering gold is described by Pliny (*NH* XXXIII 93): "The goldsmiths also use gold-solder (*chrysocolla*) of their own for soldering gold, and according to them it is from this that all the other substances with a similar green colour take the name. The mixture is made with Cyprian copper verdigris and the urine of a boy who had not reached puberty with the addition of soda; this is ground with a pestle ... in mortars ... and the Latin name for the mixture is *santerna*. It is in this way used for soldering the gold called silvery-gold" (transl. by Rackham, *Pliny, Natural History*, 71–3). The gold pieces to be soldered were coated with this glutinous material and placed on a charcoal fire. Soda probably acted as a flux while copper produced a surface alloy with the gold, which alloy had a lower melting point than pure gold.

48] Scholars have expressed doubts about the authenticity of §§ 15–16 (see Berthelot, *CAAG* III 50 n. 4). Festugière in particular considered the opposition between young and old 'alchemists' in the passage as a sign of its later composition, since such a polarity would imply a developed state of the art (*Révélation d'Hermès*, vol. 1, 225–6). However, three elements support the authenticity of these two paragraphs. First, according to the history of alchemy traced by Zosimus (*CAAG* II 213–4 and 239–46), different groups or schools were already recognizable in the most ancient tradition. For instance, the beginning of the treatise *First Book of the Final Quittance* (*CAAG* II 240 = Festugière, *Révélation d'Hermès*, vol. 1, 364–5) contrasts Democritus with the Jewish alchemists. Furthermore, according to Zosimus' *Book on Tin* (preserved only in Syriac translation; see in particular SyrC, fols. 49ʳ-50ʳ), Democritus criticized other ancient authors who hid the natural processes at the basis of any alchemical operation (Berthelot-Duval, *CMA* II 239, translated the passage as follows: "Le philosophe dit qu'ils [i.e. the ancient alchemical authors] noyèrent dans un grand océan les écrits de la science de la nature"). Second, the Syriac tradition, which preserves §§ 15–16, also hands down a similar section (otherwise lost in Greek) in the book *On the Making of Silver* (2SyrC § 5): the similarities in content and style between the two passages seem to support the authenticity of these two 'theoretical' sections which belonged to the book on gold and to the book on silver respectively (see *supra*, pp. 12–3; 23–4). Finally, three later alchemists refer to this section. According to the alchemist Christianus, in fact, Zosimus mentions ps.-Democritus' criticism against young alchemists (*CAAG* II 406,19–20; on this text, see Martelli, *Pseudo-Democrito*, 330). Synesius (§ 11, ll. 173–6) clearly refers to the beginning of *PM* § 15. Finally, a few sentences of ps.-Democritus' condemnation of the neophytes' method are cited by Olympiodorus, *CAAG* II 103, 8–14 (see in particular, ll. 9–10 = *PM* § 14, ll. 178–80 and ll. 12–4 = *PM* § 13, ll. 159–61).

49] This recipe opens the second part of the book *On the Making of Gold* (see *supra*, p. 12) which is more focused on washes (ζωμοί), that is, on the use of liquid drugs or paints (often made from plants) for colouring metals (see Berthelot, *CAAG* I 72). Similar methods are attested by the Leiden and Stockholm papyri (see *CAAG* I 59–60; Halleux, *Papyrus de Leyde*, 42), by the Syriac tradition (see, for instance, *CMA* II, XXV), and by several medieval recipe books (Berthelot, "Sur les alliages d'or," 163–5; *CMA* I 47–9).

50] Pontic rhubarb does not seem to have been used as a dye in any of the above-mentioned recipes. In accordance with the later hermeneutic tradition, which interpreted the term *Pontion rha* (Πόντιον ῥᾶ) as an allusion to the dissolving of any solid substance (see in particular Syn. § 3, ll. 31–7 and § 8, ll. 110–2), we could suppose that ps.-Democritus is already employing the term without its usual meaning. In any case, the last line of the recipe makes it clear that it was supposed to have the same property as *elydrion* (ἐλύδριον, synonymous with χελιδόνιον; see *CAAG* II 16,18), namely *Chelidonium maius* L., a herbaceous plant with yellow flowers, whose bright yellow juice was used for preparing inks and paints (*PHolm.* 64; 65; 76; 84). Some recipes (e.g. *PLeid.X.* 67; *CMA* II 277 rec. 23; Smith and Hawthorne, "Mappae Clavicula," 30 rec. 4) prescribe dissolving mineral substances (especially *misy*) in the juice extracted from this plant.

51] The recipe seems to describe how to prepare a paint made from saffron, Aristolochia (celandine) and a juice of grapes. Taylor commented that "the [...] recipe appears to deal with the tinting of a metal by means of a layer of lacquer, coloured by various plant juices, to be applied to the surface of polished metal" ("A Survey of Greek Alchemy," 130). Similar methods are attested by *PLeid.X.* 72 and Syriac recipes (*CMA* II 276 rec. 20), which also list mineral substances (in particular orpiment). In the *PM*'s recipe, it is possible to find at least an allusion to orpiment in the sign ⟨ᵬ⟩ (i.e. ἀρσενικόν), which is handed down by the manuscripts above the name 'Cilician saffron' (κρόκος Κιλίκιος).

52] The expression "our lead" (ὁ ἡμῶν μόλυβδος) does not seem to refer to ordinary lead, but to a specific alloy, perhaps similar to the 'black lead' made from stibnite and litharge mentioned in *PM* § 6. The metal, in any case, before being treated, was made *arreustos* (ἄρρευστος). Since the lead is said to be melted at the end of the recipe, in this context the adjective is likely to mean 'harder' rather than 'not fusible' (Berthelot translated it 'peu fusible'). The procedures for hardening lead – which are well attested by ancient alchemical texts (e.g. *PLeid.X.* 1) – were interpreted by later alchemists as a way of eliminating moisture from the metal. For instance, Stephanus of Alexandria, commenting on ps.-Democritus' work, wrote (**M**, fol. 30ʳ6–8; **B**, fol. 68ᵛ16–69ʳ2; **A**, fol. 63ʳ23–6 = II 232,32–5 Ideler): "And how is what is fluid by nature (τὸ φύσει ῥευστόν) going to become not fluid (ἄρρευστον)? How do you harden a liquid and unstable substance? In fact, if it loses its moisture [fluidity], it will no longer be lead." For alchemists, the moist nature of lead – emphasised by various ancient sources, such as Pliny (*NH* XXXIV 161), Galen (XII 230,6–8 K.) and Olympiodorus (*CAAG* II 93,4–5; *CAAG* II 98,9) – was counteracted and transformed by the dryness of the other substances (in the *PM*'s recipe, various kinds of earth) which were combined with the metal.

53] The form 'flower of *oichomenion*' (οἰχομενίου ἄνθος) is puzzling, since a name like οἰχομένιον is not attested elsewhere. Zuber (see *supra*, p. 192) translated it as *Orchomenii florem*, probably having in mind *orchomenion* (Ὀρχομένιον, *ThGL* IV 2264) with the meaning attested by Hp. *De ulc.* 17 [VI 422,11 Littré; see Marie-Paule Duminil, *Hippocrate*. vol. 7: *Plaies, Nature des os, Coeur, Anatomie* (Paris: Les Belles Lettres, 1998), 66 n. 76], where it refers to a powder composed of several plants growing around the lake of Orchomenus [*Kopais*, in *RE* XI/2 (1922) 1346–60]. However, in Theophrastus' account of the typical flora of this area there is no mention of plants for dyeing (*HP* IV 10,1). On the other hand,

Berthelot's suggestion of reading ōkymenion (ὠκυμένιον) is striking (*CAAG* III 8 n. 1). Such a name – perhaps related to ōkimon (ὤκιμον), 'basil' – would be a *hapax*: the form ōkiminon (ὠκιμίνου σκευασία) is attested by Diosc. I 49,1 with reference to a unguent of basil. Comparison with other passages in the *Corpus alchemicum* does not shed new light on the question, since we find slightly different spellings, such as ēchoumenion (ἠχουμένιον in *CAAG* II 7,15, where it refers to safflower blossoms) and oichoumenon (οἰχούμενον in Zos. Alch. *CAAG* II 159,19–160,1, where it is listed among the plants with yellowing properties). If these parallels convince us that the term referred to a plant, they do not help us to reconstruct its original (and correct) spelling. We cannot rule out the possibility that the term is somehow related to the forms *echomenion/ochomenion* (ἐχομένιον/ὀχομένιον) attested by many papyri (for ἐχομένιον see *P.Oxy.* IV 729,31; XIV 1689,16; XXXI 2584,12; *P.OxyHels.* XLI 15; *PSI* XIII 1330,11; *SB* XII 10780,17; for ὀχομένιον see *BGU* IV 1017,11; *P.Mert.* I 17,13 and 22; *P. Oxy.* I 101,12; *P.Ryl.* IV 683,15; *PSI* IX 1036,7 and 1070,12). Although LSJ[9] 1281 (*s.v.* ἐχομένιον) proposed the meaning of 'coriander,' we must keep in mind Grenffel and Hunt's comment on *POxy* X 1279,18 (p. 222 n. 17): "the meaning of the word, which seems only to have been found in papyri from Oxyrhynchus, is uncertain."

54] According to Berthelot's interpretation, the recipe describes once again how to prepare a liquid dyeing drug for treating lead. Similar ingredients are also used for making golden inks in *PLeid.X.* 61 and the *Corpus Syriacum* (*CMA* II 206 rec. 17).

55] On this paragraph, see *supra*, pp. 63–4 and the notes on the Syriac version (*1SyrC* § 16 nn. 29–30).

Ps.-Democritus' *Book on the Making of Silver (AP)*

1] As in *PM* § 5, ps.-Democritus here prescribes solidifying the 'mercury' extracted from different arsenic ores. Although no explanation of the procedure is given, an interesting parallel is provided by Zosimus, who explains how to sublime arsenic ores with sulphur. The excerpt entitled *On the Evaporation of Sulphur Water which Makes Mercury Solid* (Περὶ τῆς ἐξατμίσεως τοῦ θείου ὕδατος τοῦ πήσσοντος τὴν ὑδράργυρον) describes how to process orpiment (*arsenikon*) according to the books written by Jewish alchemists: "Take orpiment and whiten it as follows: flatten oily clay by making a layer as thin as *lapis specularis*; riddle it with small holes like a sieve, and fit it firmly to a vessel that must contain one part of sulphur; the amount of orpiment you want is on the sieve; after covering with another vessel and encasing [the joints of the two vessels] with clay, cook it over two days and nights; you will find white-lead [ceruse]" (VIII 20–7 Mertens). For a reconstruction of the device described by Zosimus, see Mertens, *Zosime de*

Panopolis, CLXII. If we compare this technique with the title of the excerpt, we can infer that orpiment (*arsenikon*) was also somehow assimilated to mercury (*hydrargyros*) in ps.-Democritus' recipe. Since both arsenic ores (orpiment and realgar) and mercury were considered as 'fleeing substances,' alchemists may have tried to 'fix' them by processing them with sulphur, despite the fact that mercury was the only substance to change from a liquid state to a solid one. Arsenic ores are already solid prior to their sublimation with sulphur, unless we suppose that a 'solution' made of arsenic ores and sulphur was sublimed (in his title Zosimus refers to 'sulphur water'). Both the processes, in any case, were aimed at preparing a white solid compound that was to be laid on metals (such as copper or iron) and was expected to turn them white.

2] Both copper and iron are to be treated with the solid compound made from different kinds of 'mercury.' Robert Halleux, "Nouveaux textes sur la métallurgie antique," in *Mines et fonderies de la Gaule. Université de Toulouse-Le Mirail, 21–22 novembre 1980, table ronde du CNRS* (Paris: éd. du CNRS, 1982), 200 n. 19, recognized in *AP* § 1 the most ancient written testimony of an alloy composed of metallic arsenic ("alliage utilisant l'arsenic métallique"). Besides, the treatment of copper with arsenic ores or compounds is quite common in alchemical literature (*PLeid.X.* 22; Olymp. Alch. *CAAG* II 75; *CMA* II 47 rec. 7, and 77 rec. 9). For example, one recipe in the *Mappae clavicula* reads: "Making copper white. When [copper] begins to melt, add orpiment, not prepared, but fresh" (rec. 75 in Smith and Hawthorne, "Mappae Clavicula," 38). On this process of *dealbatio aeris* in medieval alchemy, see Robert R. Steele, "Practical Chemistry in the Twelfth Century. *Rasis de aluminibus et salibus* translated by Gerard of Cremona," *Isis*, 12 (1929): 37.

3] Both copper and iron are supposed to undergo a pre-treatment, as we can infer from the participle *theiōthenti* (θειωθέντι), 'which have been purified by sulphur.' Zosimus (*CAAG* II 179,10–3) claimed that ancient readers did not correctly understand the participle, since they took it to be synonymous with *kaenti* (καέντι) 'which have been burnt.' On the contrary, according to Zosimus' interpretation ps.-Democritus here prescribes treating copper with sulphur and iron with *magnēsia* (an unidentified substance; see Halleux, *Papyrus de Leyde*, 221). Paul T. Keyser, on the other hand, took the participle to refer only to iron, and translated it as "coppered iron," that is (n. 9), "iron plated with copper by immersion in [a solution of] χάλκανθον ['copper flower']" ["Greco-Roman Alchemy and Coins of Imitation Silver," *American Journal of Numismatics*, Second Series, 7–8 (1995–1996): 213]. However we cannot rule out the possibility that different methods were employed, as can be inferred from a Syriac recipe in which iron is processed with *magnēsia* and white sulphur (*CMA* II 69 rec. 2).

4] Ps.-Democritus lists several ingredients considered capable of whitening metals. In particular, the use of roasted cadmia (*kadmia*) and white-lead (*psymythion*) is largely confirmed by ancient sources [Forbes, *Studies*, vol. 8, 265–78; Paul

T. Craddock, *Early Metal Mining and Production* (Edinburgh: Edinburgh University Press, 1995), 292–302]. Various authors explained how to produce zinc-copper alloys (brass), whose colour varies according to the percentage of zinc: in particular, Theopompus *FGrH* 115 F 112; Strab. XIII 1,56 [Robert Halleux, "L'orichalque et le laiton," *L'Antiquité Classique*, 42 (1973): 65–6], Theophr. *Lap.* 49 and [Arist.] *Mir.* 835a 11; Plin. *NH* XXXIV 4 (Healy, *Pliny the Elder*, 310–4). According to Taylor's interpretation, ps.-Democritus here describes a "copper-zinc alloy with traces of arsenic," which would have been white when containing only 40% of copper ("A Survey of Greek Alchemy," 127). Similarly, *psymythion* (sometimes together with other lead compounds, particularly litharge) was often added to copper in order to produce a lead-copper alloy, which, in a second step, was usually mixed with silver (*PLeid.X.* 82; *Corpus Syriacum* in *CMA* II 210 rec. 1).

5] *You will melt iron by adding magnēsia.* Ancients were not able to melt iron completely because of the very high temperature required for such a process (Healy, *Pliny the Elder*, 326–7; Halleux, *Le problème des métaux*, 192–3). However, a few sources seem to claim that complete fusion was sometimes (and by chance) achieved when iron was processed with certain fluxes. Aristotle, for instance, wrote that "wrought iron indeed will melt and grow soft, and then solidify again. And this is the way in which steel is made" (*Mete.* IV 6, 383a 33–5, transl. by Lee, *Aristotle, Meteorologica*, 323). Halleux explained that iron, when roasted in contact with charcoal (working as a flux), occasionally reached a completely fluid state (*Le problème des métaux*, 193–8). Similarly, we can guess that in ps.-Democritus' recipe *magnēsia* (along with sulphur) could lower iron's melting point. Indeed, other alchemical writings – *On Iron Tempering* (Περὶ βαφῆς σιδήρου; *CAAG* II 342–5), followed by *On Tempering of Indian Iron* (Βαφὴ τοῦ Ἰνδικοῦ σιδήρου; *CAAG* II 347) – described how to process iron with carbon-rich compounds (Halleux, "Nouveaux textes," 196–7), thus witnessing a substantially developed technology related to the working of this metal.

6] *Take the abovementioned volatile substance.* Berthelot (*CAAG* I 72) identified this volatile substance with mercury (its sign ☽ is preserved over the term in the Byzantine manuscripts), which would have been mixed with tin, sulphur, and other ingredients to produce a white and whitening compound. This process would be quite similar to *PLeid.X.* 5: "*Manufacture of Asem* [silver]. Tin, 12 drachmas; mercury, 4 drachmas; earth of Chios, 2 drachmas. To the melted tin, add the crushed earth, then the mercury; stir with an iron, and put [the product] in use" (transl. by Caley, "Leiden Papyrus X," 1152). A tin amalgam was produced, which was used for tinning metals [Robert Halleux, "De stagnum «étang» à stagnum «étain», contribution à l'histoire de l'étamage et de l'argenture," *L'Antiquité Classique*, 46 (1977): 557–70; see also *PLeid.X.* 26; 84]. However, we should note that ps.-Democritus is here referring to the volatile substance he mentioned earlier, that is, "the mercury that comes from orpiment or from realgar"

(*AP* § 1, l. 1; see *infra*, *Cat.* § 3 n. 4). According to this interpretation, arsenic (rather than mercury) would have been mixed with tin to produce a whitening product (see also *Corpus Syriacum* in *CMA* II 66 rec. 35; 69 rec. 3). A similar process may also be recognizable in *PLeid.X.* 8 (= *PHolm.* 3), where copper is first processed with arsenic and then mixed with tin.

7] Metals that were involved in the making of alloys or amalgams were often pretreated and purified by means of sulphur and chlorides or other salts. For instance, *PLeid.X.* 2 reads: "*Another (purification) of Tin*. Lead and white tin also purified with pitch and bitumen. They are made pure by having alum, salt of Cappodocia and stone of Magnesia thrown on their surface" (transl. by Caley, "Leiden Papyrus X," 1151; see Halleux, *Papyrus de Leyde*, 167 nn. 1 and 3).

8] *Kobathia fumes*. The term *kobathia* (κοβάθια or κωβάθια) is attested only by alchemical writings. It was considered a sulphurous substance (*CAAG* II 10,15) and its fumes were likened to 'orpiment vapours'; see the *Lexicon on the Making of Gold*, *CAAG* II 9,15 (= M, fol. 133ʳ16; B, fol. 4ᵛ7–8; V, fol. 93ᵛ14; A, fol. 20ᵛ11–12): "*kobathia* fumes are the vapour of orpiment" (I have followed the reading of B and A which hand down the sign ϐ⁹, i.e. 'orpiment,' after 'vapour'). Zosimus (*CAAG* II 188,13) explained that *kobathia* were purple, perhaps referring to the red colour of realgar (arsenic sulphide). Sulphur or arsenic vapours were not only used in dyeing techniques related to metals. For instance, Apuleius (*Met.* IX 9, 24) explicitly claimed that sulphur fumes were employed by fullers for whitening cloth.

9] This recipe describes how to produce a white alloy imitating silver. It is not surprising to find brass (*oreichalkos*) among the ingredients, since its colour could be changed by treating the alloy with various ingredients. For instance, *PLeid.X.* 83 explains how to make a white copper-zinc-arsenic alloy by treating brass with arsenic ores (Halleux, *Papyrus de Leyde*, 181 n. 6). Slightly differently, in ps.-Democritus' recipe brass is first mixed with tin and then processed with a cement composed of *magnēsia* treated with sulphur or arsenic (*kobathia*) fumes.

10] *White sulphur*. This expression could refer either to the yellowish element that has been whitened by a preliminary treatment (see Diosc. V 107) or to a different ingredient. In the *Lexicon on the Making of Gold*, 'white sulphur' is said to be equivalent to 'copper flower' (*CAAG* II 5,8f.), 'earth of Samos' (*CAAG* II 6,18), 'treated lead' (*CAAG* II 8,22), '*nitron*' (sodium carbonate, *CAAG* II 11,17) and 'alum' (*CAAG* II 13,12). According to the whitening method described by ps.-Democritus, sulphur was ground with salty solutions (urine, salt brine): we could suppose that the sulphur particles became covered by a coating of salt, thus appearing white.

11] The first part of the recipe explains how to produce a white and shiny compound by mixing sulphur, alum, salt, and orpiment. This cement was expected to change the colour and mechanical properties of several metals. The whitening of copper by means of sulphur and arsenic compounds is often attested by alchemical writings (see *supra*, n. 2). In addition, the same ingredients could also harden both tin (depriving it of its peculiar 'cry,' the crackling sound produced when a tin bar is bent) and lead (see *PLeid.X.* 1–4).

12] Whitened litharge (see *supra*, *PM* § 6 n. 29), cadmia, orpiment, and copper-iron sulphides are here mixed together to produce a kind of white cement that was expected to whiten all the metals. Similar ingredients are included in a procedure described in *PLeid.X.* 11: "*Manufacture of Asem* [silver]. Purify lead carefully with pitch and bitumen, or tin as well; and mix cadmia and litharge in equal parts with the lead, and stir until the alloy is completed and solidifies. It can be used as natural asem [silver]" (transl. by Caley, "Leiden Papyrus X," 1152). Zinc and lead oxides were melted with lead or tin in order to produce white alloys (Halleux, *Papyrus de Leyde*, 87 n. 1; see also *Corpus Syriacum* in *CMA* II 241 rec. 13).

13] Ps.-Democritus here applies the same criterion explained in *PM* § 5 (see n. 24), according to which white compounds have whitening properties and yellow compounds yellowing ones. The reference to the overly strong fire is perhaps understandable if we consider that lead, when roasted, may produce yellow or red oxides.

14] Ps.-Democritus seems to have been aware of the close connection between litharge (lead oxide) and lead. This oxide was indeed a by-product of the melting and refining of lead ores (or lead-silver ores); for instance, the ancients melted lead sulphide (galena, PbS) in contact with charcoal, thus producing metallic lead in accordance with the following reactions: $2PbS + 2O_2 \rightarrow 2PbO$ (litharge) $+ 2SO_2$; $2PbO + PbS \rightarrow 3Pb + SO_2$; $3PbO + 2C$ (charcoal) $\rightarrow 3Pb + CO + CO_2$ [Jerome O. Nriagu, *Lead and Lead Poisoning in Antiquity* (New York: John Wiley and Sons, 1983), 86]. The production of different by-products from lead was explained in the light of the supposed "moist and unstable nature" of the metal (Steph. Alch. II 232,33 Ideler). Olympiodorus (*CAAG* II 92–3) claimed that lead could assume different colours by turning white (lead acetate or carbonate) and yellow (lead oxide); all the colours were concentrated in the natural blackness of the metal, whose nature was expected to change quickly [Cristina Viano, "Les alchimistes gréco-alexandrins et le *Timée* de Platon," in *L'alchimie et ses racines philosophiques. La tradition grecque et la tradition arabe*, ed. Cristina Viano (Paris: Vrin, 2005), 98]. In ps.-Democritus' opinion, when litharge becomes infusible (deprived of its moist nature), it also loses its capacity to turn back into lead.

15] This recipe is handed down by the so-called *Chemistry of Moses* under the title of *Washes for the Making of Silver* ('Ἀργυροποιίας ζωμοί). Comparison with the Syriac translation, which preserves a theoretical section (*2SyrC* § 5) that divides the book on the making of silver into two sections (§§ 1–4 and §§ 6–10), confirms that this recipe opened the second part of the original book, dealing with liquid substances or washes (*zōmoi*). *AP* § 6 actually describes the preparation of a paint able to whiten a superficial layer of various metallic leaves (copper, lead, iron; see *CAAG* I 72). These leaves were exposed to a double treatment: they were first dipped into a saffron-alum solution, then the solution was enriched with sulphur and arsenic sulphides and smeared on the metals.

16] *White orpiment.* The term *arsenikon* (ἀρσενικόν) refers to orpiment, a yellow sulphide of arsenic (As_2S_3; see Halleux, *Papyrus de Leyde*, 208; Healy, *Pliny the Elder*, 235) described by both Pliny (*NH* XXXIV 178) and Dioscorides (V 104). Alchemical sources usually insist on its golden or yellow/red colour: e.g. *PLeid.X.* 56,2 ἀρσενικὸν χρυσίζον, 'golden-colored orpiment,' as well as *CAAG* II 75,7f.; 150,18; 310,13; *PLeid.X.* 88,8 ἀρσενικὸν χρυσοῦν, 'golden orpiment'; *PM* § 5, l. 73 ἀρσενικὸν ξανθόν, 'yellow orpiment'; *CAAG* II 363,23 ἀρσενικὸν ἐρυθρόν, 'red orpiment.' However, in this passage ps.-Democritus specifies taking the 'white orpiment' (see also *Cat.* § 3, l. 30) without explaining how to treat the yellow ore and turn it white. On the other hand, Pliny claims that "to increase its efficacy, it [orpiment] is heated in a new earthenware pot till it changes its colour" (transl. by Rackham, *Pliny, Natural History*, 257). According to Bailey's interpretation, "the action of heat on orpiment is complicated. When heated gently, the yellow As_2S_3 is transformed into a red modification of the same composition, which regains its yellow colour on cooling. On further heating, some sulphur is lost and black variety of realgar, As_2S_2, is formed. This, on cooling, changes to the familiar red sulphide (*sandaraca*) … If heated still further, oxidation sets in and sulphur dioxide and arsenic trioxide (As_2O_3) are formed. The presence of a little of the latter substance, the toxic 'white arsenic,' is probably responsible for the increase in potency referred above" (*Pliny's Chapters*, vol. 2, 207). Ancient alchemists often describe similar processes for roasting and subliming orpiment, which perhaps produced a certain amount of white arsenic oxide. Zosimus (VIII 20–7 Mertens), for instance, explains how to sublime orpiment with sulphur (see *supra*, n. 1, and Mertens, *Zosime de Panopolis*, 198 n. 8); moreover, according to Olympiodorus (*CAAG* II 75,5–76,2), orpiment was first ground with vinegar, and afterwards distilled in a specific vessel called *asympoton* (ἀσύμποτον, 'made of non-absorbent material,' LSJ⁹ 265; Berthelot, *CAAG* III 82 n. 6, supposed that such a process could produce arsenic acid). A similar procedure is also attested in *CAAG* II 318,7–15 and 391,10–21.

17] The distinction made by ps.-Democritus is not clear and the central role played by saffron in the recipe is surprising. This ingredient, in fact, is usually employed for the making of gold paints (*PLeid.X.* 38; 56; 72; *PM* § 18). We cannot exclude the possibility that the author was using a code name; according to the *Lexicon on the Making of Gold* (*CAAG* II 8,9), for instance, saffron could correspond to divine water, a highly reactive liquid able to attack the surface of metals and change their colour (see *supra*, PM § 11 n. 38).

18] The recipe describes how to produce a wash by crushing various white ingredients and dispersing them in honey: (a) white litharge; perhaps litharge treated with salt, vinegar, *nitron* or honey (see *supra*, PM § 6 n. 29); (b) bay leaves that are usually whitish; (c) 'white realgar,' perhaps arsenic oxide produced by treating realgar with the same methods used for orpiment (see *supra*, n. 16); (d) Cimolian earth, i.e. white fuller's earth (see Halleux, *Papyrus de Leyde*, 211). Some of these ingredients are employed for making silver paints in similar recipes preserved in both the Leiden papyrus (*PLeid.X.* 61; 77) and the *Corpus Syriacum*: in CMA II 208 rec. 34, sulphur, bay leaves, litharge, and soda are mixed together to make a paint with which silver letters can be written on copper. However, we must note that ps.-Democritus' recipe (in both the Greek and Syriac versions) makes no mention of the metal (or the metallic leaves) that are to be treated.

19] *Mixtures that are not thick [anousia migmata].* According to Zosimus, the adj. 'without substance' (*anousia*) refers to those sulphurous ingredients that do not resist high temperature (*CAAG* II 168,15 τὰ ἀνούσια τὰ θειώδη τὰ μὴ ὑφιστάμενα τῷ πυρί). The same meaning seems to have been already clear to ps.-Democritus, who recommended avoiding fire that could affect the properties of the drug. In addition, specific watery ingredients were mixed with this drug in order to counteract the effect of the fire. Zosimus (*CAAG* II 189,13–7) explains that the expression "water of ashes of white poplar" – a solution made by percolating water through white poplar ashes (see Zos. Alch. IV 54–5 Mertens and the commentary at p. 149 n. 13) – must not be taken in its usual sense, since he supposed that ps.-Democritus was here hinting at the "divine water composed with quicklime" (see Martelli, "Divine Water," 18–22).

20] *The abovementioned volatile substance.* Ps.-Democritus could be referring either to mercury, as the Syriac translator seem to have interpreted (see 2SyrC § 8) or to the arsenic mentioned in *AP* § 1. Both these ingredients are often used in the making of whitening washes.

21] As already noticed by Zosimus (*CAAG* II 153,1; 155,2 and 17; etc.), in this recipe the required amount of each ingredient is specified (the only example of such specificity among the ps.-Democritean recipes). Among the substances mentioned, the plant called *Persaion* (Περσαίον) poses some problems of identification. The manuscripts record two different spellings which seem to be equivalent: *persea* (περσέα A; see also Zos. Alch. *CAAG* II 164,1) and *Persaion* (Περσαίον *scil.* φυτόν), 'Persian plant' (**MBV**). The first form is attested as the name of an Egyptian plant that has not been definitively identified (André, *Les noms des plantes*, 193 *s.v.* *Persea*): scholars have proposed both the *Cordia myxa* L. and the *Mimusops laurifolia* Friis (= *M. schimperi* L.). Ancient sources (Diosc. I 129; Plin. *NH* XV 45; Orib.

I 63 Raeder in *CMG* VI/1, 1, p. 72) sometimes confused *persea* with peach and explained that this plant, even if of Persian origin, started producing sweet fruits only after having been transplanted to Egypt. The same anecdote is told by Galenus (*De semine* II 42,1–3 De Lacy in *CMG* V/3, 1, p. 164, 13–5) about *Persaion* and is likely to derive from Bolus of Mendes' writings (at least according to a *scholion* to Nic. *Ther.* 764 = 68 [55] B 300,4 D-K; see *supra*, p. 40).

Ps.-Democritus' *Catalogues (Cat.)*

1] *Mercury that comes from cinnabar.* The Ancients used to extract mercury from its natural ore (cinnabar, HgS) using two different techniques (Halleux, *Le problème des métaux*, 179–88). The first method was a cold extraction, whereby cinnabar was pounded with vinegar (or water) in a mortar with a pestle made of various metals. Theophrastus (*Lap.* 60) specified that "quicksilver is made by pounding cinnabar with vinegar in a copper mortar with a copper pestle" (transl. by Eichholz, *Theophrastus. De lapidibus*, 81). The process, also described by Pliny (*NH* XXXIII 123), has been identified by Laszlo Taracs, "Quicksilver from Cinnabar: the First Documented Mechano-chemical Reaction?," *Journal of Minerals, Metals and Materials Society*, 52 (2000): 12–3, with a 'mechanochemical reaction' (that is, the vinegar or water had a purely mechanical action, allowing for a closer physical contact between the solid ingredients that were expected to react), which probably produced a copper amalgam (Eichholz, *Theophrastus. De lapidibus*, 128). The reaction, however, would have been very slow in the opinion of Bailey, who tried to replicate the method (Bailey, *Pliny's Chapters*, vol. 1, 233). A source of heat would have accelerated the process. In this respect, it is interesting to note that later alchemists stipulate that similar processes should be carried out in the sun; Zosimus (*CAAG* II 172,16–8) specified that "all the [alchemical] books and both Chymes and Maria say: lead mortar and lead pestle. Pound the cinnabar with vinegar in the sun until mercury (νεφέλη, lit. 'cloud') is formed." In addition, **SyrC**, fol. 60ᵛ2–5 has: "*Other Section.* Take a lead mortar and pestle; put cinnabar in it and pound it with water until mercury is formed. Other section: other people take vinegar and pound [cinnabar] in the sun."

The second method was an extraction at elevated temperatures that exploited the relatively low boiling point (356 degrees C) of mercury. It was often performed simply by heating cinnabar. Vitruvius (*De Arch.* VII 8,1–4) explained that droplets of mercury evaporate when cinnabar is brought to a high temperature: see Earle Radcliffe Caley, "Mercury and its Compounds in Ancient Times," *Journal of Chemical Education*, 5 (1928): 420. Dioscorides (V 95) and Pliny (*NH* XXXIII 123) describe a more sophisticated method that involved heating cinnabar in a simple device: an iron saucer containing cinnabar was placed on a earthenware pan, covered by another vessel called an *ambix*, and heated; drops of mercury then condensed on the covering *ambix* [F. Sherwood Taylor, "The Evolution of the Still," *Annals of Science*, 5 (1945): 188; Mertens, *Zosime de Panopolis*,

CXIX]. In this case, the mercuric sulphide of the cinnabar would be reduced by the metal of the saucer; the iron combined with the sulphur, thus liberating mercury which evaporated and recondensed on the colder surface of the *ambix*, from which it could be collected. In alchemical writings, a similar distillation method seems to be mentioned by Stephanus, who wrote in his *Letter to Theodoros* (II 208,19–24 Ideler): "Truly there is moist vapour and dry vapour. For the moist is distilled using the *phanoi* which have nipples [alembics]. But dry vapour <is distilled> using the pot and bronze cover [a device similar to that described by Pliny and Dioscorides], as is the white vapour from cinnabar" [transl. by F. Sherwood Taylor, "The Alchemical Works of Stephanos of Alexandria. Translations and Commentary, Part. II," *Ambix*, 2 (1938–1940): 39 (slightly modified)]. Two methods of distillation are clearly distinguished here: one method for "dry" (non-aqueous) vapours that consisted of heating materials such as cinnabar at high temperatures in a pot with a cover, and another for "humid" vapours (low boiling-point liquids) performed at lower temperatures in alembics. The philosopher Anonymous, a Byzantine alchemist, ascribed another method for isolating mercury from cinnabar to ps.-Democritus himself (*CAAG* II 123,3–7): "This very famous philosopher [Democritus] said: 'Who does not know that the vapour of cinnabar is the mercury of which it is composed? Therefore, if anyone grinds cinnabar with oil of soda (νιτρελαίῳ), mixes them together, puts them in the double vessels (ἐν ἄγγεσιν διπλοῖς) and lights a persistent fire, he will collect the entire vapour'." Here, cinnabar is first ground with *nitron* oil (probably a thick, concentrated solution of sodium carbonate in water) and the mixture heated. The small amount of water served to bring the cinnabar and alkali salt into closer contact. Upon heating, the sodium carbonate combined with the sulphur of the cinnabar, thus freeing the mercury to evaporate at elevated temperatures and recondense in cooler parts of the "double vessel" from which it could be collected (see also *CAAG* II 208,14–6). As we shall see, a similar procedure may have been applied not only to cinnabar, but also to other minerals (see *infra*, n. 4).

2] The text has been here restored on the basis of the comparison with Syn. §§ 13–14, ll. 225–8 (see Martelli, *Pseudo-Democrito*, 378–9). Ps.-Democritus drew a distinction between two different sulphur waters (Martelli, "Divine Water," 19–22): (a) the proper water of sulphur, i.e. a solution made with sulphur alone, where the adj. 'untouched' (ἄθικτον) seems to stress that this 'water' was free of any other component; (b) a solution of sulphur and quicklime, which can probably be identified with the sulphur water described by *PLeid.X.* 87: "*The Invention of Sulphur Water.* A handful of quicklime and another of sulphur in fine powder; place them in a vessel containing strong vinegar or the urine of a small child. Heat it from below, until the supernatant liquid appears like blood. Decant this later in order to separate it from the deposit, and use" (rec. 89 in Caley, "Leiden Papyrus X," 1161). A similar distinction is also clearly mentioned by the later alchemist Christianus (*CAAG* II 399,18–400,3), while other alchemical texts claimed that 'sulphur water' understood in a general sense is a solution of the two abovementioned minerals. For instance, a passage ascribed to Zosimus by Berthelot-Ruelle (*CAAG* II 208,14–6; but see Letrouit, "Chronologie," 36 n. 93) reads: "Divine water in a non-specific sense (τὸ δὲ ἀπολελυμένον ὕδωρ θεῖον): this is the water made with two parts of quicklime and one part of sulphur, boiled in a pot, filtered and boiled once again" (see also *CAAG* II 8,9–10).

3] On the restoration of this passage, see Martelli, *Pseudo-Democrito*, 381–2. Ps.-Democritus here specifies the different metals that should be treated with the aforementioned washes, and the different results reached by treating each of them (I have given to *dia* + acc. the sense of 'for the sake of,' 'with the purpose of'). A similar statement is also handed down by the *Turba philosophorum*, in precisely the section ascribed to Democritus [chap. LXXII in Julius Ruska, *Turba philosophorum. Ein Beitrag zur Geschichte der Alchemie* (Berlin: Julius Springer, 1931), 170, ll. 10–7]: "Dimocras (Democritus M) … dixit enim: 'pone plumbum, ferrum, et albar propter aes' [; proinde reversus ait:] 'et aes nostrum propter nummos, <et> plumbum propter aurum, ac aurum propter aurum coralli, corallique aurum propter aurum ostri'" ("Democritus … indeed said: 'lay lead, iron and *Albar* for [making] copper' [then he says the other way around] 'and our copper for [making] coins, and lead for gold, and gold for gold of coral, and gold of coral for gold of purple'").

4] *Mercury that comes from orpiment, or from realgar.* It is of course impossible to extract mercury from arsenic sulphides such as orpiment (As_2S_3) or realgar (AsS). Some scholars (*CAAG* I 238f.; Halleux, "Nouveaux textes," 200 n. 19) interpret this expression as hinting at the extraction of metallic arsenic. However, the simple combustion of arsenic sulphides can hardly produce this result, because any metallic arsenic produced would oxidize very quickly during the process (as Bailey, *Pliny's Chapters*, vol. 2, 207, pointed out in his comments to Plin. *NH* XXXIV 178; see also Forbes, *Studies*, vol. 9, 177–8; Craddock, *Early Metal Mining*, 289–90). On the production of arsenic oxide by subliming orpiment or realgar, see *supra*, AP § 6 n. 16. On the other hand, we cannot exclude the possibility that ps.-Democritus tried to process orpiment and realgar with the same methods used for extracting mercury from cinnabar (see *supra*, n. 1). In this respect, the method mentioned by the philosopher Anonymous – first grinding cinnabar with *nitron* oil and then subliming it; see *CAAG* II 123,3–7 – is particularly interesting since, when applied to orpiment or realgar, it could possibly produce metallic arsenic successfully, assuming that a large enough quantity was heated in a relatively well-sealed container (like the pot with a close-fitting lid). Unfortunately ancient sources do not explicitly describe such a process. According to *AP* § 9, orpiment is triturated in a liquid that also contains *nitron*, but it is not sublimed (although Zosimus, *CAAG* II 163,22–164,5 commenting upon this passage, seems to refer to mercury). However, we can guess that it would be a logical step, after learning that *nitron* oil liberates the mercury from cinnabar, to attempt the same with orpiment. By this analogy, the liberated arsenic from orpiment might well be called 'mercury' because it is a 'dry' vapor prepared in the same way. In any case, ps.-Democritus seems to mention different kinds of oil used for treating arsenic compounds. In particular, *AP* § 2, ll. 9–10 may be read as a method for boiling arsenic oxide "with castor and radish oil," if we identify the volatile substance mentioned at the beginning of the recipe with the oxide produced by roasting orpiment or realgar [see C. Anne Wilson, "Distilling, Sublimation, and the Four Elements: The Aims and Achievements of the Earliest Greek Chemists," in *Science and Mathematics in Ancient Greek Culture*, ed. Christopher J. Tuplin and Tracey E. Rihll (Oxford: Oxford University Press, 2002), 308]. In Multhauf's opinion, if the arsenic oxide "be fused in turn with oil or gum, we obtain a black sublimate, the element itself" (Multhauf, *Origins of Chemistry*, 108; see also 230–1). In such a process (a sublimation that had to be done out of contact with fresh air), the residue of the burning of oils could perhaps act as a reducing agent to produce metallic arsenic. A similar interpretation seems to be supported by a passage in *Isis to her son Horus*: "In this way vapour rises up: take some orpiment, boil it in water, put it in a mortar and grind it with an ear of corn and with oil; put it in a casserole and close its mouth with a bowl; light a charcoal fire until the vapour rises up; do the same with realgar" (*CAAG* II 32,23–33,3 = Mertens, *Un traité gréco-égyptien d'alchimie*, 138 [text] and 122 [comment]; in Mertens' opinion, this recipe could stem from ps.-Democritus' books). The recipe, in fact, seems to describe the reduction of arsenic sulphides, with the corn and oil providing the carbon for the reduction.

5] *[Mercury that comes from] Italian stibnite.* The association between mercury and antimony ores (*stîmi*) could refer to methods for extracting the metal. Pliny (*NH* XXXIII 101) specified that the term *stibi* referred to a white and shiny petrified *spuma* (lit. 'foam') which occurred in silver mines. The ancients were able to treat such *spuma* to obtain both white antimony oxides (Sb_2O_3) and probably metallic antimony, which was confused with lead because of its colour. Pliny (*NH XXXIII* 103) and Dioscorides (V 84) explained how to roast stibnite (antimony sulphide, Sb_2S_3) under a layer of dung and charcoal: antimony oxide was first produced along with sulphur dioxide; the antimony oxide was then reduced in contact with charcoal, probably producing some quantity of antimony (Bailey, *Pliny's Chapters*, vol. 1, 214).

6] See *1SyrC* § 3 n. 7.

Synesius' commentary

1] This saying about natures was probably already known in Hellenistic times; in fact, the second century BC astrologer Nechepso explicitly claimed that one nature is overcome by another one ("natura alia natura vincitur") [fr. 28 (2) Riess, "Nechepsonis et Petosiridis Fragmenta," 379 = Firm. Mat. *Mat.* IV 22,2]. According to Firmicus Maternus' interpretation, this formula explains the influence of planets and decans on the human body and health. However, the rules of sympathy and antipathy that regulated astrological connections (planets-men-animals-plants-minerals) were probably applied to other disciplines. Ostanes (Bidez-Cumont, *Mages*, vol. 1, 175–8) is actually presented by ancient sources as an astrologer, magician (Plin. *NH* XXX 1–2), and expert in a kind of magical pharmacopeia (Plin. *NH* XXVIII 5–7), based on knowledge of the secret properties of natural substances (Bidez-Cumont, *Mages*, vol. 1, 188–98). For instance, Dioscorides' *De materia medica* and ps.-Apuleius' *Herbarium* hand down the 'magic names' of the plants used by the Persian magus (Bidez-Cumont, *Mages*, vol. 2, 299–301; Boscherini, "L'Erbario di Apuleio"). Unfortunately the *Corpus alchemicum* does not allow us to understand the extent to which such knowledge was reinterpreted within an alchemical framework. Ostanes is rarely mentioned in the Greek alchemical treatises now extant (see Bidez-Cumont, *Mages*, vol. 2, 317–56), and the only work handed down under his name, *Ostanes the Philosopher to Petasios on the Holy and Divine Art* (*CAAG* II 261–2), seems to be a later text that cannot be used to reconstruct its 'original' production.

2] A certain scepticism about the use of liquid substances (washes) in dyeing procedures is also attested (although rejected) by *The Chemistry of Moses* (*CAAG* II 305,10–1): "There are people who do not trust the utility of washes, but they do not provide any demonstration [of that] with their work; so consider the utility of washes!" Synesius may have restricted a similar criticism to just the liquid substances extracted from plants. Such a position, in fact, could explain the allegorical interpretation he provided of the plant juices listed by ps.-Democritus, that is, that their names secretly hint at procedures for dissolving solid minerals. Zosimus had already expressed some doubts about the efficacy of plant juices for colouring metals (*CAAG* II 149,20–150,4), and tried to interpret most of the vegetable and organic liquids mentioned by ps.-Democritus as references to solutions of active mineral substances (the so-called 'divine or sulphur water'; see *CAAG* II 184–5 and Martelli, "Divine Water," 9–10).

3] The importance of procedures for dissolving mineral substances is stressed by the comparison (and the opposition) between Persian and Egyptian dyeing methods [Matteo Martelli, "La tradizione tecnico-artigianale e l'influenza orientale: lo Pseudo-Democrito alchimista," in *La scienza antica e la sua tradizione*, ed. Franco Repellini and Gianni Micheli (Milan: Cisalpino, 2011), 192–212]. Synesius testifies that Democritus shared Ostanes' criticism of Egyptian techniques (see also Steph. Alch. II 213,3 Ideler), which seem to have been based on roasting base metals (ὄπτησις, i.e. cooking in contact with fire), upon which the dyeing compositions were laid (ἐπιβολαί). The Persian magician, in contrast, preferred to smear the dyeing substances on metallic leaves and heat them gradually. The same procedure is also given by the philosopher Anonymous, who saw (*CAAG* II 264,12ff.) an example of this method in *PM* § 17 and preserved the following quotation by Ostanes: "It is necessary to dip metallic leaves into washes and to smear the dyeing drug on in this way: so, he claims, they will be quickly dyed" (*CAAG* II 265,5–6; the Greek text must be read in accordance with M, fol. 91ᵛ27–8: ἐμβάπτειν δεῖ τὰ πέταλα τοῖς ζωμοῖς καὶ οὕτω ἐπιχρίειν τὸ φάρμακον· οὕτως γάρ, φησίν, εὐχείρως προσδέξεται τὴν βαφήν). A similar procedure implies the use of viscous dyeing compounds, which were probably prepared by dissolving minerals (see Zos. Alch. *CAAG* II 168,11–7; Phil. Anon. Alch. *CAAG* II 264): the 'watery' consistency of the colorants prevented them from being affected by too intense a fire (which could make them evaporate before the dyeing process was completed).

4] The verb *exydatoō* (ἐξυδατόω), 'make watery, change into water' (LSJ⁹ 599) was misunderstood by Berthelot, who translated the sentence "si tu ne les épuisses pas de leur partie liquide" (see also Garzya, *Sinesio di Cirene*, 805: "se non li privi delle parti liquide"). On the contrary, the verb refers to the opposite process, as already pointed out by the Latin translators, both Pizzimenti (*Democritus Abderita*, 12ʳ19: "in aquam convertas") and Zuber (see *supra*, p. 197: "aqueam in naturam converteris"). The process is often attested in medical texts, where it usually refers to the dilution of liquid substances (urine, blood, milk), which are made watery (e.g. Hp. *Coac.* 564,3, Littré in vol. 5, p. 712,6; Hp. *Superf.* 17,6 Lienau in *CMG* I/2,2, p. 80,4; Diosc. II 75,3; Sor. *Gyn.* II 24,6 Ilberg in *CMG* IV, p. 72,3f.; Aet. IV 6,33–5 Olivieri in *CMG* VIII/1, p. 363,10–2). In an alchemical context, the verb can refer also to solid substances, which are changed into water by being often mixed with liquids. For instance, Zosimus used the verb for describing a process wherein "a [solid] residue is properly triturated, made watery, and washed six or seven times with sweet waters" (*CAAG* II 223,19–21 Τοῦτο δὲ σκωρίδιον λείωσον καλῶς καὶ ἐξυδάτωσον καὶ ἀπόπλυνον ἑξάκις καὶ ἑπτάκις ἐν γλυκοῖς ὕδασι κτλ.; see also *CAAG* II 203,5; 432,9; 450,2).

5] The analogy between 'flowers' (ἄνθη) and dyeing 'waters' (ὕδατα) composed from minerals is also stressed in § 6, ll. 65–8, where Synesius claims that "by speaking about flowers he [i.e. ps.-Democritus] showed us that the waters are obtained from solid substances." The reference to the plants producing flowers supports the impression that, in Synesius' interpretation, minerals as well as plants were to be 'watered' in order to flourish and develop their dyeing properties. A similar analogy seems to have been developed by the alchemist Pelagius (fourth century AD; see Letrouit, "Chronologie," 46–7), who compared both the still and the related chemical procedures to a tree: in the alembic, substances could be turned into lighter and more active forms through heating and distillation; in the same way as a tree, when carefully watered and fed by the heat of the sun, can produce wonderful flowers (*CAAG* II 260,24–261,8).

6] The ps-Democritean expression 'rhubarb from Pontus' (Πόντιος ῥᾶ; see also *PM* §
17; on the specific meaning of the adj. *Pontios*, 'from Pontus' rather than 'of the sea,'
see Martelli, *Pseudo-Democrito*, 395) was allegorically interpreted by Synesius, who
folk-etymologized both the adjective and the name (see also § 16, ll. 256–8). The
term *rhā* (ῥᾶ) was read as stemming from the verb *katarrheō* (καταρρέω), 'to flow
down, stream down' (LSJ⁹ 909), and the adj. *Pontios* (Πόντιος) was related to the
currents of the sea. The expression would thus have hinted at the dissolution of minerals, which were ground up, mixed in a liquid and perhaps distilled. Analogous allegorical readings are attested by other alchemical writings. First of all, the term
'rivers' (ποταμοί) seems to have been used as a code name for mercury and
sulphur water (*CAAG* II 14,15 and 18; II 20,1). In addition, the Byzantine alchemist
Stephanus provided a detailed explanation of the same expression 'rhubarb from
Pontus' in his seventh *Lecture on Chysopoeia* (II 234,16–25 Ideler = **M**, fol. 31ʳ;
B, fol. 70ᵛ-71ʳ; **A**, fol. 64ᵛ; **V**, fols. 46ᵛ-47ʳ): "In fact rhubarb comes from Pontus;
for the sea is called 'pontos' by them [i.e. ancient alchemists]. Rhubarb from
Pontus is very valued, [it means] that the [dyeing] compound has been entirely
worked in a mortar: in fact they gave to the metal of the mortar the name of
'pontos,' since it is ocean colour, and to the 'sea' [i.e. the water] that appears in
the mortar the name of '*rha*' ... How could waters overflow without the ocean
[read ὑπάρχοντος **BAV**ᵐᵍ, rather than ὑπάρχοντα]? How could rain [read ὄμβρων
V, rather than ὄγκων] come down if clouds do not suck (ἀνασπάω) dewy rainwater
from it [i.e. the ocean]?" The alchemist clearly refers to specific devices (mortars)
employed for crushing substances and making them liquid. Moreover, the meteorological explanation at the end of the passage seems to hint at distilling processes: the
dewy rainwater and the cloud could correspond to the vapours that were 'sucked' by
the upper part of the alembic. We could read the picture handed down by **M**, fol. 10ʳ,
between the first and the second *Lectures* of Stephanus (texts that seem to have
nothing to do with the picture; see *CAAG* I 141–2) in the light of such procedures:

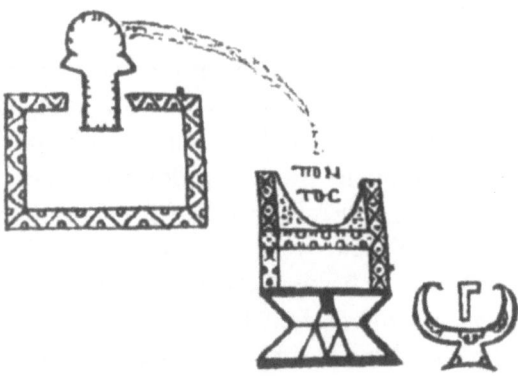

Alchemical equipment.
Source: Maricianus gr. 299, fol. 10r
Courtesy of Biblioteca Nazionale Marciana

On the right one can recognize a mortar and pestle, and in the centre a bigger mortar that is labelled with the term *pontos,* as in the abovementioned passage by Stephanus. The drawing on the left could represent a distillatory device (see *CAAG* I 142), perhaps used for distilling the substances crushed and mixed with liquid ingredients in the mortar. The lines that link the two instruments establish a connection – as in Stephanus' passage – between the 'waters' prepared in the mortar and the distilled 'waters.' A cyclical process, analogous to the meteorological phenomenon of rain, seems to be presupposed: the waters used to dilute the dry materials in the mortar can be the products of the distillation, just as the materials prepared in the mortar can be distilled.

7] As pointed out by Bidez-Cumont (*Mages*, vol. 2, 313 fr. A4b), Ostanes made Democritus swear to keep his alchemical knowledge secret. A passage handed down in **SyrC**, fol. 144^v, supports this interpretation, since the Persian magician is said to have ordered his pupils to not reveal his words (*CMA* II 326–7). In addition, in the Greek text *Isis to her son Horus* (*CAAG* II 29–8 = Mertens, *Un traité gréco-égyptien d'alchimie,* 131–2), the Egyptian goddess is forced by the angel *Amnaël* to swear a complicated and holy oath (analysed by Mertens, "Une scène d'initiation alchimique") before receiving the revelation about *chrysopoeia* (see also *CAAG* II 27,5–7; *CMA* II 320).

8] Both 'whitening' and 'yellowing' were considered to depend upon the duration and intensity of heating. Ps.-Democritus specified that the same compound, if properly heated, would turn white, but, if heated with too strong a fire, could turn yellow (*AP* § 5). The same idea is expressed by the Byzantine alchemist Christianus with reference to the distillation of eggs, which could produce either a white or a yellow water depending on the intensity of the fire (*CAAG* II 405–6). According to a passage perhaps ascribable to Zosimus, the increasing intensity of heat could make white substances turn yellow (*CAAG* II 208,9–11). In light of this information, we must interpret the term *anazōopyrēsis* (ἀναζωοπύρησις, from ἀναζωπυρέω, lit. 'rekindle, light up again,' LSJ⁹ 104) as referring to a procedure by which fires were rekindled and strengthened, perhaps by means of bellows (Zos. Alch. *CAAG* II 147,16; 182,13). We cannot exclude the possibility that this method was also meant to 'rekindle' and regenerate the burned substances. For instance, according to Stephanus, copper ashes were regenerated by the action of heat and air which could instil a soul (ψυχή) and a vital spirit (ζωτικὸν πνεῦμα) into it again [II 210,6–20 Ideler; see Maria K. Papathanassiou, "L'oeuvre alchimique de Stéphanos d'Alexandrie: structure et transformations de la matière, unité et pluralité, l'énigme des philosophes," *Chrysopoeia*, 7 (2000–2003): 19–21].

9] Synesius here draws a false etymology for the expression 'dog's milk' (see Stéphanidès "Notes sur les textes chymeutiques," 315). He reads the genitive *kynos* (κυνός, 'of dog') as stemming from the adj. *koinos* (κοινός), 'common, ordinary.' In alchemical contexts this adjective is often associated with the names of various substances, such as 'ordinary silver' (ἄργυρος κοινός, *CAAG* II 19,5; 130,4; 156,1), 'ordinary lead' (μόλυβδος κοινός, *CAAG* II 37,13 and 15; 93,14; 94,8), 'ordinary vinegar' (ὄξος κοινόν, *CAAG* II 12,5). As pointed out by Papathanassiou ("L'oeuvre alchimique de Stéphanos," 18–9) with reference to Stephanus' writings, similar specifications make it clear that alchemists employed – at least in such cases – substances that were also used by practitioners in other fields (e.g. pharmacy and medicine; see Zos. Alch. *CAAG* II 217,19–21). In addition, these ingredients were probably processed by means of distilling devices: the verb *anapherō* (ἀναφέρω), in fact, is often attested with reference to the vapours that 'rise up' through the alembic (Moyses Alch. *CAAG* II 301,4 ἀνένεγκαι διὰ τοῦ ἀμβίκου; Steph. Alch. II 217,17 Ideler τῆς ἄνω ἀναφερομένης αἰθάλης; II 224,8 ἡ ἀναφερομένη ἀτμίς; II 241,8 τὸ θεῖον ὕδωρ τὸ ἐκ τοῦ συνημμένου ὀργάνου ἀναφερόμενον).

10] Synesius is here paraphrasing the text of Ps.-Dem. *Cat.* § 2, ll. 18f., by playing with the similarity between the verbs *metalloioō* (μεταλλοιόω), 'to change, to transform' (LSJ⁹ 1114) and *metalleuō* (μεταλλεύω), 'to mine' (in particular metals; see Halleux, *Le problème des métaux*, 22). The same association is attested by Stephanus in his third *Lecture on Chrysopoeia* (II 209,6–8 Ideler): "So also the bodies, being made metallic (μεταλλευόμενα, lit. 'being mined') and being changed (μεταλλοιούμενα) from the contrary nature, become by a certain method level and aetherial" (transl. by Taylor, "The Alchemical Works of Stephanos," 39–41). Taylor interpreted the participle *metalleuomena* as referring to metallic bodies (the same use is attested in Olympiodorus' commentary on *Meteorology* IV; see Viano, *La matière des choses*, 163ff.), while Papathanassiou ("L'oeuvre alchimique de Stéphanos," 20–1) read it with the ordinary meaning of 'being mined.' Despite this difference, it is likely that neither Synesius nor Stephanus had in mind only the extraction of minerals (in particular metallic minerals) from the mine, but also the purifying treatments undergone by such ores.

11] According to the ps.-Democritus' passage quoted by Synesius, the internal, hidden nature of the *sōmata* (σώματα, lit. 'bodies') could be extracted by means of *hydrargyros* (ὑδράργυρος, 'mercury'), since mercury attracts any kind of body. The wide range of meanings that the two terms assume in alchemical literature makes it impossible to give a single technical explanation. On the one hand, ps.-Democritus could refer to mercury's property of easily amalgamating with metallic bodies; the later alchemist Olympiodorus referred to Synesius' dialogue by commenting: "All the ancients know that [mercury] is white, 'fleeting' and unstable [lit. 'without substance'], but that it can absorb any fusible body and attract it to itself" (*CAAG* II 90,21–91,1). This mercury-metal amalgam was then distilled, perhaps with the aim of purifying the amalgam and making it whiter (thus extracting mercury again; see *The Book of Crates* in *CMA* III 55). On the other hand, the term *sōmata* may have referred to any kind of white solid substance, such as the

Chian earth, *asteritēs* and white cadmia listed in the previous lines (similar ingredients are distilled in order to produce a white sulphur water in Zos. Alch. IV 68–75 Mertens). In this case, 'mercury' could refer to any kind of liquid substance with which such ingredients were mixed before being distilled.

12] Synesius here provides the description of a distillatory device taken from a ps.-Democritean text which has not survived (Taylor, "The Evolution of the Still,", 195–7; see also *AP* § 10). It must be considered, along with Maria's passages quoted by Zosimus, as one of the earliest references to a true alembic, composed of different parts that are given specific names, but which do not match Maria's nomenclature [Zos. Alch. II 1–5; III 1–27 Mertens; see Mertens, *Zosime de Panopolis*, CXVI-CXXX; Matteo Martelli, "Greek Alchemists at Work: 'Alchemical Laboratory' in the Greco-Roman Egypt," *Nuncius*, 26 (2011): 291–305].

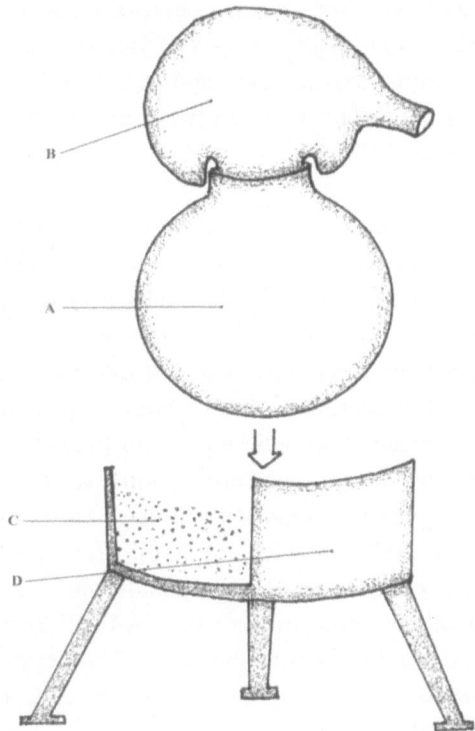

(A) The lower part of the device is called the *bōtarion* (βωτάριον); a term attested only in alchemical literature, which must be read as the diminutive of *bōtion* (βωτίον), synonymous with *stamnion* (σταμνίον; see Hsch β 1407,1 L.), i.e. a glass or ceramic vessel for wine (Pollux *Onom.* VI 14,3 Bethe).
(B) The upper part corresponds to a second vessel provided with a *mastarion* (μαστάριον), i.e. a breast-shaped protuberance. The *Corpus alchemicum* preserves various references to breast-shaped devices (*CAAG* II 199,5; 210,12; 275,12; 278,12; 291,13 etc.), amongst which a passage by Zosimus presents strong

similarities with Synesius' description (Zos. Alch. IX 7–18 Mertens): "Then mix them [eggs] together, and taking other eggs, and breaking them, put them in the vessel ... and luting all around the *ambix* [i.e. upper part of the alembic] and the breast-shaped cup [*mastarion*] with its receiver ... give it to be heated" (transl. by Taylor, "The Evolution of the Still," 198). In Synesius' passage, however, there is no reference to the receivers.

(C–D) A cauldron (λεβής) containing warm ashes was put under the device, which was gently heated in order to ensure a gradual evaporation of the contents. The part in contact with the ashes is the *bōtarion*. An anonymous commentator – perhaps in order to elucidate the rare term – added the note (afterwards included into the text) "that is a *kērotakis*." This last term is problematic and could refer to either a particular device for colouring metallic leaves by means of sulphur vapours (Mertens, *Zosime de Panopolis*, CXXX-CLII; Taylor, "A Survey of Greek Alchemy," 130–5) or just the specific source of heat used for carrying out similar operations (Martelli, *Pseudo-Democrito*, 416–8; "Greek Alchemists at Work," 304–8). Later alchemical texts seem to consider *bōtarion* and *kērotakis* as synonymous (*CAAG* II 164,22–165,7), probably because neither was placed in direct contact with the fire, but was set on a layer of heated ashes.

13] The process called *sēpsis* (σῆψις), 'fermentation,' referred to any procedure whereby a solid or liquid substance was treated with reactive liquors (often vinegar) and changed in colour. In *CAAG* II 23,1 it is related to the *iōsis* (ἴωσις) process, a way of treating metals to produce a kind of 'rust.' Copper, for instance, was processed with vinegar in order to produce green copper acetates (Diosc. V 91; Plin. *NH* XXXIV 110f.). Olympiodorus explicitly claimed: "*sēpsis* cannot happen without a liquid substance" (*CAAG* II 90,9). In our text, the fermented ingredient is liquid (namely the product of the distillation) and must be macerated by adding another liquid substance: the philosopher Anonymous explains that Synesius intended to prepare the dyeing compound "by adding one liquid to another one" (*CAAG* II 440,6–9). After processing, the liquor probably changed colour: it is called, in fact, 'Aminaios wine,' i.e. a wine often associated by alchemists with yellow liquid substances (*CAAG* II 19,18; 339,16). This liquid, after being macerated, could of course be used for macerating other substances: Synesius himself seems to refer to similar procedures at § 13, l. 208, where gold is macerated in an unspecified liquor.

14] On this expression, see *supra*, Cat. § 3, n. 4. The alchemical signs (⟁ and ⟁) that appear in manuscripts with the meaning of *arsenikon* (ἀρσενικόν, 'orpiment') are unusual (see also § 16, l. 267; § 19, l. 307), since this ingredient is usually referred to by the symbols ℞, ℞ or ⅄ (Martelli, *Pseudo-Democrito*, 423). It is difficult to justify this different shape (a different inclination); however, we should note that the signs ⟁ and ⟁ seem to have been used especially when *arsenikon* was associated with mercury: see A, fol.17ᵛ4 (= L, fol. 3ᵛ22; *CMAG* VIII 956), where the substance called 'mercury of orpiment' (ὑδράργυρος ἀρσενικοῦ) is indicated by the symbol ⟁⟁.

15] After mentioning the different kinds of *hydrargyros* (ὑδράργυρος, 'mercury') listed by ps.-Democritus, Synesius (§§ 9–11) deals with two main points: the nature of this ingredient and its interaction with the other ingredients in dyeing processes. Two pairs of polar concepts must be kept in mind in order to better understand Synesius' explanation: (a) *sōma-psychē* (σῶμα-ψυχή), 'body-soul,' in which the second term could be replaced by *pneuma* (πνεῦμα), 'spirit' (see § 15, l. 248); and (b) *hylē-eidos* (ὕλη-εἶδος), 'matter-form.' With regard to the nature of *hydrargyros*, Synesius first insists on the opposition between the unity and the plurality of 'mercury.' *Hydrargyros* could in fact refer either to mercury proper, i.e. the metal extracted by processing cinnabar in various ways (see *supra, Cat.* § 2, n. 1), or to any kind of volatile substance produced by treating other minerals, such as orpiment, using similar methods. According to the first pair, any product of a distilling or subliming process was identified with the 'soul' of the distilled/sublimed ingredient; a Syriac passage ascribed to Hippocrates (a misunderstanding for Democritus in Berthelot-Duval's opinion) reads: "Orpiment has a body and a soul; its soul is the 'mercury' [ܠܐܢ, *'nono*, lit. 'cloud'] that rises up from it, when it is liquefied and distilled; its body is the heavy and dense part" (*CMA* II 42,2–4). The passage recalls ps.-Democritus' expression "mercury that comes from orpiment" and makes it clear that ancient alchemists identified 'mercury' with any kind of 'soul' (spirit, vapour) extracted from different substances. Another passage from the *Corpus Syriacum* (**SyrC**, fol. 58ᵛ4-8) is more explicit: "On the fact that mercury comes from all the bodies; since in his opinion mercury is the unique [substance] that comes from all the bodies, Pebychius [an ancient alchemist also quoted in *Syn. Alch.* § 10, l. 163] stopped at the market and shouted: all the bodies are mercury." In this way, mercury was identified with the common constituent of all bodies, their *hylē* according to the matter-form pairing. Synesius (§ 10, l. 150) states that the *hylai* (matter-constituents) of all bodies are their souls, since it was considered possible to extract from all bodies a kind of *psychē* (distilled vapour; see Zos. Alch. *CAAG* II 250,13–251,6) in which alchemists recognized a type of 'mercury.'

16] The verb *diorganizomai* (διοργανίζομαι) must be translated as 'treated *dia organōn* (διὰ ὀργάνων),' 'treated by means of specific devices.' Especially in the excerpt *On the Four Elements* (*CAAG* II 337–42), the verb refers to treatments performed by means of distillatory devices (339, 4 and 14–6). A similar meaning must also be read in Synesius' context, which clearly refers to the distillation to which mercury (and the substances mixed with it) was exposed (see also § 7, ll. 91–106).

17] Mercury's interactions with other ingredients are discussed here. In dyeing processes, *hydrargyros* acts as the vehicle that absorbs the colour of the body with which it interacts, and then transfers that colour to the metallic body to be dyed. In the explanation of this process, the body–soul and matter–form pairs (see *supra*, n. 15) seem to partially overlap. The 'soul' extracted from the processed bodies was considered their active and dyeing element; Stephanus explicitly claimed: "Copper as well as a man has a soul and a body; it is necessary to make the matter of its body (τὴν ὕλην τοῦ σώματος) orphaned, so that what we have is its spirit (πνεῦμα), that is, its dyeing element ... What is its soul and which is its body? The soul is its thin part extracted by treatment, i.e. its dyeing spirit (τὸ βαπτικὸν πνεῦμα)" (II 13–8 Ideler). He went on to explain that bodies had to be crushed and dissolved in order for their dyeing spirit to be extracted. These procedures were performed, according to Synesius' explanation, by means of *hydrargyros*. This was first mixed with the body from which a specific dyeing property had to be extracted; a property identified by Synesius with the colour, the *eidos* (form) or the *psychē* (soul) of the body. In this phase, mercury was compared to wax that can absorb any colour, as clearly explained by Hermes' quotation about white or yellow wax honeycombs (Festugière, *Révélation d'Hermès*, vol. 1, 247; see also Phil. Christ. Alch. *CAAG* II 420,5–13). Though *hydrargyros* (like any distilled product) was identified with a *psychē*, at this stage it also operated as a body in order to receive the colour extracted from another body. In this sense Synesius defined it as *hypostatikē*, (ὑποστατική) 'substantial, giving a support to'; an adjective that generaly pertains to the four metallic bodies that do not evaporate when heated (Zos. Alch. *CAAG* II 148,8–9). In the second phase of the procedure, in contrast, *hydrargyros* was considered to act as a *psychē* ('soul') or *pneuma* ('spirit') and was called *anypostatos* (ἀνυπόστατος), 'unsubstantial,' since it represented the dyeing element that penetrated one of the four metallic bodies (*tetrasōmia*; see *infra*, n. 18) and reshaped its *hylē* ('matter') by giving it a new colour. In this sense it was said to be *katochimos* (κατόχιμος), i.e. 'possessed, held in possession' (LSJ⁹ 930; see also Martelli, *Pseudo-Democrito*, 431–2) by the metallic bodies which received the new *eidos* ('form') brought in by the 'mercury,' and were in turn changed by it.

18] The term *tetrasōmia* (τετρασωμία) may refer both to the four metallic bodies (Olymp. Alch. *CAAG* II 96,6: "the *tetrasōmia* is the four [metallic] bodies"; they are associated with the four feet of the *ouroboros* snake in *CAAG* II 22,1–3) and to their common *hylē* ('matter'; see *CAAG* II 235,5–6: "the *hylē* of the [metallic] bodies is called *tetrasōmia*"). This second meaning lies at the basis of the long comparison between the activities of craftsmen and alchemists. Both work upon specific *hylē* (matter) in order to give it an *eidos* (form): carpenters work wood, stonemasons work stone, and alchemists work *hydrargyros*. 'Mercury,' however, was considered *hylē* in a double sense: (a) it is the matter that absorbs the colours of the other bodies which alchemists mix with it; (b) it is also the common constituent, the common *hylē*, of all the bodies (see *supra*, n. 15). This second meaning is essential for a correct understanding of the second step of the procedure: the mixing of the coloured *hydrargyros* with the metallic bodies to be dyed. In this phase, the metallic bodies (the *tetrasōmia*) represent the *hylē* that must be dyed by 'mercury.' On the other hand, 'mercury,' since it is the common constituent of these bodies, can mingle perfectly with them (it is said to master metallic bodies and to be mastered by them; see

Olymp. Alch. *CAAG* II 96,6–14, and Viano, "Les alchimistes gréco-alexandrins," 99–100). In this way it can change them by introducing a new *eidos* for which it acts as vehicle.

19] See *supra*, n. 15

20] This statement is interesting if read alongside Zos. Alch. II 1–20 Mertens, where the alchemist explains how to make mercury yellow by mixing it with sulphur vapours (see *supra*, PM § 4 n. 23). Zosimus (ll. 15–7) wondered how it was possible that mercury, "that is white actually and potentially," became yellow when fixed with white vapours. Synesius, however, distinguishes between the colour that 'mercury' has actually and the colour it has potentially. On the one hand, it is silver in colour, that is, white (λευκή); on the other hand, since it can absorb any colour, it can become, for instance, yellow (ξανθή) if mixed with yellow substances such as sulphur.

21] Synesius is here drawing a false etymology for the term *magnēsia* (μαγνησία) by relating the first part *magn-* to the verb *meignumi* (μείγνυμι), 'to mix, to mingle,' and the second part, -*ēsia* to *ousia* (οὐσία), 'substance' (see Stéphanidès, "Notes sur les textes chymeutiques," 320). The expression would thus mean that 'mercury' (mentioned immediately before the body of *magnēsia*) had to be mixed with all the solid substances, i.e. all the bodies.

22] The association between *chrysokolla* (χρυσόκολλα), 'malachite' (a green copper carbonate) and *batrachion* (βατράχιον), 'lit. ranunculus,' is confirmed by the *Lexicon on the Making of Gold*, CAAG II 6,6. The identification of the second term is not easy, even if according to a recipe in the *Corpus Syriacum* (CMA II 31 rec. 2) it should refer to a golden compound (it is said to be like 'gold flower') made by mixing orpiment and copper and heating them in a pot closed with a clay lid.

23] The colour of malachite and *batrachion* form the background to Synesius' discussion of chromatic differences between green and yellow. In order to understand the passage, we must bear in mind that the adj. *klōros* (χλωρός) could refer to different shades [Marie-Hélène Marganne, "Le système chromatique dans le Corpus Aristotélicien," *Études Classiques*, 46 (1978): 198ff.], from 'green' to 'yellow-green' and 'yellow,' thus overlapping with *ōchros* (ὠχρός), 'pale yellow, sallow,' that was usually employed with reference to skin complexion (see LSJ⁹ 2042). This overlap is well attested in ancient sources [Maria Fernanda Ferrini, *Pseudo Aristotele. I Colori. Edizione critica, traduzione e commento* (Pisa: ETS, 1999), 105] and especially in medical literature (e.g. Gal. *In Hipp. Acut. comm.* XV 544,10–1 Kühn; *In Hipp. Epid. VI comm.* XVII/1 929,7 K.). Synesius stresses this link between *klōros* and *ōchros*, and introduces at the end the colour of ochre (the *eidos* of *ōchrotēs*), which he relates to a golden colour [see Alberta Lorenzoni, "*Eustazio: paura 'verde' e oro 'pallido' (Ar. Pax 1176, Eup. fr. 253 K.-A., Com. adesp. frr. 390 e 1380A E.),*" *Eikasmos*, 5 (1994): 139–63].

24] A chromatic criterion still guides Synesius' reading of ps.-Democritus' catalogue, where the commentator emphasises the yellow ingredients – whose names are either masculine or feminine – listed by the ancient alchemist. They refer to the production of yellow and yellowing compounds, made by macerating gold or, presumably, other yellow substances.

25] A second criterion of classification is, in Synesius' opinion, the alternation between dry and moist ingredients. Here the commentator recalls the interpretation of the adj. *Pontikos* (Ποντικός), 'from Pontus,' which has already been used in the expression 'Pontus rhubarb' (see *supra*, n. 6).

26] On this distinction between two kinds of 'sulphur water', see *supra*, *Cat.* § 2 n. 2.

27] By mentioning the feminine name *hē chalkanthos* (ἡ χάλκανθος) in the same form attested in ps.-Democritus' catalogue (§ 2, l. 8), Synesius tacitly corrects the mistake made by Dioscorus, who instead used the neuter form (l. 233 χάλκανθον ξανθόν; on the alternation between the two forms, see LSJ⁹ 1972).

28] The term *exischnōsis* (ἐξίσχνωσις, 'thinning, refining,' LSJ⁹ 595) has been introduced on the basis of comparison with other passages (e.g. Zos. Alch. *CAAG* II 169,13–14; Pelag. Alch. *CAAG* II 260,4–5 and 9), where the term overlaps with *exiōsis* (ἐξίωσις), i.e. different techniques for removing rust. The verb *exischnaino* (ἐξισχναίνω or ἐξισχνόω, 'to make dry, whiter,' LSJ⁹ 595) may in fact refer to living bodies that get thinner and emaciated (*ThLG* III 1313; D.C. IV 17,11 = Const. VII Porph. *Sent.* 414,29ff.; Eus. *Ps.* in *PG* XXXIII 349,50–3; Them. *Or.* I 10a-b; *Suda* σ 133,3–4 Adler). It is not surprising that alchemists used a similar analogy when referring to their procedures for scraping metals clean and removing their rust.

29] On the association between 'flowers,' 'souls' and 'spirit,' see *supra*, nn. 5 and 15.

30] On the expression 'rhubarb from Pontus,' see *supra*, n. 6. Here Synesius again seems to tacitly correct Dioscorus (see *supra*, n. 27). In fact, while the priest used the common form *rha Pontikon* (ῥᾶ Ποντικόν), Synesius quotes the name of the substance in exactly the rare spelling attested by ps.-Democritus (*rha Pontion*).

31] On the expression 'dog's milk,' see *supra*, n. 9. Different kinds of milk are often mentioned in alchemical recipes, such as dog's milk (*PHolm.* 11; 13; 61), cow's milk (*PHolm.* 12; 18; 153), and goat's milk (*PHolm.* 114). These names were also used as code names hinting at mercury, sulphur, and divine water (see Zos. Alch. *CAAG* II 154,17f.; *CAAG* II 6,14; *CMA* II 6,19). Synesius also referred to some physical properties of dog's and donkey's milk, which cannot withstand high temperatures. This claim could be based on the ancient classification of different kinds of milk, according to which the densest was cow's milk (since it curdles), and the lightest were the milks of camel, horse, and donkey (Ar. *HA* III 20, 521b 26–33; Galen. *De alim. facul.* III 14,3f. Helmreich in *CMG* V/4,2, p. 345). Since it lacks casein, donkey's milk was expected to evaporate very easily (see also Plin. *NH* XXVIII 158).

32] On this section, see *supra*, *Cat.* § 2 n. 3.

33] The term *anakampsis* (ἀνακάμψις, lit. 'a bending back,' LSJ⁹ 107) seems to have been used by Synesius with reference to a cyclic process according to which a body could be first 'killed' (its 'soul' was detached from it) and reduced to black ashes, then revitalized and restored by giving a new colour/spirit to it. Mercury of course played a central role in this process, as the vehicle that brought in the new spirit (see *supra*, n. 17). Stephanus used the term with a similar meaning, by comparing the restoration of copper (first reduced to ashes, then receiving a new vital spirit; see Papathanassiou, "L'oeuvre alchimique de Stéphanos," 18–21) with the elements (II 210,20–24 Ideler): "And likewise all the elements have creations, destructions, changes and restorations from one to another (ἀνακάμψεις). So also copper, being burnt and restored with oil of roses and being expelled, after it has undergone this many times, becomes without stain, better than gold" (transl. by Taylor, "The Alchemical Works of Stephanos," 41).

34] The meaning and the terms of this comparison are not clear. The term *magnēsia* is followed by the name of two different ingredients, the first one yellow/red (gold coral) and the second one white (Italian antimony). We cannot rule out the possibility that Synesius wanted to stress again the two main colours involved in alchemical procedures. Moreover, if the second quotation seems to stem from the beginning of ps.-Democritus' catalogue for the making of silver (*Cat.* § 2, l. 27, *magnēsia* or Italian antimony), the first one does not, as we would have expected, match the catalogue for the making of gold (*Cat.* § 1, ll. 2–3), where we find *chrysokolla* instead of *chrysokorallos* ('gold coral'). Though we cannot exclude a mistake in the transmission of the text – which perhaps originally had the term *chrysokolla* as suggested by Pizzimenti (1573, 18ʳ9), who translated it as "corpus magnesiae crysocollam" – I have preferred to keep the reading of the manuscripts. *Chrysokorallon*, in fact, is *lectio difficilior*, and it would be difficult to explain how the easier *chrysokolla* has been misread and replaced with the rare *chrysokorallos*.

35] The expression "a single colour" probably hints at gold, as defined by Zos. Alch. II 147–8 Mertens: "a unique species derived from many species" (see Viano, "Les alchimistes gréco-alexandrins," 101).

36] The verb *hydrargyrizō* (ὑδραργυρίζω) is a *hapax*, attested also by some later alchemists (Philos. Christ. Alch. *CAAG* II 274,3; 279,17; 420,9; Philos. Anon. Alch. *CAAG* II 439,18–20). Synesius is here playing with a *figura etymologica*, using a verb that clearly hints at the distillation of mercury, the operation that, according to § 7, ll. 92–106, comes before the maceration of the distilled substance.

Notes on the Syriac texts

*On the correspondences between the Greek and the Syriac names of ingredients, see the indexes at the end of the volume. Specific cases are discussed in the following notes. When supported by the lexica, I have provided a vocalized transliteration of the Syriac terms (according to Western pronunciation; I opted for a simplified system of transliteration, where I did not mark the length of the vowels). Otherwise, only the consonants have been transliterated.

First book (*1SyrC*)

1] The Cambridge manuscript hands down a version of the recipe which is slightly different from both the Greek and the SyrL translation. In particular the central part (ll. 3-4) introduces two melting pots (see *infra*, n. 4), in which mercury seems to be treated with the ingredients listed in the first lines of the recipe. This procedure is supposed to produce a 'red rust' used for treating silver and gold; there is no mention, however, of the white drug that, according to the other two versions of the recipe (*PM* § 5, l. 70; **SyrL**, l. 3), was employed for whitening copper. On the other hand, we cannot exclude the possibility that the term *Hermēs* was used to translate the Greek *chalkos* (χαλκός), 'copper' (see *infra*, nn. 14 and 17). In the last sentence (l. 7), *Hermēs* seems to indicate the metal (the copper, according to the Greek version, see *PM* § 5, l. 74) that was made shiny by means of mercury (*ḥalbo*, lit. 'milk'; **SyrC** reads *b-ḥalo*, which I have corrected to *ḥalbo* by comparison with **SyrL**).

2] I have corrected ܣܛܘܡܐ (*sṭomo*) – usually attested as the transcription of the Greek *stomōma* (στόμωμα), 'steel' (*ThSyr* II 2601; the spellings ܣܛܘܡܐ, *sṭo'mo*, and ܐܣܛܘܡܐ, *'esṭo'mo*, are attested as well: see *B.B.* I 222,24 and II 1328,9) – into ܣܛܝܡܐ (*sṭimo*). This last form, in fact, is also used in the following recipe (§ 2, l. 3) for transliterating the Greek *stîmi* (στῖμι; its more common transcription is however ܐܣܛܝܡܝ, *stimi*: see *ThSyr* II 2599 and *B.B.* II 1331,1).

3] Note that the Greek adj. *xanthos* (ξανθός), 'yellow' – that usually qualifies gold and golden-yellow substances (such as sulphur in this case) – is regularly translated with ܣܘܡܩ (*sumoq*), 'red, reddish' (*ThSyr* II 2665; the Syriac adj. usually corresponds to the Greek πυρρός or ἐρυθρός).

4] The term ܟܘܢܣ (*kwns*) seems to be a transcription, not attested elsewhere, of the Greek *chōnos* (χῶνος) or *choanos* (χόανος), 'melting pot' (*CMA* II 263). According to *B.B.* I 879,6, the spelling ܟܘܢܘܣ (*kunos*; see *ThSyr* I 1708) had already been recorded by the lexicographer Ḥenaišo bar Serošway (ninth century) with the same meaning of بوطق (*būṭaq*, 'melting pot'). Though it would be possible to standardize the SyrC reading into ܒܘܛܩܐ, we must consider that other forms are attested as well: the London manuscripts, in fact, usually have ܚܘܢ or ܚܘܢܐ, transcriptions of the Greek form χώνη/χοάνη: see Rubens Duval, "Notes de lexicographie syriaque et arabe," *Journal asiatique*, n.s. 2 (1893): 317. I have therefore preferred to keep the SyrC reading and to propose the possible correction only in the critical apparatus.

5] The rare Greek term *chrysokorallos* (χρυσοκόραλλος), lit. 'gold coral,' seems to have been translated (or transliterated) into Syriac as ܟܪܘܣܘܩܘܠܝܘܢ (*krwswqwlywn*), a form that is attested again at § 4, l. 3 where it corresponds to the Greek *chryso-konchylion* (χρυσοκογχύλιον; see *PM* § 8, l. 95). Indeed, the Syriac term *krwswqwlywn* is used to render two different Greek terms, both probably referring to a kind of purple gold. Its interpretation is unclear, especially in regard to the second part *qwlywn*, which does not seem to coincide either with *korallos* or with *konchylion*. In fact, the first Greek term (including the form *korallion*) is usually translated as ܟܣܢܐ (*kesno, ThSyr* I 1787) or ܩܘܪܠܝܘܢ (*qwrlywn, ThSyr* II 3734; *B.B.* I 911 *s.v.* ܟܣܢܐ); the second is transliterated as ܩܢܩܘܠܐ (*qnqulo, ThSyr* II 3667). The Syriac *krwswqwlywn* may somehow be related to the more common ܟܪܘܣܘܩܘܠܐ (*krwswqwl'*), i.e. the transcription of the Greek word *chrysokolla* (χρυσόκολλα). This last Syriac term (*ThSyr* I 1816), in fact, could cover the two meanings of the corresponding Greek word (*chrysokolla*, both malachite and an alloy for soldering gold) in addition to the less common meaning of 'coral' (see also *B.B.* I 918,13 *s.v.* ܟܪܘܣܘܩܘܠܐ). It is therefore not surprising that at § 4, l. 3 the rare word ܟܪܘܣܘܩܘܠܝܘܢ seems to have been explained in terms of the more common ܟܪܘܣܘܩܘܠܐ, which was probably a marginal note that has at some point been incorporated into the text.

6] The Syriac tradition hands down two slightly different versions of the recipe. Compared to **SyrL**, the **SyrC** version (a) does not give the second name of the *pyritēs* (i.e. *argyritēs*) at the beginning of the recipe; (b) introduces 'wax' among the ingredients used for processing the mineral; and (c) does not specify the colour of the drug (which was expected to be yellow/red according to *PM* and **SyrL**). On the other hand, compared to the Greek text, both Syriac versions mention Samos rather than Coptos at the end of the recipe.

7] The term ܪܬܝܢܐ (*rtyn'*) could be read as a rare spelling of ܪܗܛܢܐ (*rehṭno*), ῥητίνη, 'resin' (*ThSyr* II 3837; the form ܪܝܬܝܢܝ, *riṭini*, is attested as well; see *ThSyr* II 3897; Duval, "Notes de lexicographie syriaque," 335). However, comparison with the Greek version complicates the issue, since Byzantine manuscripts hand down two corrupt readings, i.e. νίθεως (*nitheōs*, **MV**) and νύθου (*nythou, sic* **BA**). Berthelot-Ruelle, on the basis of Hsch. ν 689 L. (νυθόν· ἄφωνον, σκοτεινόν), have accepted the second form with the meaning of 'grey' (the adj. would qualify the term 'litharge'); yet we would have then expected the feminine (νυθῆς). On the other hand, 'resin' does not fit very well into the context, in which several mineral substances are listed, unless we consider the term to be a cover name. In addition, an entry in the *Lexicon on the Making of Gold* (*CAAG* II 11,17) reads: "*nitron* (νίτρον, 'soda') as the white sulphur that makes copper 'shadowless'; the same effect is produced by *aphronitron* (ἀφρόνιτρον, lit. 'soda foam') and *rhytinē* earth (ῥυτίνη γῆ **BA**; ῥιτίνη γῆ **M**)." The adj. **rhytinos* (ῥύτινος/ῥίτινος) is not attested elsewhere, and Berthelot was not sure whether to relate it to the term 'resin' (*rhytinē*, ῥυτίνη) or to the verb 'to flow' (*rheō*, ῥέω). In any case, we cannot rule out that the term behind the Syriac form ܪܛܝܢܐ (*rtyn'*) referred to a kind of earth or mineral, whose name could be somehow related to the term *rhitheon* (ῥίθεον), attested by ps.-Democritus' *Catalogues* as a name for the red *nitron* (νίτρον πυρρόν; Ps.-Dem. Alch., *Cat.* § 3, l. 31).

8] Although using the same ingredients as the Greek recipe, the Syriac version seems simpler: it does not mention the result (gold-dust in the Greek) obtained after the first washing process and it prescribes dyeing only silver (and not gold) at the end of the procedure.

9] The Syriac expression ܫܩܠ ܐܘܩܘܢܘܡܝ (*šqal 'wqwnwmy'*), 'to take a treatment,' translates the Greek verb *oikonomeō* (οἰκονομέω), 'to treat, to process.' However, we must observe that the Syriac translator often preferred to make explicit which kind of a 'treatment' was implied by the Greek verb. In this recipe, for instance, the second occurrence of the Greek verb (*PM* § 7, l. 84) has been translated by the Syriac verb ܫܘܓ (*šog*), 'to cleanse (*Aph.* 'to wash');' see also § 2, l. 1 and § 7, l. 1 where *pa'el* form of the verb ܒܫܠ (*bašel*), 'to cook' corresponds with the more generic *oikonomeō* of the Greek recipes (*PM* § 6, l. 76; § 11, l. 108).

10] Where the Greek text has the adj. *Attikos* (Ἀττικός), 'Attic,' both **SyrC** and **SyrL** hand down the term ܐܢܛܠܝܐ (*'anṭalia*), which usually refers to the city of *Attaleia* (Ἀττάλεια, i.e. the modern Antalya; see *ThSyr* I 269). This variant is difficult to explain, especially if we consider that the name of the ingredient 'Attic ochre' is usually translated with expressions such as ܐܘܟܪܐ ܕܡܢ ܐܛܝܩܝ (*'okro d-men 'aṭiqi*, 'ochre from Attica'; Mm. 6.29, fol. 25ʳ20), ܐܘܟܪܐ ܗܘ ܕܐܛܝܩܝ (*'okro haw d-'aṭiqi*, 'ochre of Attica'; Mm. 6.29, fol. 25ʳ14) and ܐܘܟܪܐ ܐܛܝܩܝ (*'okro 'aṭiqi*, 'Attic ochre'; Mm. 6.29, fol. 30ᵛ13 and 20; fol. 75ᵛ21). On the basis of these examples, it might be possible to correct the **SyrC** reading to ܐܛܝܩܐ (see also *ThSyr* I 132) or ܐܛܝܩܝܐ.

11] In the Syriac recipe (handed down only by **SyrC**) two lines of the Greek text are missing (*PM* § 8, ll. 91–3: οὐ τὸν λίθον λέγω – στυπτηρία ἐξιπωθείσῃ).

12] On the relationship between the term ܟܪܘܣܘܩܘܠܝܘܢ (*krwswqwlywn*) and the more common ܟܪܘܣܘܩܘܠܐ (*krwswqwl'*), see *supra*, n. 5.

13] The Greek term *kinnabaris* (κιννάβαρις), 'cinnabar,' is usually transliterated either as ܩܢܒܪܝܣ (see also *supra*, § 1, l. 7) or as ܩܝܢܒܪܝܣ (*qinabaris*; see *ThSyr* II 3658); the **SyrC** reading ܩܝܒܪܝܣ (*qybrys*) is not, however, registered by the lexica. Although it could easily be corrected into ܩܝܢܒܪܝܣ (*qinabaris*), I have preferred to keep the form handed down by the manuscript, which is similar to the spelling attested by *B.B.* II 1709,6: ܩܒܪܝܣ ܐܒܘ ܩܣܡ ܗܩܕܠܐ زنجفر, "*Qbrys*. Cinnabar [ar. *zunǧufr* or *zinǧafr*] according to alchemists" (see also *B.B.* II 1775,13). The same word *zunǧufr*, written with Syriac letters (ܙܘܢܓܦܪ), has been set down in the margin of **SyrC**.

14] Here the Syriac recipe is again different from the Greek: (a) it does not include vinegar among the ingredients for whitening cinnabar; and (b) at the end of the recipe there is no mention of the copper, which was expected to turn into electrum according to the Greek version. On the contrary, **SyrC** introduces the term *Hermēs*, which was supposed to absorb half of an unspecified 'dye,' possibly referring either to the dyeing compound whose production is described in the first part of the recipe or, perhaps, to the gold that was added in the dyeing process (see *PM* § 9 n. 35). We should note that the planet 'Hermes' was associated with mercury or quicksilver in the list of planets/metals edited by Berthelot-Duval, *CMA* II 6. Such lists were not, however, stable and the same planet could also be associated with other metals: (a) tin, in the list handed down by **M** fol. 6ʳ (*CMAG* VIII 4) and in Olympiodorus' commentary *in Mete*. III, 6 (p. 267,5 Stüve); (b) copper, in the writings of the astrologers Vettius Valens (*CCAG* II 92,5) and Rhetorius (*CCAG* VII 221,21); (c) electrum, in the list handed down by **A** [*CAAG* II 25, where mercury is also included; see Arthur Ludwich, *Maximi et Ammonis carminum de actionum auspiciis reliquiae. Accedunt anecdota astrologica* (Leipzig: Teubner, 1877), 121] and in the Syriac text *De causa causarum* (see Duval, "Notes de lexicographie syriaque," 297); and (d) iron, in Celsus (see Orig. *Cels*. VI 22,16-8).

15] The *incipit* of the Syriac recipe (ܟܕܡܐ ܠܗ ܦܪ ܩܠܘܡܕܐ) is unclear, especially when compared with the Greek version, which reads (*PM* § 10, ll. 102f.): Τὴν δὲ Κυπρίαν καδμίαν, τὴν ἐξωσμένην λέγω (λεύκαινε κτλ.), "[Whiten] Cyprian cadmia, I mean the one that has been forced out [of its ores]." Behind the Syriac ܩܠܘܡܕܐ ܠܩܕܪܐ (*l-qedro qa'dmia*; the *lomad* introduces an accusative in Syriac) I am tempted to recognize the Greek 'Cyprian cadmia' (accusative), although the term ܩܕܪܐ (*qedro*), 'pot' (*ThSyr* II 3499), does not match the Greek adj. *Kyprios* (Κύπριος), 'Cyprian.' Moreover, even if we correct ܩܕܪܐ (*qedro*) to ܩܦܪܘܝܐ (*qproyo*) or ܩܘܦܪܘܝܐ (*quproyo*), the expression ܦܪ ܠܗ remains difficult to interpret and does not seem to correspond to the second part of the Greek sentence. In fact, if we read ܦܪ (*qore*) as the *pe'al* participle, third person masc. sing., from ܩܪܐ (*qro*, 'to call'), we should translate the whole sentence as: "he (?) gives to the *qedro* the name of cadmia." However, in this case, the subject of the participle would remain unspecified; in addition, no further sources seem to confirm such an identification between cadmia and *qedro* (see, for instance, *B.B.* I 535,14 and II 1687,15). On the other hand, if we interpret ܦܪ (*qorro*) as the *pe'al* participle, third person fem. sing., from ܩܪ (*qar*, 'to grow cold, to cool'), we could take cadmia as the subject of the sentence and translate: "the cadmia grows cold," etc.; however, the accusatives ܠܩܕܪܐ (*l-qedro*) and ܠܗ (*loh*, feminine) would hardly fit into such a sentence. In conclusion, since neither solution is satisfactory, I have preferred to add a *crux desperationis* to the sentence.

16] 'White lead' rather than a more generic 'white earth' is mentioned at the beginning of the Syriac translation of this recipe, which describes how to make a copper-lead alloy (see *supra*, *PM* § 12 n. 43). Both terms refer to an artificial substance made by mixing various ingredients: the Syriac recipe lists only white-lead, *helkysma* and *magnēsia*, while the Greek text also includes Italian stibnite and litharge.

17] In two cases the Syriac recipe introduces the term *Hermēs* (the planet usually associated with mercury) in places where, according to the Greek text, we would have expected 'copper.' First, at l. 4 we read 'rust of *Hermēs*' (ܚܣܝܐ ܕܗܘ ܘܗܪܡܣ) instead of the Greek "the rust that has been scraped off" (*PM* § 12, l. 122: ἰὸς ξυστός); and second, at l. 6, "*Hermēs* and lead" are listed as the components of the alloy called *molybdochalkos* (lead-copper) in Greek. Since planet-metal associations were quite flexible in this early period (see *supra*, n. 14), we cannot rule out the possibility that 'Hermes' here refers to a solid metal. However, the possible meaning 'copper' remains quite dubious: the Syriac recipe, in fact, mentions the metal twice under the names of ܢܚܫܐ (*nḥošo*) and ܐܦܪܘܕ ('*aprod* = Ἀφροδίτης, i.e. the planet Venus).

18] The adj. *arreustos* (ἄρρευστος), 'not fusible' (LSJ[9] 247) and *atrētos* (ἄτρητος), 'not perforable' (LSJ[9] 273), which qualify the lead-copper alloy in the Greek recipe (*PM* § 12, ll. 124f.), do not correspond to the Syriac ܪܟ (*ro'ak*) 'soft' and ܢܗܝܪܐ (*nahiro*), 'bright.' The Syriac translator may not have correctly interpreted the Greek recipe, or have had a different text at his disposal.

19] Where the Greek text (*PM* § 13, l.136) prescribes applying the dyeing 'drug' to copper or silver, the Syriac recipe introduces the term *Hermēs*, which is difficult to read as 'mercury' in this context. We would have expected, in fact, a reference to the solid metal to be dyed. On the problems arising from planet-metal associations, see *supra*, n. 14. Moreover, the Syriac version does not preserve the second part of the Greek recipe (*PM* § 13, ll. 138–42).

20] As in the Greek version, a theoretical discussion of the power of natures and the methods of physicians introduces the second section of the *Book on the Making of Gold*, which is more focused on the use of liquid ingredients (washes). Although the Syriac texts cover the same main points as the Greek one, their readings are often slightly different: such a gap is difficult to explain, since it could be due either to a different Greek text behind the Syriac translation, or to specific choices of the translator, who could have tried to interpret and partially rework the text he had at his disposal.

21] The plur. adj. ܚܟܝܡܐ (*ḥakime*, 'wise, skillful, expert') is difficult to interpret. If it refers to 'natures,' it may be translated in two slightly different ways: either "they [i.e. natures] are expert of wonder(s)," or "they are wonderful experts" (with *ḥakime* as a nominal adjective). Both translations sound rather odd, as already pointed out by Berthelot-Duval, who proposed a quite free interpretation (*CMA* II 269): "Elles sont l'oeuvre des sages admirables." However, comparison with the Greek text suggests a possible solution. *PM* § 15, ll. 155–6, in fact, reads: "O you who are prophets with me [*symprophētai*], I know that you are not incredulous, but rather open to wonder." If we interpret *ḥakime* ('wise men') as an attempt to translate the Greek *symprophētai* (see also *3SyrC* § 1 n. 1 and *2SyrC* § 5, l. 1), we may suppose a lacuna in the Syriac text: a copyist perhaps omitted a verb such as ܚܟܡܝܢ (*ḥokmin*, 'they know'), susceptible to omission by a *saut du même au même*. The Syriac text, so restored, seems to better correspond to the Greek version: the translator perhaps simplified the Greek passage and just specified that 'wise men' recognized the greatness of natures.

22] The translation of the term *rha* (ῥᾶ/ῥῆον, 'rhubarb') – usually ܪܐܘܢ (*r'wn*) in Syriac (*ThSyr* II 3781) – is unclear. The Syriac term ܦܝܣܝܢ (*pysyn*) was interpreted by Berthelot-Duval (*CMA* II 269) as an unusual transcription of the Greek *pissa* (πίσσα) 'pitch' (its usual spelling is ܦܝܣܣܐ, *pissa* or ܦܝܣܐ, *pisa*; see *ThSyr* II 3120). On the other hand, we must note that the word 'rhubarb' seems to have been a key term in ps.-Democritus' recipes, at least given the importance attached to it by later commentators (e.g. Syn. Alch. §§ 3 and 16). If we do not want to suppose a misreading of such an important term, we could interpret the form ܦܝܣܝܢ (*pysyn*) in the light of the term ܦܣܐܘܣ (*ps'ws*; see *SyrLex. Suppl.* 269), which is attested by **SyrL** (*CMA* II 5, 9) with the meaning of 'red rhubarb.'

23] The Syriac translator usually refers to honey (ܕܒܫܐ, *debšo*) rather than to 'cerate' (*kērōtē*, κηρωτή, a wax-based paste) for specifying the consistency of the wash whose preparation is described in the recipe.

24] The Syriac recipe does not prescribe mixing rhubarb with celandine (see *PM* § 17, ll. 194-6), but recommends leaving the metallic leaf to cool in the shade (a detail not mentioned by the Greek text).

25] Two metals are prescribed to be treated with the saffron-based wash; however, where the Greek text mentions silver and copper leaves, the Syriac recipe has copper as the first metal and *Hermēs* as the second. Since it is not possible to laminate mercury, it is likely that in this recipe the planet 'Hermes' refers to a solid metal (see *supra*, n. 14).

26] The mention of God introduced by the Syriac translator could betray the Christian *milieu* in which he was working.

27] The Syriac tradition preserves the two anomalous forms ܣܘܪܛܘܣ (*swrṭws*; **SyrC**) and ܣܘܛܘܣ (*swṭws*; **SyrL**) in the place of the Greek *pyrites* (πυρίτης). Berthelot-Duval (*CMA* II 21 n. 8) interpreted the **SyrL** reading as a transliteration of the Greek *anthos* (ἄνθος), lit. 'flower,' which is attested by *PM* in the following line (§ 15, l. 211). On the other hand, especially on the basis of the **SyrC** reading, it would be possible to read the form as a corruption of ܦܘܪܝܛܝܣ, *puriṭis* (= gr. *pyrites*). However, the presence of this ingredient in the recipe is not certain, if we consider that all other substances listed here by ps.-Democritus are plants. I have therefore preferred to keep the reading handed down by the manuscript tradition.

28] The form ܟܡܘܢܝܘܢ (*kmwnywn*) seems to be the transliteration of the Greek *oichomenion* (οἰχομένιον). On the other hand, **SyrL** has the reading ܟܡܘܢܝܩܘܢ (*kmwnyqwn*), which has been recorded by *LexSyr. Suppl.* 166 with the explanation of '*oechomene* [basil?]' (on the basis of *CMA* II 21). As already noted (see *supra*, *PM* § 19 n. 53), this interpretation is based on the assumption that the Greek term handed down by *PM* was somehow linked to the terms *akinos* (ἄκινος) or *ōkinon* (ὤκιμον), 'basil.'

29] This paragraph is different from the Greek version in many respects. First of all, in *PM* (§ 20, l. 213) ps.-Democritus introduces the name of the alchemist Pammenes, who is supposed to have taught Egyptian priests about a method probably to be identified with the technique illustrated in the preceding recipe (*PM* § 19). In fact, according to Zosimus (*CAAG* II 148, 15f.), ps.-Democritus mentioned Pammenes in regard to the working of lead, which is actually the topic of *PM* § 19. In the Syriac passage, however, there is no mention of Pammenes and the subject of the very first line is not explicit. It is not clear, in fact, to what the pron. ܗܕܐ (*hode*) refers. I have implied the term ܟܝܢܐ (*kyono*), 'nature,' which is mentioned immediately before the pronoun. Yet the sentence "since it overcomes nature" could be considered as the translation of the ending of *PM* § 19 (l. 212 ἡ φύσις τὴν φύσιν κρατεῖ), wrongly attached to the following paragraph. If this is the case, we could find a certain similarity between the Syriac ܗܕܐ ܗܝ ܪܒܬܐ ('this is greatest') and the Greek text preserved by Byzantine manuscripts (αὕτη ἡ Παμμένου ἐστί): the superlative 'greatest' (ܗܝ ܪܒܬܐ) could be explained by supposing that the translator misunderstood the name *Pammenes* (Παμμένης) and read it as παμμεγέθης (*pammegethes*), 'very great, immense.' A similar correction was proposed for the Greek text as well: see Hammer Jensen, *Die älteste Alchimie*, 88, and *RE* XVIII/2 (1942), 1633.

30] The end of the Syriac paragraph does not follow the Greek: it does not include the comparison with the healing properties of a few ingredients (human excrement, quicklime and Rhamnus, *PM* § 20, ll. 216–23), nor it does not make the topic of the second book explicit. On the contrary, the last few lines of the Syriac text expand the idea of the secrecy and the value of the unique substance that was supposed to be effective in any alchemical operation.

Second book (*2SyrC*)

1] The two metals that were supposed to be treated with 'mercury' are different in the Syriac recipe, where *Hermēs* and 'alum' are listed instead of copper and iron (first purified with sulphur). On the polysemy of the term *Hermēs*, which could refer to several metals, see *1SyrC* § 5 n. 14. On the other hand, the term 'alum' could stem from a misinterpretation of the Greek *sidērō theiothenti* (σιδήρῳ θειωθέντι); however, 'alum' seems to correspond to the mineral with which the metal was to be treated (before receiving 'mercury') rather than to the metal on which 'mercury' was laid.

2] In the second part of the Syriac recipe there is no mention of the Greek term *magnēs* (μάγνης), 'magnetite'; the Syriac text, in fact, omits the Greek expression "or a pinch of magnetite" (*AP* § 1, l. 7: ἢ μαγνήτος βραχύ). In addition, in the following sentence (ll. 7-8: "For magnetite has affinity with iron") the term *magnēs* is replaced by the Syriac ܡܓܢܣܝܐ (*magnisia*), a transcription of the Greek *magnēsia* (μαγνησία). Such an omission may be explained by considering the similarity between the two Greek names *magnēs* and *magnēsia*, especially when transcribed into Syriac letters, such as ܡܓܢܣܝܐ (*magnisia*) and ܡܓܢܛܝܣ (*magnaṭis*; see *ThSyr* II 2006). We cannot exclude the possibility that *magnaṭis* was originally also in the Syriac recipe.

3] The translation of the second recipe differs substantially from the Greek text, where a greater number of ingredients are involved in the process. The Syriac version, in fact, includes only the substances mentioned in the first sentence of the Greek version (*AP* 2, ll. 9–10): (a) 'mercury,' (b) alum, and (c) simple oil (while the Greek text specifies "castor or radish oil").

4] The Syriac expression ܐܢܠ ܕܨܒܐ may be read in different ways: (1) ; as *tenono d-ṣob'o*, lit. 'the dyeing smoke' (see *CMA* II 24 n. 1); (2) as *tenono d-ṣabo'o*, lit. 'smoke of the dyer' (maybe 'fumes used by dyers'); (3) as *tenono d-ṣbo'o*, 'smoke of a pigment/dye' (i.e. 'fumes produced by [burning] a pigment'). It seems to be an attempt to translate the Greek *kapnos tōn kobathiōn* (καπνὸς τῶν κοβαθίων). The rare term *kobathia* (κοβαθία or κωβαθία; see *supra*, *AP* § 3 n. 8), which is never attested in the *Corpus alchemicum Syriacum*, at least according to Bethelot-Duval's index (*CMA* II 381), was perhaps misunderstood by the translator.

5] The Syriac sentence ܘܗܘܘ ܙܕܩܐ ܡܢ ܗܘ ܡܐ ܕ ܡܚܣܡܟ, "you must add part of this [i.e. copper?] while you are stirring," seems to be incomplete: there is no mention, in fact, of the tin that is involved in the process as described by *AP* § 3, ll. 25-6: "and melt it by adding gradually one ounce of previously purified tin; stir with your hands, etc." (ἐπιβάλλων κατ᾽ ὀλίγον κασσιτέρου προκαθαρισθέντος οὐγγίαν μίαν, καθύπο χεῖρα κινῶν). In particular, the syntagma ܡܢ ܗܘ (*men hono*) is anomalous, since we would have expected the simpler form ܡܢܗ (*meneh*) if the translator meant to specify that just a part of copper (the pron. ܗܘ seems to refer to this metal) had to be added. We cannot rule out the possibility that the Syriac text is somehow corrupt at this point. Unless we suppose a more substantial lacuna, it is perhaps possible to partially restore it by reading the name of the tin – ܐܢܟ ('*onko*) or ܙܘܣ (*zews*; see *ThSyr* I 1104; *CMA* II 2-3) – instead of the pronoun ܗܘ (*hono*).

6] The Syriac recipe is composed of two sections that originally belonged to two different Greek recipes. The incipit ("take white unburnt sulphur") corresponds to *AP* § 4, l. 31, i.e. to a recipe that describes the making of a sulphur-based compound to be used for treating different metals (copper, iron, lead). The rest of the Syriac recipe is a translation of *AP* § 5, in which the author prescribes processing litharge with several ingredients in order to prepare a white drug supposedly able to whiten base metals.

7] The Syriac form ܦܡܛܝܣ (*pmtys*) – which seems to be *vox nihili* – has been corrected into ܦܘܪܝܛܝܣ (*puriṭis*) on the basis of *AP* § 5, ll. 41f. (τρῖψον σὺν θείῳ, ἢ καδμίᾳ, ἢ ἀρσενικῷ, ἢ πυρίτῃ κτλ.). Berthelot-Duval translated the passage (*CMA* II 271): "avec de la céruse [ψιμύθιον?]"; however the Greek term *psimythion* is usually transcribed as ܦܣܝܡܝܬܝܢ (*psimitin*; see *ThSyr* II 3188 and Duval, "Notes de lexicographie syriaque," 327).

8] This paragraph, which marks the end of the first section of the book on the making of silver, is lost in Greek. However, its main topic, a discussion of the contrast between the unity and plurality of substances, is a feature proper to other passages by ps.-Democritus (see, in particular, *PM* §§ 4, 15–6, 20 and the corresponding Syriac translations). We must suppose that the book on silver-making, as well as the one on gold-making, was originally divided into two sections – namely, the sections on solid substances (*2 SyrC* §§ 1–4 ≈ *AP* §§ 1–5) and on liquid drugs or washes (*2SyrC* §§ 6–9 = *AP* §§ 6–9) – by a theoretical paragraph which has been preserved only in the Eastern tradition.

9] Instead of "copper, lead, and iron," metals that are involved in the process according to the Greek version (*AP* § 6, ll. 55–6), the Syriac recipe mentions only *Hermēs*, which is likely to refer not to mercury, but to a solid metal (see *1SyrC* § 5 n. 14).

10] The Syriac recipe is less precise in describing the method by which metallic leaves are to be treated: it does not, in fact, mention either the new vessel into which the leaves are placed, or the kind of fire on which they are to be heated for one day.

11] At this point SyrC has ܘܗܘܡܐ ܐܘ̈ ܡܡ ܚܕ̈ ܐܘܙܦܐ ܦܟܠܝ̈ܗ ܘܣ ܚܠܐ, "Rub them on the surface and leave the half of one side behind." The sentence is not clear, and Berthelot-Duval tried to interpret it by translating: "Enduis-en la lame extérieure-ment sur une seule face et laisse l'autre intacte" (*CMA* II 272). However, if we compare the Syriac with the Greek text – which clearly prescribes using just one half of the dyeing drug – we could suppose that the expression ܦܠܓܗ (*pelgeh*), 'a half of it,' refers to the drug rather than to the metallic leaf. According to this interpretation I have corrected ܚܠܐ ܝܣ, (*d-ḥad gabo*) 'of one side,' as ܚܣܠ ܚܠܐ, (*l-ḥad gabo*) 'on one side': the author would thus prescribe leaving half of the drug aside and keeping it for the second part of the process. It is, however, difficult to find the point at which the translator refers again to this reserved amount of drug: at l. 4 the expression ܚܣܪ ܡܥ ܚܬܢܣ (*b-ḥad men kyonin*), 'in one of the natures,' sounds quite bizarre, especially when compared with the Greek τὸ τοῦ φαρμάκου λείψανον, 'the rest of the drug.'

12] The central part of the recipe is different from the Greek version, since the Syriac one does not mention either honey or the metallic body that is to be whitened: it simply specifies heating the drug on live coals (a detail missing from *AP* § 8).

Again by Democritus (*3SyrC*)

1] According to Greek alchemical sources, in a few passages of his work Democritus is supposed to have mentioned the people to whom he addressed his teaching: see, in particular, *CAAG* II 427,3–6, where kings, priests, and prophets are explicitly mentioned (see also *CAAG* II 158,3). In the translation of the first book on the making of gold (*1SyrC* § 11, l. 4; see n. 20), the Syriac term ܚܟܝܡܐ (*ḥakimo*, lit. 'wise man') seems to correspond to the Greek term *symprophētēs* (συμπροφήτης; *PM* § 15, l. 155).

2] The third Syriac book is composed of four paragraphs (*3SyrC* §§ 1–4) dealing with precious stones, and three paragraphs (*3SyrC* §§ 5–7) dealing with purple dyeing: topics probably covered by the original last two books of ps.-Democritus. After a brief introduction (*3SyrC* § 1), three texts on the dyeing of stones open with the same formula "here is for you," a formula that is also attested in § 1 (l. 3). The very first line of this paragraph is, however, problematic, since the text does not make explicit the subject implied by the verb ﺠﻤﺴ (*mayte*, lit. 'he makes come, brings forward'). If the name of Democritus, the supposed author of the following sections, is implied, we should consider the sentence a later interpolation, perhaps introduced by the person who epitomized the original books. This paragraph, in any case, does not seem to fit with the beginning of ps.-Democritus' book on stones as quoted by the Byzantine alchemist Christianus (*CAAG* II 395,2ff.; see *supra*, p.15): "Since I have just now finished my account related to my second treatise and I am going to present a complete exposition of the methods [for the making] of stones, I come to my third treatise by premising something useful for my writing, that is: sulphurous substances are mastered by sulphurous substances, and liquid substances by the corresponding liquid substances."

3] The Byzantine alchemist called the Philosopher Anonymous quotes the ps.-Democritean passage (*CAAG* II 122,4–17; text to be corrected on the basis of M, fol. 84ʳ11–26) which underpins the first part of the Syriac paragraph (*3SyrC* § 2, ll. 1–10). However, the Greek text is slightly different. On the one hand it preserves the name of the mysterious stone, which is said to be called *aphroselēnos* (see *supra*, PM § 5 n. 22); on the other hand, it does not describe the complete process of the stone, but it stops after mentioning the milk of a jenny or goat. In addition, instead of the term *Hermēs* (l. 9; on its different possible meanings, see *1SyrC* § 5 n. 14), the Greek quotation has 'copper' (*chalkos*; see *CAAG* II 122,13–4: καὶ αὕτη μόνη λευκαίνει τὸν χαλκόν, "and only this one [i.e. cinnabar vapour] whitens copper").

4] Various spellings of the name of this substance are attested by alchemical texts: *komaron* (κόμαρον; attested also in the masculine), *kom[m]ari* (κόμ[μ]αρι), and *kōmaris* (κώμαρις). A few recipes from *PHolm.* (85, 87, 97) describe how to dilute this ingredient; for example, *PHolm.* 97 reads: "The dissolving of *kommari*. To dissolve *kommari*. Grind tartar with quicklime, put it in a small dish and stir it. Pour the clear water in another vessel, put ground comarum in it, stir it and it will give up its color at once. Then let is clarify until the following day and you will find purple" (rec. 92 in Caley, "The Stockholm Papyrus," 991; see also *CMA* II 329, § 3; *PHolm.* 87 specifies at the end: "[The solution of *komari*] is also useful as a preliminary coating for every stone"). In addition, we know that ps.-Democritus dealt with the dyeing properties of this substance thanks to a later recipe book handed down by the mss. *Parisini gr.* 2325 (B; see fols. 160ᵛ1–173ᵛ8) and 2327 (A; see fols. 147ʳ1–159ʳ5) and published by Berthelot-Ruelle (*CAAG* II 350–64) under the title of *Deep Tinctures of Stones, Emeralds, Rubies and Jacinths from the Book Taken out from the Sancta Sanctorum of Temples*. Although scholars have identified *komari* as a dye extracted from a plant (Halleux, *Paryrus de Leyde*, 218), according to this recipe book (*CAAG* II 350,8; 358,25–6) it refers to a mineral or stone called τάλκ (*talk*; i.e. Greek transcription of the Arabic طلق, *ṭalq*, 'talc'). A similar opinion seems to have been defended by ps.-Democritus himself, who claimed (*CAAG* II 356,13–4): "Consider the *komaron* as the stone" (τὸ γὰρ κόμαρον νόμιζε τὸν λίθον; see also *CAAG* II 357,13–5).

5] This sentence is attributed to ps.-Democritus in a passage of the abovementioned recipe book, which reads (*CAAG* II 357,10–2): "But Democritus, coming to the *komari*, asserts by the following words: 'besmear whatever stone you want, crush it and it will become a pearl'" (ἀλλὰ γὰρ ὁ Δημόκριτος, ἐπὶ τῆς κομάρεως ἐλθών, κατηγορεῖ φάσκων· ἐπίχριε ὅσον βούλει λίθον, λειώσας αὐτόν, καὶ ἔσται μαργαρίτης). On the basis of this comparison, we cannot rule out the possibility that the Syriac ܡܪܡܪܝܛܝܣ (*mrmrytys* = gr. μαρμαρῖτις, 'like marble') – not attested by the lexica – is a mistake for the Greek *margaritēs*, 'pearl' (**SyrL** has ܡܪܝܛܝܣ, *mrytys*, at this point), which would represent the final result of the process (see *CMA* II 26 n. 3).

6] This recipe – not preserved in the Byzantine tradition – employs a difficult nomenclature that is not always understandable. An allegorical description of the process, which probably involves *Decknamen*, opens the text. In particular, the term ܝܩܕܟܘܢ (*yqdkwn*), which I have proposed to translate as 'snake' by following Berthelot-Duval's interpretation (*CMA* II 273), is neither attested by lexica nor matches the transcription of any Greek word, although the mention of 'his bones' (l. 2) could suggest that it refers to a living being (see *CAAG* II 22,12–6 which directs that the flesh be stripped off a snake; similar allegories involving human beings are attested by Zos. Alch. X-XII Mertens). It is impossible to give a sure interpretation of the term. It might be related to the claim of few later alchemists (Phil. Anon. Alch. *CAAG* II 121,10–5; 125,5; 263,5–6; Steph. Alch. II 236,35 Ideler), who often refer to the currents of the Nile as the place to find a specific stone that was supposed to have a *pneuma*. The so-called philosopher Anonymous (*CAAG* II 121,14–5) claimed that alchemists had to take the 'heart' (*kardia*, καρδία) out of the stone, and this stone is explicitly described as a snake (or *ouroboros*) by the iambic poem (II 331,26–332,23 Ideler = vv. 135–57 Goldschmidt) ascribed to the alchemist Theophrastus (sixth/seventh century?; see Letrouit, "Chronologie," 82–3): "Though not a stone, it yet is made a stone / from metal, having three hypostases, / for which the stone is prized and widely known; / yet all the ignorant search everywhere / as though the prize were not close by at hand. / Deprived of honor yet the stone is found / to have within a sacred mystery, / a treasure hidden and yet free to all. / A dragon springs therefrom which, when exposed / in horse's excrement for twenty days [dubious text; see Goldsmith's edition] / devours his tail till naught thereof remains. / This dragon, whom they Ouroboros call, / is white in looks and spotted in his skin / ... He swims and comes unto a place within / the currents of the Nile; his gleaming skin / and all the bands which girdle him around / are bright as gold and shine with points of light." Translation by Charles A. Browne, "The Poem of the Philosopher Theophrastos upon the Sacred Art: A Metrical Translation with Comments upon the History of Alchemy," *Scientific Monthly*, 11 (1920): 204.

7] The three ingredients mentioned in this sentence are difficult to interpret. The first term, ܐܦܪܣܠܝܘܢ ('*prslywn*) could be read as a misspelling for ܐܦܪܣܠܡܘܢ (see *B.B.* I

267,26), the transcription of the Greek *aphroselēnos* (ἀφροσέληνος). However, although this ingredient probably played a role in making stones (see *CAAG* II 350,8; 357f.), in this recipe it seems to specify the composition of the vessel rather than its contents. The two following terms are more problematic, since it is unclear whether they refer again to the vessel in which the stones are put, or to the ingredients by which the stones are processed. Neither word, in fact, is attested by the lexica. If referring to the vessel, the first expression ܩܡܘܠܐܕܒ (*b-d-qmwl'*) could be interpreted as "in [the vessel] of *qmwl'*," where *qmwl'* might be a misspelling for ܩܡܘܠܝܐ (*qimolia*; see *ThSyr* II 3603), 'Cimolian earth' (= gr. Κιμωλία; even though this material seems not to be suitable for a vessel). For the second term, I have not been able to identify any Greek term to account for the spelling.

8] Among the ingredients that compose the dyeing wash, the substance called ܩܠܦܝܣܘܢ ܕ ܦܝܠܘܢ (*pilon d-qlpyswn*) is difficult to identify. While it seems possible to recognize the Greek term *pēlos* (πηλός), 'mud or dregs,' in the Syriac ܦܝܠܘܢ (*pilon*; see *ThSyr* II 3105), the second term ܩܠܦܝܣܘܢ (*qlpyswn*) remains uncertain.

9] The last three recipes of the collection deal with purple dyeing, the topic of ps.-Democritus' book *On Purple* (Περὶ πορφύρας), from which they probably stem. The first two recipes describe a dyeing process which involves *phykos* (φῦκος; ܦܘܩܘܣ, in Syriac), a dyeing alga mentioned both in *PM* (§ 2, l. 24) and in the Leiden and Stockholm papyri (see Halleux, *Papyrus de Leyde*, 233). According to the first recipe, *phykos* was steeped in water containing a bit of lime. The term ܟܠܫܐ (*kelšo*) may translate both the Greek *titanos* (τίτανος), 'lime' and *asbestos* (ἄσβεστος), 'quicklime.' Solutions of alkaline substances are often mentioned by the above mentioned papyri for washing wool (Halleux, *Papyrus de Leyde*, 44). *PHolm.* 135, for instance, reads: "Cold-dyed purple. Pulverize quicklime in cistern water. Pour the lye off and mordant what you wish therein from morning until evening. Then rinse it out in fresh water [and] colour it in the first place in an extract of *phykos*. Then put in vitriol in addition" (rec. 129 in Caley, "The Stockholm Papyrus," 996; translation slightly modified).

10] Both vinegar and alum are mentioned by the Leiden and Stockholm papyri with reference to *phykos*-based dyeing processes (*PHolm.* 46; 96; 100; 113; 114; 130). Both ingredients were used as mordants for fixing dyes on wool: for instance, *PHolm.* 100 explicitly mentions *phykos* pre-treated with alum (φῦκος στυπτήριον); *PHolm.* 113 reads similarly: "Dyeing with *phykos*. To dye with *phykos*. Wash the wool as is previously described. For a mina of wool take 4 chus of urine and a half a mina of alum. Mix these, and at the same time make a fire beneath them until they boil up. Put the wool in and stir incessantly, but when the wool sinks down and the liquor subsides then rinse the wool out. Boil in drinking water three times as much *phykos* as the weight of the wool, take the *phykos* out, put the wool in and stir up uniformly until the wool becomes soaked. Then pulverize a quarter of a mina of vitriol for each mina of wool and mix them. Stir up incessantly and thereby make the wool uniform. Then take it out, rinse out and let the wool dry as in other cases" (rec. 108 in Caley, "The Stockholm Papyrus," 994; translation slightly modified). In our recipe the mordant is added to the dyeing bath itself.

11] The form ܐܣܝܛܘܣ ('srṭws), not attested by the lexica, could be interpreted as a misspelling for ܐܣܛܪܘܣ ('sṭrws), which is registered by *ThSyr* I 303 with the meaning of 'a stone that comes out of the sea' (ܡܢ ܟܐܦܐ ܕܢܦܩܐ ܡܢ ܝܡܐ); however, this definition is attested by *B.B.* I 228,19 under the term ܐܣܛܪܝܣܘ). The word might recall the Greek *asteritēs* (ἀστερίτης, *scil.* γῆ, often attested by the *Corpus alchemicum*; see, in particular, *AP* § 2, l. 15), which, according to Photius (*Bibl. cod.* 190, 153b 22ff.) and to the lexicon *Suda* (ι 333 A.), refers to a stone hidden inside a fish called Pan. Mertens related the term to the word *astēr* (ἀστήρ), which would refer to a Samian clay (LSJ⁹ 261; Plin. *NH* XXXV 191; Diosc. V 172): see *CMA* II 7,19f. (transl. p. 14) and 302 § 4; *B.B.* I 508,22.

12] The form ܗܓܝܘܢ (*hgywn*) seems to correspond to ܐܓܝܘܢ ('*gywn*; see *SyrLex. Suppl.* 3), 'wool' (see the Greek ἔριον). The Syriac translation of the ninth book of Galen's *On Simple Drugs*, which is handed down by **SyrC**, reads at fol. 127ᵛ15 (see also *CMA* II 305): ܐܓܘܢ ܗܝ ܕܐܝܬ ܥܡܪܐ, "'*gwn* (perhaps to be read ܐܓܘܢ) that is the wool."

13] The last recipe does not specify which kind of dye was employed. In addition, the term ܪܓܡܢ (*rgmn*) is dubious, even if it could be interpreted as a transliteration of the Greek *eregma* (ἔρεγμα) or *eregmos* (ἐρεγμός), 'flour, crushed corn' (LSJ⁹ 684), a term usually transcribed as ܐܪܓܡܘܣ ('*rgmws*; see *ThSyr* I 369). The use of 'crushed corn' in dyeing processes seems to be attested by *PHolm.* 137: "Dissolving of *phykos*. Take and wash *phykos*, air it and lay it aside. Then take and cook crushed corn (ἐρεγμόν) in considerable water. When it is cooked then mix *phykos* with the water from the crushed corn. When you let the *phykos* become cold together this, then you dissolve it in this manner" (rec. 131 in Caley, "The Stockholm Papyrus," 997; translation slightly modified).

Appendix

TABLE 1.

COMPARISON BETWEEN V AND **MBA**

V	MBA
1 1ʳ5-7ʳ16 *Ἐκ τῶν Δημοκρίτου φυσι-κῶν καὶ μυστικῶν*	Second part of the work *Φυσικὰ καὶ μυστικά*, (= **M**, fols. 68ʳ8-71ʳ6, **B**, fols. 10ᵛ20-17ʳ16, **A**, fols. 25ᵛ25-29ᵛ3)
2 7ʳ17- 10ᵛ *Περὶ ἀργύρου*	Work entitled *Περὶ ἀσήμου ποιήσεως* in **M** (fols. 71ʳ7-72ᵛ8) and *Περὶ ποιήσεως ἀσήμου* in **B** (fols. 17ʳ16-20ʳ18) and **A** (fols. 29ᵛ4-31ʳ22)
3 33ᵛ13-35ᵛ16 *Ἐκ τῶν Δημοκρίτου περὶ πορφύρας φυσικῆς*	First part of the work *Φυσικὰ καὶ μυστικά* (**M**, fols. 66ʳ7-68ʳ8, **B**, fols. 8ᵛ10-10ᵛ20 and **A**, fols. 24ʳ5-25ᵛ25)

TABLE 2.

NATURAL AND SECRET QUESTIONS: GREEK AND SYRIAC TRADITION

Greek mss. (*PM* §§ 5-20)	London Syriac mss. = **SyrL** (*CMA* II 10-2)	Cambridge Syr. ms. = **SyrC** (*1SyrC* §§ 1-16)
PM § 5 Λαβὼν ὑδράργυρον	ܡܚܣ ܚܣܝܠ (*CMA* II 10, ll. 4-11)	ܡܚܣ ܣܚܟܝܠ (*1SyrC* § 1 = fol. 90ᵛ2-12)
6 Πυρίτην ἀργυρίτην	ܩܘܙܢܝܡܗܣ ܐܘܢܚܘܢܝܡܗܣ (10, ll. 12-6)	ܚܐܦܐ ܘܡ ܩܙܗܢܝܡ (§ 2 = fol. 90ᵛ12-7)
7 Πυρίτην οἰκονόμει	ܡܚܣܢܝܡܗܣ ܘܩܘܙܢܝܡܗܣ (10, ll. 17-20)	ܡܚܣܢܝܡܗܣ ܘܩܘܙܢܝܡܗܣ (§ 3 = fols. 90ᵛ17-91ʳ4)
8 Τὸν Κλαυδιανὸν λαβών	*omittit*	ܟܚܠܩܗܘܣܘܡܗܣ ܚܚܝܒܘܢܝ (§ 4 = fol. 91ʳ4-9)
9 Τὴν κιννάβαριν λευκήν	*omittit*	ܣܚܙܢܝܡܗܣ ܚܚܝ ܣܘܙܘܠܠ (§ 5 = fol. 91ʳ9-14)
10 Τὴν δὲ Κυπρίαν καδμίαν	*omittit*	ܚܡܘܙܠ ܡܐܘܡܚܠ (§ 6 = fol. 91ʳ14-20)
11 Τὸν ἀνδροδάμαντα οἰκονόμει	ܐܝܘܙܘܘܡܚܘܘܣܗܣ ܘܡܚܠ ܚܡܠܠ (10, l. 21–11, l. 5)	ܐܝܘܙܘܘܡܚܘܘܣܗܣ ܘܡܚܠ ܚܡܠܠ (§ 7 = fol. 91ʳ20-91ᵛ10)
12 Λαβὼν γῆν λευκήν	ܡܚܣ ܐܚܙܐ ܣܘܙܘܐ (11, ll. 5-12)	ܡܚܣ ܐܚܙܐ ܣܘܙܘܐ (§ 8 = fols. 91ᵛ10-92ʳ2)
13 Τῷ θείῳ τῷ ἀπύρῳ	ܚܠܡܝ ܐܩܘܙܘܝ (11, ll. 13-5)	ܚܠܡܝ ܐܩܘܙܘܝ (§ 9 = fol. 92ʳ2-7)
14 Χρυσόκολλαν τὴν τῶν Μακεδόνων	ܚܙܡܩܩܣܘܠܠ ܚܘܗ ܘܡܚܣܘܡܝܣ ܠܘܣܘܡܝ (11, ll. 15-9)	ܚܙܡܩܩܣܘܠܠ ܚܘܗ ܘܚܡܚܣܘܡܝܣ ܠܘܣܘܡܝ (§ 10 = fol. 92ʳ7-14)

Continued

TABLE 2.

CONTINUED

	Greek mss. (*PM* §§ 5-20)	London Syriac mss. = **SyrL** (*CMA* II 10-2)	Cambridge Syr. ms. = **SyrC** (*1SyrC* §§ 1-16)
15	Ὦ φύσεις φυσέων δημιουργοί	*omittit*	ܐܘ ܟܝܢܐ ܡܫܟܚܢܐ. ܠܗܘܢ ܕܟܝܢܐ (§ 11 = fol. 92r14-92v7)
16	Οὗτοι δὲ ἀκρίτῳ καὶ ἀλόγῳ ὁρμῇ	ܘܕܝܢ ܐܝܟ ܕܦܪܘܫ ܡܢ ܡܬܚܫܒܢܘܬܐ (1, l. 8-2, l. 2 = *1SyrC* § 12, ll. 4-12)	<ܗܢܘܢ> ܕܝܢ ܡܢ ܥܠܬ ܘܠܐ ܢܝܫܐ (§ 12 = fols. 92v7-93r3)
17	Λαβὼν τὸ Πόντιον ῥᾶ	*omittit*	ܣܒ ܦܢܛܝܩ ܕܡܢ ܦܢܛܘܣ (§ 13 = fol. 93r3-12)
18	Δέξαι κρόκον Κιλίκιον	*omittit*	[ܘ]ܣܒ ܟܘܪܟܡܐ ܕܡܢ ܩܝܠܝܩܝܐ (§ 14 = fol. 93r13-93v3)
19	Λαβὼν μόλυβδον τὸν ἡμῶν	ܣܒ ܐܒܪܐ ܕܝܠܢ (11, l. 19–12, l. 2)	ܣܒ ܐܒܪܐ ܗܘ ܕܝܠܢ (§ 15 = fol. 93v4-12)
20	Αὕτη ἡ <ἀγωγὴ τοῦ> Παμμένους	ܡܢ ܕܝܢ ܠܟܝܢܐ ܕܟܠ ܗܘܐ ܫܘܠܡܐ ܠܗܢܐ ܟܬܒܐ (12, ll. 2-5)	ܡܢ ܕܝܢ ܠܟܝܢܐ ܕܟܠ ܗܘܐ ܫܘܠܡܐ ܠܗܢܐ ܟܬܒܐ (§ 16 = fols. 93v12-94r3)

TABLE 3.

ON THE MAKING OF SILVER: GREEK AND SYRIAC TRADITION

	Greek mss. (*AP* §§ 1-10)	London Syriac mss. = **SyrL** (*CMA* II 12-3)	Cambridge Syr. ms. = **SyrC** (*2SyrC* §§ 1-10)
AP § 1	<Λαβὼν> ὑδράργυρον τὴν ἀπὸ τοῦ ἀρσενικοῦ	ܣܐܡܐ ܕܡܬܥܒܕ ܒܐܪܣܢܝܩܘܢ (*CMA* II 12, ll. 6-11)	ܣܐܡܐ ܗܘ ܕܡܬܥܒܕ ܒܐܪܣܢܝܩܘܢ (*2SyrC* § 1 = fol. 94r6-94v1)
2	Λαβὼν τὴν προγεγραμμένην νεφέλην	ܣܒ ܠܟ. ܥܢܢܐ ܕܟܬܝܒ (12, ll. 11-5)	ܣܒ ܠܟ ܥܢܢܐ ܕܟܬܝܒ (§ 2 = fol. 94v1-7)
3	<Λαβὼν> μαγνησίαν λευκήν	ܣܒ ܡܓܢܝܣܝܐ ܚܘܪܬܐ (12, ll. 15-21)	ܣܒ ܕܝܢ ܡܓܢܝܣܝܐ ܚܘܪܬܐ (§ 3 = fol. 94v7-18)
4	Λαβὼν θεῖον τὸ λευκόν	*omittit*	ܣܒ ܠܬܐܘܢ ܐܦܘܪܘܢ ܚܘܪܐ
5	Τὴν δὲ λευκανθεῖσαν λιθάργυρον	*omittit*	(§ 4 = fols. 94v18-95r9)[1]
	omittit	*omittit*	ܢܣܒ ܠܗܘܢ ܚܢܢ ܗܟܝܠ ܐܘ ܫܪܝܪܐ (§ 5 = fol. 95r10-95v3)
6	Λαβὼν κρόκον Κιλίκιον	ܣܒ ܟܘܪܟܡܐ ܩܝܠܝܩܝܐ (pp. 12, l. 22–13, l. 5)	ܣܒ ܟܘܪܟܡܐ ܕܡܢ ܩܝܠܝܩܝܐ (§ 6 = fol. 95v3-13)
7	Δέξαι λευκὴν τὴν λιθάργυρον	*omittit*	ܗܒ ܠܗܘܢ ܕܝܢ ܠܠܝܬܐܪܓܘܪܘܢ (§ 7 = fols. 95v13-96r3)
8	Λαβὼν τὴν προγεγραμμένην νεφέλην	*omittit*	ܣܒ ܕܝܢ ܣܐܡܐ (§ 8 = fol. 96r3-9)
9	Δέξαι ἀρσενικοῦ οὐγγίαν μίαν	*omittit*	ܗܒ ܠܗܘܢ ܐܪܣܢܝܩܘܢ. ܐܘܢ ܚܕܐ (§ 9 = fol. 96r9-16)
10	Ἀπέχετε πάντα	ܗܐ ܡܫܡܠܝܢ ܐܢܘܢ ܟܠܗܘܢ (13, ll. 6-8)	ܗܐ ܡܫܡܠܝܢ ܐܢܘܢ ܟܠܗܘܢ (§ 10 = fols. 96r16-96v2)

[1] In **SyrC** two recipes preserved by the Byzantine tradition (*AP* §§ 4-5) have been summarized in a single recipe (see *2SyrC* § 4 n. 6).

TABLE 4.

PS.-DEMOCRITEAN EXCERPTS INCLUDED IN *THE CHEMISTRY OF MOSES*

The Chemistry of Moses (CAAG II 300-5)	Ps.-Democritus' PM and AP
1. *CAAG* II 306,14 Περὶ ἀργυροποιίας	= AP § 1
2. *CAAG* II 306,15-307,16 four catalogues of dyeing substances: (a) Ὕλη χρυσοποιίας (Ϟποιίας **A**) (b) Ὕλη ζωμῶν. Ζωμοί. Τὰ δὲ ἐν ζωμοῖς ἐστι ταῦτα (c) Ὕλη ἀργυροποιίας (Ϟποιίας **A**) (d) Ταῦτα ἄνθη προτετίμηνται παρὰ τῶν προγενεστέρων κτλ. This section is closed by the following sentence: Ταῦτα παρὰ τοῦ προειρημένου διδασκάλου μεμαθηκὼς ἠσκούμην ὅπως ἀκούσω τὰς φύσεις. Ἡ φύσις γὰρ τὴν φύσιν νικᾷ, καὶ ἡ φύσις τὴν φύσιν κρατεῖ.	= PM § 2, II. 27-33 = PM § 3 (first and last sentences)
3. *CAAG* II 307,18 Οἰκονομία πυρίτου²	= PM § 7
4. *CAAG* II 307,19 Οἰκονομία πυρίτου ἀργυρίτου	= PM § 7

² The same title is also attested by the Syriac tradition that introduces the translation of *PM* § 7 with the specification: ܡܥܒܕܢܘܬܐ ܕܦܘܪܝܛܐ, "The making of pyrite" (see *1SyrC* § 3, l. 1).

TABLE 5.

COMPARISON BETWEEN PS.-DEMOCRITUS'S *CATALOGUES* INCLUDED IN *THE CHEMISTRY OF MOSES*
(TABLE 4, N. 2) AND SYNESIUS'S COMMENTARY

	CHEMISTRY OF MOSES	SYNESIUS' COMMENTARY	INGREDIENTS
(a)	**Ὕλη χρυσοποιίας**		
1.	Ps. Dem. Alch. *Cat.* § 1, l. 2	Syn. Alch. § 13, II. 202-3³	ὑδράργυρος ἡ ἀπὸ κινναβάρεως
2.	*Cat.* § 1, l. 2	§ 13, l. 203⁴	σῶμα μαγνησίας
3.	*Cat.* § 1, l. 3	§ 13, l. 203⁵	χρυσόκολλα
4.	*Cat.* § 1, l. 4	§ 13, II. 204-6	κλαυδιανόν
5.	*Cat.* § 1, l. 4	§ 13, II. 204-6	ἀρσενικὸν ξανθόν
6.	*Cat.* § 1, l. 4	§ 13, l. 209	καδμία
7.	*Cat.* § 1, l. 4	§ 13, l. 209	ἀνδροδάμας
8.	*Cat.* § 1, l. 5	§ 13, II. 211-4	στυπτηρία
9.	*Cat.* § 1, l. 5	§ 13, II. 215-6	θεῖον ἄπυρον
10.	*Cat.* § 1, l. 5	§ 13, l. 218	πυρίτης
11.	*Cat.* § 1, l. 6	*omittit*	ὤχρα Ἀττική
12.	*Cat.* § 1, l. 6	§ 13, II. 221-3	σινωπὶς Ποντική
13.	*Cat.* § 1, l. 6	§ 13, l. 224	θεῖον ὕδωρ ἄθικτον
14.	*Cat.* § 1, II. 7-8	§ 14, II. 225-6	θεῖον ὕδωρ τὸ δι' ἀσβέστου
15.	*Cat.* § 1, l. 8	§ 14, l. 230	θείου αἰθάλη
16.	*Cat.* § 1, l. 8	§ 14, l. 232	σῶρι ξανθόν
17.	*Cat.* § 1, l. 8	§ 14, l. 233	χάλκανθος ξανθή
18.	*Cat.* § 1, l. 8	§ 14, l. 233	κιννάβαρις

Continued

TABLE 5.

CONTINUED

	Chemistry of Moses	Synesius' commentary	Ingredients
(b)	Ὕλη ζωμῶν		
1.	Cat. § 2, l. 11	§ 6, ll. 66- 7; § 15, l. 242	κρόκος κιλίκιος
2.	Cat. § 2, ll. 11-2	§ 6, l. 67; § 15, l. 242	ἀριστολοχία
3.	Cat. § 2, l. 12	§ 15, l. 243	κνήκου ἄνθος
4.	Cat. § 2, l. 12	omittit	ἐλύδριον
5.	Cat. § 2, l. 12	§ 15, ll. 243-8	ἄνθος ἀναγαλλίδος
6.	omittit	§ 16, ll. 254-8[6]	ῥᾶ Πόντιον
7.	Cat. § 2, l. 13	§ 16, l. 262	κυανός
8.	Cat. § 2, l. 13	§ 16, l. 262	χάλκανθος
9.	Cat. § 2, l. 13	§ 16, l. 265	κόμμι ἀκάνθης
10.	Cat. § 2, l. 14	omittit	ὄξος
11.	Cat. § 2, l. 14	§ 6, l. 69; § 16, l. 265	οὖρον ἄφθορον
12.	Cat. § 2, l. 14	omittit	ὕδωρ θαλάσσιον
13.	Cat. § 2, l. 14	§ 6, l. 70; § 16, l. 266	ὕδωρ ἀσβέστου
14.	Cat. § 2, l. 15	§ 6, l. 70; § 16, l. 266	ὕδωρ σποδοκράμβης
15.	Cat. § 2, l. 15	§ 6, l. 70	ὕδωρ φέκλης
16.	Cat. § 2, l. 15	§ 6, l. 71; § 16, l. 266	ὕδωρ στυπτηρίας
17.	Cat. § 2, l. 15	§ 16, l. 267	ὕδωρ νίτρου
18.	Cat. § 2, ll. 15-6	§ 16, l. 267	ὕδωρ ἀρσενικοῦ
19.	Cat. § 2, l. 16	§ 16, l. 267	ὕδωρ θείου ἀθίκτου
20.	Cat. § 2, l. 16	§ 17, l. 275-6	οὖρὸς γάλακτος ὀνείου
21.	Cat. § 2, ll. 16-7[7]	§ 6, l. 72; § 17, ll. 270 and 275	ἀπὸ κυνὸς γάλα
(c)	Ὕλη ἀργυροποιίας		
1.	Cat. § 3, l. 26	§ 8, l. 125; § 19, ll. 304-5	ὑδράργυρος ἡ ἀπὸ ἀρσενικοῦ
2.	Cat. § 3, l. 26	§ 8, l. 126	(ὑδράργυρος ἡ ἀπὸ) σανδαράχης
3.	omittit	§ 19, l. 305	(ὑδράργυρος ἡ ἀπὸ) θείου
4.	Cat. § 3, l. 26	§ 19, l. 305	(ὑδράργυρος ἡ ἀπὸ) ψιμυθίου
5.	Cat. § 3, l. 27	§ 19, l. 305	(ὑδράργυρος ἡ ἀπὸ) μαγνησίας
6.	Cat. § 3, l. 27	§ 19, ll. 305-6	(ὑδράργυρος ἡ ἀπὸ) στίμμεως
7.	Cat. § 3, l. 29	§ 7, l. 88	γῆ Χία
8.	Cat. § 3, l. 29	§ 7, l. 89	καδμία λευκή
9.	Cat. § 3, l. 29	§ 7, l. 89	γῆ ἀστερίτης

[3] See also Syn. Alch. § 5, ll. 43-4; § 8, l. 125; § 11, l. 168; § 19, ll. 306-7. [4]See also Syn. Alch. § 11, l. 181-2. [5]See also Syn. Alch. § 12, ll. 188-91. [6]See also Syn. Alch. § 3, ll. 31-7. [7]According to the text preserved by *The Chemistry of Moses*, after listing the abovementioned ingredients, ps.-Democritus insisted on the dyeing properties of the liquid substances and specified which base metals were to be processed with these ζωμοί. This second part of the catalogue is also often commented upon by Synesius: Ps.-Dem. Alch. *Cat.* § 2, ll. 18-20 = Syn. Alch. § 6, ll. 74-7; Ps.-Dem. Alch. *Cat.* § 2, ll. 21-3 = Syn. Alch. § 18, ll. 293-5.

TABLE 6.

ALCHEMICAL SIGNS RECORDED IN THE CRITICAL APPARATUS

	M	V	B	A
ἄργυρος	☾	☾	☾	☾
ἀρσενικόν	ꞵ Λ	ꞵ	ꞵ Λ	ꞵ Λ
ἀφροσέληνος	ἀφρο☾	ἀφρο☾	ἀφρο☾	ἀφρο☾
δραχμή (?)	Ɣ	Ɣ	Ɣ	Ɣ
ἡμέρα	6	6	6	6
ἥμισυ	ς̄	ϟ	ϟ	ϟ
θεῖον	X	X	X	X
θεῖον (ἄθικτον)	ϻ	ϻ	ϻ	ϻ
κασσίτερος	☿ ♉	☿ ♉	☿	☿
κιννάβαρι	⊙	⊙	⊙	⊙
λαβών	λ^α			
λιθάργυρος	λι☾ λιθαρ☾	λι☾	λι☾	λι☾ λιθαρ☾
μαγνησία	μ͂Γ μ͂γ	μ͂Γ μ͂γ	μ͂Γ μ͂γ	μΓ
μέρος	μ^ε	μ^ε		
μόλυβδος	♄ μ^λ	♄ ♄	♄ ♄	♄ ♄
μολυβδόχαλκος	μ̂♀	μ̂♀	Ι♀	μ♀
νίτρον	Ñ	Ñ	Ñ	Ñ
οὐγγία	Ꞁο	Ꞁο	Ꞁο	
ὄξος				⌖
πέταλον	□	□	□	□
ποίει	Π^οι	Π^οι		
πυρίτης				Π^υ
σανδαράχη	ꞵ	ꞵ	ꞵ	ꞵ
σίδηρος	♂	♂	θ̄	θ̄
στυπτηρία (σχιστή)	✳	✳	✳ σχιστή ✳ σχιστή et s.l. ✳	✳ σχιστή ✳ σχιστή et s.l. ✳
τρῖβε/τρῖψον	@	@	@	@
ὑδράργυρος	☽	☽	☽	☽
ὕδωρ	ῠ̂	ῠ̂	ῠ̂	ῠ̂
ὕδωρ θαλάσσιον	≈	≈	≈	≈
χάλκανθον (-ος)			☿̽	☿̽
χαλκός	♀ fort. ♀	♀ fort. ♀	♀	♀
χρυσόκολλα	♉	♉	♉	♉

Continued

TABLE 6.

CONTINUED

	M	V	B	A
χρυσός	♃	♃	♃	♃
signum incertum			◎	◎
signum incertum			♀	
signum incertum	♍	♍		

Index of the Greek names of substances and relevant terms

For the name of each substance I have provided an English translation and, when possible, the corresponding Syriac name. Since the Syriac translation does not always correspond to the Greek text preserved in the Byzantine tradition, the Syriac equivalent has been given only when supported by at least one passage in which the Greek and the Syriac texts overlap. Otherwise, a possible Syriac translation (attested in passages no longer extant in Greek) is suggested within parentheses.

ἄγχουσα (Λαοδικηνή) — alkanet
Ps.-Dem. Alch. *PM* § 2, l. 25

ἄκανθα (κόμμι) — acacia (gum)
Ps.-Dem. Alch. *Cat.* § 2, l. 13 ‖ Syn. Alch. § 16, l. 265

ἅλμη — brine — ܡܢܐ ܘܡܚܠܐ
Ps.-Dem. Alch. *PM* § 9, l. 98; § 11, l. 111; § 12, l. 119 ‖ *AP* § 3, l. 18; § 4, l. 32; § 6, l. 54

ἅλς/ἅλας — salt — ܡܠܚܐ
Ps.-Dem. Alch. *PM* § 16, l. 174 ‖ *AP* § 4, l. 32; § 9, l. 81

ἅλς Καππαδοκικός — Cappadocian salt
Ps.-Dem. Alch. *PM* § 13, l. 139 ‖ *Cat.* § 3, l. 31

ἄμπελος (χυλός) — grape juice — ܣܥܪܐ ܘܚܩܦܠܐ
Ps.-Dem. Alch. *PM* § 18, l. 198

ἀναγαλλίς (ἄνθος) — flower of the Anagallis
Ps.-Dem. Alch. *Cat.* § 2, l. 12 ‖ Syn. Alch. § 15, ll. 243, 245, 246, 247

ἀνδροδάμας — *androdamas* — ܐܒܪܘܕܡܘܣ
Ps.-Dem. Alch. *PM* § 11, l. 108 ‖ *Cat.* § 1, l. 4 ‖ Syn. Alch. § 13, ll. 209, 210

ἄνθος — flower (dye)
Ps.-Dem. Alch. *PM* § 2, l. 27 ‖ Syn. Alch. § 6, ll. 66, 67; § 15, ll. 243, 246, 248; § 16, l. 263
ἄνθος τὸ τῆς Ἀχαίας — Achaean flower (called λακχά) Ps.-Dem. Alch. *PM* § 2, l. 29
ἄνθος τὸ τῆς ἀνωτέρας Συρίας — the flower from the upper Syria (called κόγχος) Ps.-Dem. Alch. *PM* § 2, l. 33
ἄνθος τὸ τῆς Συρίας — the Syrian flower (called ῥίζιον) Ps.-Dem. Alch. *PM* § 2, l. 30
ἄνθος θαλάσσιον — sea flower
Ps.-Dem. Alch. *PM* § 2, l. 25
see *s.vv.* ἀναγαλλίς, οἰχομένιον, κρόκος, κνῆκος, χαλκός

ἀργυρίτης (cf. *s.v.* πυρίτης)

ἄργυρος — silver — ܣܐܡܐ
Ps.-Dem. Alch. *PM* § 5, l. 71; § 7, l. 88; § 8, l. 94; § 9, l. 100; § 10, l. 105; § 11, l. 113; § 13, l. 136; § 14, l. 148; § 18, l. 199 ‖ *AP* § 10, l. 85 ‖ *Cat.* § 2, l. 22 ‖ Syn. Alch. § 1, l. 13; § 8, ll. 110, 111, 117

ἀριστολοχία (-εία) — Aristolochia — ܐܪܝܣܛܠܟܝܐ
Ps.-Dem. Alch. *PM* § 18, l. 201; § 19, l. 209 ‖ *Cat.* § 2, l. 11 ‖ Syn. Alch. § 6, l. 67; § 15, l. 242

ἀρσενικόν — orpiment — ܐܪܣܢܝܩܘܢ
Ps.-Dem. Alch. *PM* § 5, l. 69; § 8, l. 93 ‖ *AP* § 1, l. 1; § 5, l. 42; § 9, l. 79 ‖ *Cat.* § 3, l. 26 ‖ Syn. Alch. § 8, l. 126; § 19, ll. 305, 307
ἀρ. ἐκστραφέν — turned inside out orpiment — ܐܪܣܢܝܩܘܢ ܕܡܬܗܦܟ
Ps.-Dem. Alch. *AP* § 1, l. 4
ἀρ. λευκόν — white orpiment
Ps.-Dem. Alch. *AP* § 6, l. 58 ‖ *Cat.* § 3, l. 30
ἀρ. ξανθόν — yellow orpiment — ܐܪܣܢܝܩܘܢ ܣܘܡܩܐ
Ps.-Dem. Alch. *PM* § 5, l. 73 ‖ *Cat.* § 1, l. 4 ‖ Syn. Alch. § 12, l. 199; § 13, l. 205
ὕδωρ ἀρσενικοῦ — orpiment water
Ps.-Dem. Alch. *Cat.* § 2, l. 16 ‖ Syn. Alch. § 16, l. 267

ἄσβεστος — unslaked lime, quicklime — ܟܠܚܐ
Ps.-Dem. Alch. *AP* § 8, l. 74 ‖ *Cat.* § 1, l. 8 ‖ Syn. Alch. § 14, l. 227
ἄσβ. οἰκονομηθεῖσα — treated quicklime
Ps.-Dem. Alch. *PM* § 20, l. 220
ἄσβ. στακτή — filtrate of quicklime
Ps.-Dem. Alch. *AP* § 9, l. 82
ὕδωρ ἀσβέστου — quicklime water (ܡܝܐ ܟܠܚܐ in *3SyrC*)
Ps.-Dem. Alch. *Cat.* § 2, l. 14 ‖ Syn. Alch. § 6, l. 70; § 14, ll. 226, 227; § 16, l. 266

ἄσημος — silver
Ps.-Dem. Alch. *AP* tit.

ἀφροσέληνος (-ον) — moon foam — ܐܦܪܘܣܠܝܢܘܢ (ܐܦܪܘܣܠܝܢܐ in *3SyrC*)
Ps.-Dem. Alch. *PM* § 5, l. 68 ‖ *AP*, § 2, ll. 15, 16
ἀφ. ὑαλοῦν — moon foam of glass
Ps.-Dem. Alch. *Cat.* § 3, l. 32

βατράχιον — ranunculus (name of χρυσόκολλα)

Ps.-Dem. Alch. *Cat.* § 1, l. 3 ‖ Syn. Alch. § 12, ll. 188, 190

βρύον (θαλάσσιον) — seaweed
Ps.-Dem. Alch. *PM* § 1, ll. 5, 6, 12

γάλα — milk — ܚܠܒܐ
γάλα ὄνειον (οὐρός) — (the serous part of) jenny' milk (ܚܠܒܐ ܕܐܬܢܐ in *3SyrC*)
Ps.-Dem. Alch. *Cat.* § 2, l. 16 ‖ Syn. Alch. § 17, ll. 275, 276
γάλα κυνός — dog's milk
Ps.-Dem. Alch. *Cat.* § 2, l. 17 ‖ Syn. Alch. § 6, l. 71; § 17, ll. 271, 275

γῆ — earth
(γῆ) ἀστερίτης — *asteritēs*
Ps.-Dem. Alch. *AP* § 2, l. 15 ‖ *Cat.* § 3, l. 29 ‖ Syn. Alch. § 7, l. 89
(γῆ) Κιμωλία — Cimolian earth (ܩܡܘܠܝܐ, in *3SyrC*) Ps.-Dem. Alch. *AP* § 7, l. 66 ‖ *Cat.* § 3, l. 29
γῆ λευκή — white earth (ܐܪܥܐ ܚܘܪܬܐ, 'white lead' in *1SyrC*) Ps.-Dem. Alch. *PM* § 12, l. 117
γῆ Πάρου — Parian earth
Ps.-Dem. Alch. *PM* § 19, l. 207
γῆ Χία — Chian earth — ܚܝܐ
Ps.-Dem. Alch. *PM* § 19, l. 207 ‖ *AP* § 2, l. 15 ‖ *Cat.* § 3, l. 29 ‖ Syn. Alch. § 7, l. 88
see *s.v.* Σινωπίς

δάφνη (φύλλα) — bay leaves (ܕܦܢܐ ܕܛܪܦܐ, 'shrubbery leaves' in *2SyrC*) Ps.-Dem. Alch. *AP* § 7, l. 66

δρόσος — dew
Ps.-Dem. Alch. *PM* § 12, l. 120

ἔλαιον — oil — ܡܫܚܐ
Ps.-Dem. Alch. *PM* § 9, l. 97; § 16, l. 182
see *s.v.* κίκινον and ῥαφάνινον

ἕλκυσμα — *helkysma* — ܗܠܩܘܣܡܐ
Ps.-Dem. Alch. *PM* § 12, l. 117

ἐλύδριον — celandine — ܗܠܘܕܪܝܘܢ
Ps.-Dem. Alch. *PM* § 17, l. 195; § 18, l. 201; § 19, l. 209 || *Cat.* § 2, l. 12

ἐρέα — wool (ܥܡܪܐ and ܐܪܝܘܢ in *3SyrC*)
Ps.-Dem. Alch. *PM* § 1, ll. 8, 14, 19

ἐρυθρόδανον (’Ιταλικόν) — (Italian) madder
Ps.-Dem. Alch. *PM* § 2, l. 25

ζωμός — wash — ܙܘܡܐ
Ps.-Dem. Alch. *PM* § 1, ll. 9, 11; § 16, l. 184; § 17, l. 192; § 18, l. 197; § 19, l. 210 || *AP* § 6, ll. 55, 61; § 7, l. 70; § 8, l. 75; § 9, l. 83 || *Cat.* § 2 tit. and l. 1 || Syn. Alch. § 2, l. 22; § 6, l. 66; § 8, l. 108; § 15, l. 242

ἤλεκτρον — electrum
Ps.-Dem. Alch. *PM* § 9, l. 101 || *Cat.* § 2, l. 23

θάλασσα — seawater — ܡܝܐ ܕܝܡܐ (and ܡܝ ܝܡܐ)
Ps.-Dem. Alch. *PM* § 7, l. 84; § 11, l. 108; § 12, l. 119 || *AP*, § 6, ll. 54, 63
see *s.v.* ὕδωρ θαλάσσιον

θεῖον — sulphur — ܟܒܪܝܬܐ, ܬܝܘܢ, ܐܬܝܘܢ
Ps.-Dem. Alch. *PM* § 8, l. 93; § 16, l. 182 || *AP* § 1, ll. 6, 7; § 2, l. 11; § 4, l. 38; § 5, l. 41 || *Cat.* § 1, l. 7 || Syn. Alch. § 13, l. 224; § 14, l. 229; § 19, l. 305
θεῖον ἄκαυστον — incombustible sulphur
Ps.-Dem. Alch. *Cat.* § 1, l. 5 || Syn. Alch. § 13, ll. 216, 217
θεῖον ἄπυρον — unburnt sulphur — ܐܬܝܘܢ ܐܦܘܪܝܢ, ܬܝܘܢ ܐܦܘܪܘܢ, ܐܬܝܘܢ ܐܦܘܪܘܢ
Ps.-Dem. Alch. *PM* § 5, l. 68; § 7, l. 87; § 9, l. 99; § 11, l. 114; § 13, l. 132; § 14, l. 147; § 19, l. 212 || *AP* § 6, l. 58; § 8, l. 77 || *Cat.* § 1, l. 5 || Syn. Alch. § 5, l. 51; § 13, l. 215
θεῖον λευκόν — white sulphur (ܐܦܘܪܘܢ ܚܘܪܐ ܐܬܝܘܢ, ‘unburnt white sulphur’ in *2SyrC*)
Ps.-Dem. Alch. *AP* § 4, l. 31
αἰθάλη θείου — sulphur vapour
Ps.-Dem. Alch. *AP* § 3, l. 20 || *Cat.* § 1, l. 8 || Syn. Alch. § 14, l. 230
καπνὸς θείου — sulphur fume — ܬܢܢܐ ܕܬܝܘܢ
Ps.-Dem. Alch. *AP* § 3, l. 20
ὕδωρ θεῖον — divine water
Syn. Alch. § 5, l. 50; § 7, l. 102
ὕδωρ θεῖον ἄθικτον — untouched divine water — ܡܝܐ ܕܐܠܗܐ
Ps.-Dem. Alch. *PM* § 11, l. 113 || *Cat.* § 1, l. 6
ὕδωρ θείου ἄθικτον — untouched water of sulphur
Syn. Alch. § 13, l. 224
ὕδωρ θείου
Syn. Alch. § 16, l. 267
ὕδωρ θείου ἀθίκτου
Ps.-Dem. Alch. *Cat.* § 2, l. 16

ἰός — rust — ܫܘܚܬܐ, ܚܠܘܕܐ
Ps.-Dem. Alch. *PM* § 13, l. 141 || Syn. Alch. § 14, l. 238
ἰὸς ξυστός — scraped off rust (ܚܠܘܕܐ ܗܘ ܕܗܪܡܝܣ, ‘rust of *Hermēs*’ in *1SyrC*) Ps.-Dem. Alch. *PM* § 12, l. 122
see *s.v.* χαλκός and χρυσός

ἰσάτις — woad
Ps.-Dem. Alch. *PM* § 2, ll. 32, 34

καδμία (-εία) — cadmia — ܩܕܡܝܐ, ܩܐܕܡܝܐ
Ps.-Dem. Alch. *AP* § 5, l. 42 || *Cat.* § 1, l. 4 || Syn. Alch. § 13, ll. 209, 210
καδ. Κυπρία — Cyprian cadmia
Ps.-Dem. Alch. *PM* § 10, l. 102
καδ. λευκή — white cadmia — ܩܕܡܝܐ ܚܘܪܬܐ
Ps.-Dem. Alch. *AP* § 8, l. 74 || *Cat.* § 3, l. 29 || Syn. Alch. § 7, l. 89

κυανός — azurite — ܟܘܢ
Ps.-Dem. Alch. *PM* § 12, l. 124 || *Cat.* §
2, l. 13; § 3, l. 32 || Syn. Alch. § 16, l. 262
κυ. ψωρώδης — azurite that easily peels off
Ps.-Dem. Alch. *PM* § 13, l. 133

λακχά — *lakcha*
Ps.-Dem. Alch. *PM* § 1, ll. 16, 18, 19; §
2, l. 29
see *s.v.* ἄνθος

λάπαθον — dock
Ps.-Dem. Alch. *PM* § 1, ll. 16, 17

λέκιθος (see *s.v.* ᾠόν)

λιθάργυρος — litharge — ܠܕܐܪ/ܘܓܪܘܢ,
ܣܘܪܓܪܐ (*SyrL*)
Ps.-Dem. Alch. *PM* § 6, l. 80 (μέλαν) ||
AP § 5, l. 50
λιθ. λευκανθεῖσα — whitened litharge
Ps.-Dem. Alch. *AP* § 5, l. 41
λιθ. λευκή — white litharge —
ܣܘܐܬ/ ܠܕܐܪ/ܘܓܪܘܢ ܣܘܐܬ (ܣܘܐܬ/ܛܘܪܘܡܘ in *SyrL*)
Ps.-Dem. Alch. *PM* § 6, l. 78; § 12,
l. 119 || *AP* § 7, l. 65 || *Cat.* § 3, l. 30

λίθος — stone — ܟܐܦܐ
Ps.-Dem. Alch. *PM* § 8, ll. 91-2 (Κλαυ-
διανόν); § 11, l. 115 (gloss) || *Cat.* § 1,
l. 3 (βρατράχιον) || Syn. Alch. § 1,
l. 13; § 10, l. 154; § 12, l. 188 and §
13, l. 191 (βρατράχιον)

μάγνης — magnetite
Ps.-Dem. Alch. *AP* § 1, l. 7

μαγνησία — *magnēsia* — ܡܓܢܣܝܐ
Ps.-Dem. Alch. *PM* § 12, l. 118 || *AP* § 1,
l. 6; § 8, l. 74 || *Cat.* § 3, l. 27 || Syn.
Alch. § 19, l. 305
μαγ. λευκανθεῖσα — whitened *mag.* —
ܡܓܢܣܝܐ ܘܡܚܘܪܬܐ

Ps.-Dem. Alch. *AP* § 1, l. 3; § 3, l. 28
μαγνησία λευκή — white *magnesia* —
ܡܓܢܣܝܐ ܚܘܪܬܐ
Ps.-Dem. Alch. *AP* § 3, l. 18 || *Cat.* § 3,
l. 32
σῶμα μαγνησίας — body of *magnēsia* —
ܩܘ ܘܓܫܡܐ ܕܡܓܢܣܝܐ
Ps.-Dem. Alch. *PM* § 5, l. 67 || *Cat.* § 1,
l. 2 || Syn. Alch. § 11, ll. 179, 181, 184; §
13, l. 203; § 19, ll. 313, 314

μάρμαρον (-ος) — marble
Ps.-Dem. Alch. *AP* § 4, l. 34

μάρμαρος — gleaming — ܡܪܡܪܘܢ
Ps.-Dem. Alch. *PM* § 8, 90

μέλι — honey — ܕܒܫܐ
Ps.-Dem. Alch. *PM* § 9, l. 97 || *AP* § 7, l. 66
μέλι λευκότατον
Ps.-Dem. Alch. *AP* § 8, l. 75

μήνη (see *s.v.* ἄργυρος)
Ps.-Dem. Alch. *PM* § 17, l. 187

μίσυ — *misy* — ܡܣܘܝ
Ps.-Dem. Alch. *PM* § 9, l. 98; § 13, ll.
133, 139; § 14, l. 147 || *AP* § 8, l. 73
μίσυ ὀπτόν — roasted *misy*
Ps.-Dem. Alch. *Cat.* § 3, l. 30
μίσυ ὠμόν — raw *misy*
Ps.-Dem. Alch. *Cat.* § 3, l. 30

μόλυβδος — lead — ܐܒܪ/, ܐܒܐ and ܐܒܪܘܢ
(*SyrL*)
Ps.-Dem. Alch. *PM* § 6, l. 78; § 19, ll.
206, 211 || *AP* § 4, l. 37; § 5, ll. 51,
52; § 6, l. 56 || *Cat.* § 2, l. 23

μολυβδόχαλκος — lead-copper alloy —
ܐܒܪܘܢ ܘܐܒܪ
Ps.-Dem. Alch. *PM* § 12, l. 125 || *Cat.* §
2, l. 23

πυρ. λευκανθείς — whitened pyrite (ܐܠܒܣܛܪܘܢ ܕܐܬܚܘܪ, 'whitened alabaster' in *2SyrC*) Ps.-Dem. Alch. *AP* § 1, l. 5

ῥᾶ — rhubarb — ܦܩܣܡ (?)
Ps.-Dem. Alch. *PM* § 17, ll. 194, 196; § 19, l. 210
ῥᾶ Ποντικός — rhubarb from Pontus
Syn. Alch. § 16, l. 254
ῥᾶ Πόντιος — rhubarb from Pontus — ܦܩܣܡ ܕܡܢ ܦܢܛܘܣ
Ps.-Dem. Alch. *PM* § 17, l. 186 || *Cat.* § 2, l. 13 || Syn. Alch. § 3, ll. 31, 33; § 16, l. 255

ῥάμνος — Rhamnus plant
Ps.-Dem. Alch. *PM* § 20, l. 222

ῥαφάνινον (ἔλαιον) — castor oil (ܡܫܚܐ, 'oil' in *2SyrC*)
Ps.-Dem. Alch. *PM* § 10, l. 104 || *AP* § 2, l. 10

ῥίζιον — little root
Ps.-Dem. Alch. *PM* § 2, l. 30 (see *s.v.* ἄνθος)

ῥίθεον — ?
Ps.-Dem. Alch. *Cat.* § 3, l. 31

ῥόδιον (τὸ 'Ιταλικόν) — pomegranate (? or rose extract)
Ps.-Dem. Alch. *PM* § 2, l. 26

σανδαράχη — realgar — ܐܢܕܪܘܟܝܢ (ܣܢܕܪܘܟܝܢ in *SyrL*)
Ps.-Dem. Alch. *PM* § 8, l. 93 || *AP* § 1, l. 1; § 4, l. 33; § 6, l. 57 || *Cat.* § 3, l. 26 || Syn. Alch. § 8, l. 126
σαν. ἄπυρος — unburnt realgar — ܐܢܕܪܘܟܝܢ ܕܠܐ ܢܘܪܐ
Ps.-Dem. Alch. *AP* § 1, l. 5
σαν. λευκή — white realgar
Ps.-Dem. Alch. *AP* § 7, l. 66
σαν. οἰκονομηθεῖσα — processed realgar — ܣܢܕܪܘܟܝܢ ܕܐܬܕܒܪܬ (*SyrL*)
Ps.-Dem. Alch. *PM* § 5, l. 73

σιδηρίτης — name of the πυρίτης ἀργυρίτης (see *s.v.*) — ܣܝܕܪܝܛܝܣ
Ps.-Dem. Alch. *PM* § 6, l. 76

σίδηρος — iron — ܦܪܙܠܐ
Ps.-Dem. Alch. *PM* § 20, l. 217 || *AP* § 1, ll. 6, 8; § 4, l. 36; § 6, l. 56
σίδ. θειωθείς — iron purified with sulphur
Ps.-Dem. Alch. *AP* § 1, l. 2
σκωρία σιδήρου — iron slag
Ps.-Dem. Alch. *PM* § 1, l. 2

σινωπίς (Ποντική) — earth of Sinope from Pontus
Ps.-Dem. Alch. *Cat.* § 1, l. 6 || Syn. Alch. § 13, ll. 221, 222

σκώληξ — worm
σκώληξ ὁ πορφύριος — purple worm
Ps.-Dem. Alch. *PM* § 2, l. 26
σκώληξ ὁ τῆς Γαλατίας — Galatian worm
Ps.-Dem. Alch. *PM* § 2, l. 29

σποδοκράμβη (ὕδωρ) — cabbage ashes (= ܟܪ̈ܒܐ ܕܩܘܩܡ? in *3SyrC*)
Ps.-Dem. Alch. *Cat.* § 2, l. 15 || Syn. Alch. § 6, l. 70; § 16, l. 266

σποδός — ash — ܩܛܡܐ
σποδὸς λευκίνων ξύλων (ὕδωρ) — ashes of white poplar (water of) — ܡܝ̈ܐ ܕܐܝܟ ܒܗܘܢ ܩܛܡܐ ܕܩܝ̈ܣܐ
Ps.-Dem. Alch. *AP* § 7, l. 69

στῖμι — stibnite — ܣܛܝܡܐ, ܟܘܚܠܐ
Ps.-Dem. Alch. *PM* § 11, l. 112 (μελανία)
στῖμι 'Ιταλικόν — Italian stibnite — ܣܛܝܡܐ ܐܝܛܠܝܩܝܢ, ܟܘܚܠܐ ܐܝܛܠܝܩܘܢ,
ܣܛܝܡܐ ܐܝܛܠܝܩܘܢ

Ps.-Dem. Alch. *PM* § 5, l. 68; § 6, l. 78; § 12, l. 118 || *Cat.* § 3, l. 27 || Syn. Alch. § 19, ll. 305, 314

(στῖμι) Κοπτικόν — Coptic stibnite — ܡܩܘܡܩܡ ܘܡܢ (ܠܩܡܩܕ)
Ps.-Dem. Alch. *PM* § 6, l. 79

στῖμι Χαλκηδόνιον — Chalcedonian stibnite
Ps.-Dem. Alch. *PM* § 11, l. 110

στυπτηρία — alum — ܐܠܩܘܗܘܦܝܐ, ܪܘܙܐ, ܐܠܩܘܗܘܦܝܐ
Ps.-Dem. Alch. *PM* § 9, l. 98; § 13, l. 139; § 14, l. 147; § 19, l. 207 || *AP* § 2, l. 10; § 4, l. 32; § 8, l. 72
στ. ἡ ἀπὸ Μήλου — alum from Milos
Ps.-Dem. Alch. *PM* § 5, l. 69
στ. ἐξιπωθεῖσα — dried alum
Ps.-Dem. Alch. *PM* § 8, l. 92 || Syn. Alch. § 13, ll. 211, 213
στ. σχιστή — scissile alum (just ܐܠܩܘܗܘܦܝܐ, 'alum' in *2SyrC*) Ps.-Dem. Alch. *AP* § 3, l. 19; § 9, l. 81
στ. ταπεινωθεῖσα — 'humbled' alum
Ps.-Dem. Alch. *Cat.* § 1, l. 5
στ. ξανθή — yellow alum — ܐܠܩܘܗܘܦܝܐ ܡܩܘܡܩܠ
Ps.-Dem. Alch. *PM* § 7, l. 87
ὕδωρ στυπτηρίας — alum water
Ps.-Dem. Alch. *Cat.* § 2, l. 15 || Syn. Alch. § 6, l. 71; § 16, l. 266

συκάμινον (χυλός) — mulberry (juice)
Ps.-Dem. Alch. *AP* § 9, l. 81

σφέκλη — wine dregs — ܡܩܘܕܠܩܡ (ܐܩܘܕܠܩܡ)
Ps.-Dem. Alch. *AP* § 3, l. 23

σῶρι — *sōri* — ܡܩܘܢ, ܡܩܘܢܗܐܘܩܡ
Ps.-Dem. Alch. *PM* § 9, l. 98; § 13, ll. 132, 133 || Syn. Alch. § 14, l. 234
σῶρι ξανθόν — yellow *sōri*

Ps.-Dem. Alch. *Cat.* § 1, l. 8 || Syn. Alch. § 14, l. 232

τερεβινθίνη (ῥητίνη) — terebinth (resin) — ܐܩܘܚܩܡ ܘܚܠܩܡ
Ps.-Dem. Alch. *PM* § 10, l. 104

τίτανος — lime — ܠܩܝܠ (ܚܠܩܡ in *SyrL*)
Ps.-Dem. Alch. *PM* § 8, l. 93
τίτ. ὀπτός (-ή) — roasted lime — ܚܠܩܡ ܡܩܘܦܠ (*SyrL*)
Ps.-Dem. Alch. *PM* § 5, l. 69 || *AP* § 5, l. 44 || *Cat.* § 3, l. 32

ὑδράργυρος — mercury — ܐܠܚܩܡ, ܐܠܩܚ, ܐ̈ܪܝ (?), ܐܠܩܘܡܩܘܙܐ (?)
Ps.-Dem. Alch. *PM* § 5, ll. 67, 75; § 16, l. 177; § 18, l. 204 || *AP* § 2, l. 16 || Syn. Alch. § 5, l. 51; § 7, ll. 92, 93; § 8, l. 123; § 9, ll. 127, 137; § 10, 160 (φιλοτεχνουμένη); § 11, ll. 168, 183; § 12, l. 185; § 19, ll. 309, 310, 320
ὑδρ. ἡ ἀπὸ ἀρσενικοῦ — mercury that comes from orpiment — ܐܠܚܩܡ ܘܡܩܚܠܩܚܣ ܚܠܘܙܣܡܩܡܘ,
Ps.-Dem. Alch. *AP* § 1, l. 1 || *Cat.* § 3, l. 26 || Syn. Alch. § 8, l. 125; § 19, ll. 305, 307
ὑδρ. ἡ ἀπὸ θείου — mercury that comes from sulphur
Syn. Alch. § 19, l. 305
ὑδρ. ἡ ἀπὸ κινναβάρεως — mercury that comes from cinnabar
Ps.-Dem. Alch. *Cat.* § 1, l. 2 || Syn. Alch. § 5, l. 43; § 8, l. 125; § 11, l. 168; § 13, l. 202; § 19, l. 306
ὑδρ. ἡ ἀπὸ μαγνησίας — mercury that comes from *magnēsia*
Ps.-Dem. Alch. *Cat.* § 3, l. 27 || Syn. Alch. § 19, l. 305
ὑδρ. ἡ ἀπὸ σανδαράχης — mercury that comes from realgar — ܐܠܚܩܡ ܘܡܩܚܠܩܚܣ ܚܠܩܒܘܙܩܚ

Ps.-Dem. Alch. *AP* § 1, l. 1 || *Cat.* § 3, l. 26 || Syn. Alch. § 8, l. 126

ὑδρ. ἡ ἀπὸ στίμεως Ἰταλικοῦ — mercury that comes from Italian stibnite
Ps.-Dem. Alch. *Cat.* § 3, l. 27 || Syn. Alch. § 19, l. 305

ὑδρ. ἡ ἀπὸ ψιμυθίου — mercury that comes from white lead
Ps.-Dem. Alch. *Cat.* § 3, l. 26 || Syn. Alch. § 19, ll. 305, 307

ὑδρ. λευκή — white mercury
Syn. Alch. § 11, ll. 171, 172

ὑδρ. ξανθή — yellow mercury
Syn. Alch. § 11, ll. 170, 173

ὕδωρ — water — ܡܝܐ
Ps.-Dem. Alch. *PM* § 1, ll. 6, 12, 20 || *AP* § 10, l. 86 || Syn. Alch. § 5, l. 49; § 6, ll. 66, 68; § 7, l. 101; § 15, l. 247

ὕδωρ ἀέριον — rainwater
Ps.-Dem. Alch. *PM* § 12, l. 120

ὕδωρ θαλάσσιον — seawater — ܡܝܐ ܕܝܡܐ
Ps.-Dem. Alch. *PM* § 1, l. 18; § 11, l. 110 || *AP* § 3, l. 19 || *Cat.* § 2, l. 14; (see *s.v.* θάλασσα)

see *s.vv.* ἀρσενικόν, ἄσβεστος, θεῖον, νίτρον, σποδοκράμβη, σποδός, στυπτηρία, φέκλη

φάρμακον — drug — ܣܡܐ, ܣܡܡܐ
Ps.-Dem. Alch. *PM* § 13, l. 135; § 15, ll.159, 165; § 17, ll. 189, 191; § 20, ll. 216, 219 || *AP* § 2, l. 13; § 3, l. 27; § 4, l. 34; § 6, l. 57; § 7, ll. 67-8; § 8, 78 || Syn. Alch. § 2, l. 26

φέκλη (ὕδωρ) — water of lees
Ps.-Dem. Alch. *Cat.* § 2 l. 15 || Syn. Alch. § 6, ll. 70, 73

φῦκος — seaweed (ܦܘܩܘܣ in *3SyrC*)
Ps.-Dem. Alch. *PM* § 2, l. 24 (called ψευδοκογχύλιον)

φυλάνθιον (τὸ δυτικόν) — *phylanthion* (?)
Ps.-Dem. Alch. *PM* § 2, l. 26

χάλκανθον (-ος) — copper flower — ܡܚܠܡܝ, ܡܚܠܡܝܠܘ
Ps.-Dem. Alch. *PM* § 9, l. 99; § 13, ll. 132, 138, 141 || *Cat.* § 2, l. 13 || Syn. Alch. § 14, l. 234; § 16, l. 262

χάλκανθος (-ον) ξανθή (-όν) — yellow copper flower
Ps.-Dem. Alch. *Cat.* § 1, l. 8 || Syn. Alch. § 14, l. 233; § 16, l. 261

χάλκανθον χλωρόν — green copper flower
Ps.-Dem. Alch. *PM* § 13, l. 134

χαλκίτης — *chalkitēs*
Ps.-Dem. Alch. *PM* § 12, l. 124 (ܐܘܗܪܘ, 'copper' in *1SyrC*)

χαλκός — copper — ܐܘܗܪܘ, ܐܘܗܪܘܒܠܝ, ܣܡܐ
Ps.-Dem. Alch. *PM* § 5, l. 70; § 9, l. 100; § 12, l. 126; § 13, l. 136; § 14, l. 148; § 16, l. 175; § 18, ll. 199, 200 || *AP* § 3, l. 29; § 4, l. 35; § 6, ll. 55, 62; § 10, l. 85 || *Cat.* § 2, ll. 21, 22; § 3, l. 28 || Syn. Alch. § 10, l. 158; § 14, l. 235

χαλ. ἀσκίαστος — 'shadowless' copper — ܐܘܗܪܘ ܕܐܣܡܣܡܠܘ (*SyrL*), ܐܘܗܪܘ ܡܣܡܣܡܠܘ (*SyrL*)
Ps.-Dem. Alch. *PM* § 5, ll. 71, 74; § 12, l. 130

χαλ. θειωθείς — copper purified with sulphur
Ps.-Dem. Alch. *AP* § 1, l. 2

χαλ. κεκαυμένος — burnt copper — ܣܡܐ ܝܡܘܡܠܐ
Ps.-Dem. Alch. *PM* § 12, l. 123

χαλ. ὑπόλευκος (ὀρείχαλκος) — white copper (brass)
Ps.-Dem. Alch. *AP* § 3, l. 24

ἄνθος χαλκοῦ (see *s.v.* χάλκανθον) — copper flower

Index of the Syriac names of substances and relevant terms

For the name of each substance I have provided an English translation and, when possible, the corresponding Greek name. Since the Syriac translation does not always correspond to the Greek text preserved by the Byzantine tradition, the Greek equivalent has been proposed only when supported by at least one passage in which the Greek and the Syriac texts overlap. Otherwise I have proposed the possible Greek equivalent within parentheses.

οἶνος αὐστηρός – robust wine – ܚܡܪܐ ܥܙܝܙܐ
1SyrC § 7, l. 1

κνῆκος – safflower – ܡܘܪܕܐ
1SyrC § 15, l. 2
flower of safflower – ܘܪܕܐ ܦܩܚܐ
1SyrC § 14, l. 1

τίτανος – lime – ܟܠܫܐ
1SyrC § 1, l. 2

πέταλον (or φυλλόν) – leaf – ܛܪܦܐ
1SyrC § 14, ll. 2 (ܛܪܦܐ ܕܢܚܫܐ, 'copper leaves'), 3
(ܛܪܦܐ ܕܗܪܡܣ, 'leaves of *Hermēs*') ||
2SyrC § 6, l. 5; § 7, l. 2
see *s.v.* ܚܕܪ

a tange of briars, shrubbery – ܚܕܪ
2SyrC § 7, l. 2 (ܛܪܦܐ ܕܚܕܪ, 'shrubbery leaves')

snake (?) – ܚܘܝܐ
3SyrC § 4, l. 1

lead – ܐܒܪܐ
1SyrC § 8, l. 6; *1SyrL* § 2, l. 3 (?)

λίθος – stone – ܟܐܦܐ
1SyrC § 2, l. 1 (ܟܐܦܢܝ̈ܐ) || *3SyrC* § 1, l. 3; § 2,
ll. 1, 3, 9, 13; § 3, l. 6; § 4, l. 6

θεῖον – sulphur – ܟܒܪܝܬܐ
1SyrL § 1, l. 1 (ܕܝܡܐ ܟܒܪܝܬܐ, 'sea sulphur')

στῖμι – stibnite – ܣܘܦܐ
1SyrC § 7, l. 3
στῖμι Ἰταλικόν – Italian stibnite – ܣܘܦܐ ܐܝܛܠܩܝܐ
1SyrL § 1, l. 1; § 2, l. 2

juice (= χυλός) – ܥܨܪܐ
3SyrC § 4, ll. 8 (ܥܨܪܐ ܕܣܡܣܘܩܐ, 'juice of comfrey'), 9 (ܥܨܪܐ ܕܟܪܒܐ ܕܒܪܐ, 'juice of wild cabbage')

σανδαράχη – realgar – ܙܪܢܝܟܐ
1SyrC § 1, l. 6; § 4, l. 2 || *2SyrC* § 6, l. 3; § 7, l. 2
ܙܪܢܝܟܐ ܕܠܐ ܢܘܪܐ
σαν. ἄπυρος – realgar without fire –
2SyrC § 1, l. 5

ὄξος – vinegar – ܚܠܐ
2SyrC § 3, l. 2 (ܚܡܝܨܐ ܚܠܐ); § 4, l. 3; § 8, l. 2;
§ 9, l. 3 || *3SyrC* § 4, ll. 2, 6; § 6, l. 2
ὀξάλμη – vinegar and salt – ܚܠܐ ܘܡܠܚܐ
1SyrC § 3, ll. 1–2; § 7, ll. 2, 3–4
ὀξύμελι – vinegar and honey – ܚܠܐ ܘܕܒܫܐ
1SyrC § 3, l. 2 || *2SyrC* § 4, l. 2
ὄξος δριμύτατον – sharp vinegar – ܚܠܐ ܚܪܝܦܐ
1SyrC § 15, ll. 3–4
vinegar of citrons – ܚܠܐ ܕܐܬܪܘܓܐ
3SyrC § 4, l. 2

ܚܠܒܐ
ὑδράργυρος or νεφέλη – mercury (lit. milk) –
1SyrC § 1, ll. 2, 7; § 14, l. 7 || *2SyrC* § 1, l. 2
(ܚܠܒܐ [...] ܡܢ ܐܪܣܢܝܟܘܢ = ὑδρ. ἀπὸ ἀρσενικοῦ;)
(ܚܠܒܐ [...] ܡܢ ܙܪܢܝܟܐ = ὑδρ. ἀπὸ σανδαράχης);
§ 2, l. 1 (= gr. νεφέλη); § 8, l. 1 (= gr. νεφέλη);
§ 10, l. 2
(= gr. νεφέλη) || *3SyrC* § 2, l. 7

γάλα – milk – ܚܠܒܐ
2SyrC § 9, l. 2 (ܚܠܒܐ ܕܬܘܪܬܐ, 'cow's milk');
3SyrC § 2, ll. 10–1
(ܚܠܒܐ ܕܐܬܢܐ, 'jenny's milk' = γάλα ὄνειον in
CAAG II 122; ܚܠܒܐ ܕܥܙܐ, 'goat's milk' = γάλα
αἴγειον)

chickpea – ܚܡܨܐ
3SyrC § 4, l. 3

οἶνος – wine – ܚܡܪܐ
1SyrC § 13, l. 4; § 14, l. 2 || *2SyrC* § 6, l. 7
χυλος τῆς ἀμπέλου – wine of vines – ܚܡܪܐ ܕܓܦܬܐ
1SyrC § 13, l. 1

red soda (νίτρον πυρρόν?) – ܣܘܡܩܐ ܢܬܪܘܢ
1SyrC § 15, l. 1

σῶρι – *sôri* – ܣܐܘܪܝ
1SyrC § 9, l. 1

σιδηρίτης – *siderites* – ܣܝܕܪܝܛܝܣ
1SyrL § 2, l. 1

ἄργυρος – moon/silver – ܣܗܪܐ
1SyrC § 1, l. 5; § 3, l. 4; § 4, l. 3; § 5, l. 4;
§ 6, l. 4; § 7, l. 6; § 13, l. 2 (?) ‖ 2SyrC §
1, l. 1; § 5, l. 2; § 10, l. 1

comfrey (= σύμφυτον) – (ܣܘܡܦܘܛܘܢ)
3SyrC § 4, l. 9

(πυρίτης in *PM*) – ? – ܣܘܪܝܩܘܣ
1SyrC § 15, l. 2

σῶρι – *sôri* – ܣܘܪܝ
1SyrC § 5, l. 3

στυπτηρία – alum – ܣܛܘܦܛܪܝܐ
1SyrC § 1, l. 2; 1SyrL § 1, l. 2

στῖμι – stibnite – ܣܛܝܡܐ
1SyrC § 1, l. 1 (ܣܛܝܡܐ ܐܝܛܠܝܩܐ, 'Italian stibnite' =
στῖμι Ἰταλικόν); § 2, ll. 2–3 (ܣܛܝܡܐ ܐܝܛܠܝܩܐ,
'Italian stibnite' = στῖμι Ἰταλικόν) and ll. 3–4
(ܣܛܝܡܐ ܡܢ ܣܡܘܣ, 'stibnite from Samos')

φάρμακον – drug – ܣܡܐ
1SyrC § 2, l. 4

σανδαράχη – realgar – ܣܢܕܪܟܝ
1SyrL § 1, l. 6 (ܣܢܕܪܟܝ ܕܐܬܐܠܝܬ, 'processed realgar' =
σανδαράχη οἰκονομηθεῖσα)

σφέκλης – wine dregs – ܣܦܩܠܣ
2SyrC § 3, l. 4

ὕδωρ – water – ܡܝܐ
2SyrC § 5, l. 9; § 6, l. 5 ('pure water'); § 7,
l. 4; § 10, l. 2
‖ 3SryC § 2, ll. 13–4; § 5, ll. 2–3;
§ 6, l. 3; § 7, l. 2
θάλασσα – seawater – ܡܝܐ ܕܝܡܐ
1SyrC § 3, l. 2
ܡܝܐ ܕܝܡܐ
θάλασσα and ὕδωρ θαλάσσιον – seawater –
1SyrC § 7, ll. 1, 3; § 8, l. 2 ‖ 2SyrC § 3, l. 2;
§ 6, l. 1; ‖3SyrC § 6, l. 4
ἄλμη – water and salt – ܡܝܐ ܘܡܠܚܐ
1SyrC § 5, l. 2 ‖ 2SyrC § 3, l. 1

amalgam (= μάλαγμα?) – ܡܠܓܡܐ
2SyrC § 3, l. 5

ἄλας – salt – ܡܠܚܐ
1SyrC § 12, l. 8
see s.v. ܝܡܐ and ܡܝܐ

μάρμαρος (-ον) – gleaming/marble – ܡܪܡܪܘܢ
1SyrC § 4, l. 1 ‖ 3SyrC § 2, l. 11 (ܡܪܡܪܘܢ)

marble? (better 'pearl' = ܡܪܓܢܝܬܐ) – ܡܪܡܪܝܬܐ
3SyrC § 3, l. 5

χολή – bile – ܡܪܪܬܐ
1SyrC § 6, l. 2 (ܡܪܪܬܐ ܕܬܘܪܐ, 'bile of a bull' = χολή
μοσχεῖα)

ἔλαιον – oil – ܡܫܚܐ
1SyrC § 5, l. 1; § 10, l. 3 ‖ 2SyrC § 2, l. 1
κίκινον – radish oil – ܡܫܚܐ ܕܦܩܠܐ
1SyrC § 6, l. 3

χαλκός – copper – ܢܚܫܐ
1SyrC § 10, l. 1 (?)‖ 2SyrC § 3, l. 5
χαλκὸς κεκαυμένος – burnt copper – ܢܚܫܐ ܩܠܝܐ
1SyrC § 8, l. 4
νίτρον – soda – ܢܬܪܘܢ
2SyrC § 9, l. 1

κροκόμαγμα – saffron-sauce – ܡܢܙܪܘܟܣ
1SyrC § 15, l. 3

flour? (= ἐρεγμός) – ܪܓܡ
3SyrC § 7, l. 4

ῥητίνη – resin? – ܪܛܝܢ
1SyrC § 2, l. 2; *1SyrL* § 2, l. 2

ἰός – rust – ܚܠܘܕܐ
1SyrC § 1, l. 4; § 10, l. 1 (<ܕܢܚܫܐ> ܚܠܘܕܐ,
'copper rust') ‖ *2SyrC*
§ 3, l. 7; § 9, ll. 4–5

ἰός – rust – ܚܘܣܪ
1SyrC § 8, l. 4 (ܚܘܣܪ ܕܗ ܪܡܙܘܣ, 'rust of
Hermēs')

ἥλιος/χρυσός – sun/gold – ܫܡܫܐ
1SyrC title; § 1, l. 5; § 3, l. 5; § 4, l. 3;
§ 5, l. 4; § 6, l. 4;
§ 7, l. 6; § 8, l. 3 (sun); § 9, l. 4; § 13, l. 6
(sun); ‖ *2SyrC* § 10, l. 1

θεῖον – sulphur – ܬܠܝܐ
1SyrC § 12, l. 12 ‖ *2SyrC* § 1, l. 6
θεῖον ἄπυρον – unburnt sulphur – ܬܠܝܐ ܕܠܐܘܩܕ
1SyrC § 15, ll. 5–6 ‖ *2SyrC* § 6, l. 3; § 8, l. 4
θεῖον ἄπυρον – unburnt sulphur – ܬܠܝܐ ܕܠܐܘܙܪ
1SyrC § 3, l. 3; § 7, l. 6
white unburnt sulphur – ܬܠܝܐ ܕܠܐܘܙܪ ܚܘܪܐ
2SyrC § 4, l. 1
ὕδωρ θεῖον (ἄθικτον) – sulphur water –
ܡܝܐ ܕܬܠܝܐ
1SyrC § 7, l. 5

θεῖον – sulphur – ܬܠܝ
θεῖον ἄπυρον – unburnt sulphur – ܬܠܝ ܕܠܐܘܙܪ
1SyrC § 5, l. 3; § 9, l. 1; § 10, l. 4
red (= ξανθόν?) sulphur – ܬܠܝ ܣܘܡܩܐ
1SyrC § 1, l. 2; *1SyrL* § 1, l. 2
καπνός θείου – sulphur fumes – ܬܢܢ ܕܬܠܝ
2SyrC § 3, l. 3

σποδός – ash – ܩܛܡܐ
2SyrC § 7, l. 5 (ܡܬܩܪܐ ܩܛܡܐ ܗܘ ܕܚܐܒܐ ܘܐܝܠ ܩܛܡܐ = ὕδωρ
σποδοῦ ξυλῶν)

κυανός – azurite – ܩܘܢܐ
1SyrC § 8, l. 5

κίτρον – citron – ܩܛܪܘܢ
2SyrC § 3, l. 2 (ܚܠܐ ܕܩܛܪܘܢ, 'vinegar of citrus')‖
3SyrC § 4, l. 2 (ܕܩܛܪܘܢ)

κιννάβαρις – cinnabar – ܩܢܒܪܝܣ
1SyrC § 5, l. 1

κιννάβαρις – cinnabar – ܩܢܒܪܝܣ
1SyrL § 1, l. 7 (ܩܢܒܪܝܣ ܕܐܬܗܦܟܬ, 'transformed
cinnabar' = κιννάβαρις ἐκστραφεῖσα)

κιννάβαρις – cinnabar – ܩܢܒܪܝܣ
3SyrC § 2, l. 8

Κλαυδιανόν (-ός) – *klaudianon* – ܩܠܘܕܝܢܘܢ
1SyrC § 4, l. 1
ܩܠܘܣܢܘܢ – ?
3SyrC § 4, l. 7

χάλκανθος (-ον) – copper flower – ܩܠܩܢܬܘܣ
1SyrC § 5, l. 3; § 9, l. 1

χάλκανθος (-ον) – copper flower – ܩܠܩܢܬ
1SyrC § 9, l. 2

Cimolian earth (?) – ܩܡܘܠܐ
3SyrC § 4, l. 4

flour – ܩܡܚܐ
2SyrC § 7, l. 2

κιννάβαρις – cinnabar – ܩܢܒܪܝܣ
1SyrC § 1, l. 7 (ܩܢܒܪܝܣ ܣܘܡܩܐ, 'red cinnabar')

κασία – cassia – ܩܣܝܐ
1SyrC § 14, l. 7

οὖρον – urine – ܬܝܢܐ

1SyrC § 3, l. 2; § 7, ll. 2–3 ‖ *2SyrC* § 9, l. 3

οὖρον δαμάλεως – urine of a bull – ܕܬܘܪܐ, ܬܝܢܐ

1SyrC § 10, l. 2

καπνός – smoke – ܬܢܢܐ

2SyrC § 3, ll. 2, 4 (ܕܚܕܐ, ܬܢܢܐ = καπνός κοβαθίων?)

see *s.v* ܩܒ

Bibliography

Abt, Theodor, and Salwa Fuad, *The Book of Pictures by Zosimus of Panopolis. Corpus Alchemicum Arabicum* II.2 (Zurich: Living Human Heritage, 2011)

Adler, William, and Paul Tuffin, *The Chronography of George Synkellos, A Byzantine Chronicle of Universal History from the Creation* (Oxford: Oxford University Press, 2002)

Amigues, Suzanne, *Théophraste, Recherches sur les plantes*, 5 vols. (Paris: Les Belles Lettres 1988–2006)

André, Jacques, *Pline l'Ancien. Histoire naturelle, livre XXI* (Paris: Les Belles Lettres, 1969)

André, Jacques, *Les noms de plantes dans la Rome antique* (Paris: Les Belles Lettres, 1985)

Aufrère, Sidney H., *L'univers minéral dans la pensée égyptienne*, 2 vols. (Cairo: IFAO, 1991)

Bailey, Kenneth C., *The Elder Pliny's Chapters on Chemical Subjects*, 2 vols. (London: Edward Arnold & Co, 1929–32)

Beck, Lily Y., *Pedanius Dioscorides of Anazarbus, De Materia Medica* (Hildesheim: Olms-Weidmann, 2005)

Beretta, Gemma, *Ipazia d'Alessandria* (Rome: Editori Riuniti, 1993)

Beretta, Marco, *The Alchemy of Glass: Counterfeit, Imitation, and Transmutation in Ancient Glassmaking* (Sagamore Beach, MA: Science History Publications, 2009)

Berthelot, Marcellin, "Des origines de l'alchimie et des œuvres attribuées à Démocrite d'Abdère," *Journal des Savants*, 49 (1884): 517–27

Berthelot, Marcellin, *Les origines de l'alchimie* (Paris: Georges Steinheil, 1885)

Berthelot, Marcellin, "Sur les alliages d'or et d'argent et sur les recettes des orfèvres au temps de l'Empire Romain et du Moyen Âge," *Annales de chimie et de physique*, 22 (1891): 145–72

Berthelot, Marcellin, "Papyrus de Leyde," *Mémoires de l'Académie des sciences de l'Institut de France*, 49 (1906): 266–307

Berthelot, Marcellin, and Charles-Émile Ruelle, *Collection des anciens alchimistes grecs*, 3 vols. (Paris: Georges Steinheil, 1887–88)

Berthelot, Marcellin, and Rubens Duval, *La chimie au Moyen-Âge*, vol. 2: *L'alchimie syriaque* (Paris: Imprimerie Nationale, 1893)

Berthelot, Marcellin, and Octave Victor Houdas, *La chimie au Moyen-Âge*, vol. 3: *L'alchimie arabe* (Paris: Imprimerie Nationale, 1893)

Betz, Hans Dieter, *The Greek Magical Papyri in Translation, Including the Demotic Spells* (Chicago: University of Chicago Press, 1986)

Bidez, Joseph, *Vie de Porphyre, le philosophe néo-platonicienne* (Ghent: van Goethem, 1913)

Bidez, Joseph, and Franz Cumont, *Les mages hellénisés*, 2 vols. (Paris: Les Belles Lettres, 1938)

Black, Matthew, and Albert-Marie Denis, *Pseudepigrapha Veteris Testamenti Graece.* Vol. 3. *Apocalypsis Henochi Graece*, ed. M. Black; *Fragmenta Pseudepigraphorum Graeca*, ed. A.M. Denis (Leiden: Brill, 1970)

Bladel, Kevin van, *The Arabic Hermes: from Pagan Sage to the Prophet of Science* (Oxford and New York: Oxford University Press, 2009)

Boeren, Petrus C. van, *Codices Vossiani chymici* (Leiden: Bibliotheca Universitatis Leidensis, 1975)

Browne, Charles A., "The Poem of the Philosopher Theophrastos upon the Sacred Art: A Metrical Translation with Comments upon the History of Alchemy," *Scientific Monthly*, 11 (1920): 193–214

Böhmer, Harald, and Recep Karadag, "Farbanalytische Untersuchungen," in *Die Textilien aus Palmyra*, ed. Andrea Schmidt-Colinet, Annamarie Stauffer and Khālid Al-As'ad (Mainz: Philipp von Zabern, 2000), 82–90

Boscherini, Silvano, "L'Erbario di Apuleio e i precetti dei profeti," *Galenos*, 1 (2007): 113–8

Burkhalter, Fabienne, "La production des objets en métal (or, argent, bronze) en Égypte hellénistique et romaine à travers les sources papyrologiques," in *B.C.H. suppl. 33: Commerce et artisanat dans l'Alexandrie hellénistique et romaine. Actes du colloque d'Athènes, 11–12 décembre 1988*, ed. Jean-Yves Empereur (Athens École française d'Athènes, 1998), 125–33

Caley, Earle Radcliffe, "The Leiden Papyrus X: An English Translation with Brief Notes," *Journal of Chemical Education*, 3 (1926): 1149–66

Caley, Earle Radcliffe, "The Stockholm Papyrus: An English Translation with Brief Notes," *Journal of Chemical Education*, 4 (1927): 979–1002

Caley, Earle Radcliffe, "Mercury and its Compounds in Ancient Times," *Journal of Chemical Education*, 5 (1928): 419–24

Cardon, Dominique, *Natural Dyes. Sources, Tradition, Technology and Science* (London: Archetype Publications, 2007)

Conybeare, Frederick C., James Rendel Harris, and Anne S. Lewis, *The Story of Ahikar from the Aramaic, Syriac, Arabic, Armenian, Ethiopic, Old Turkish, Greek and Slavonic Versions,* second edition enlarged and corrected (Cambridge: Cambridge University Press, 1913)

Craddock, Paul T., "Gold in Antique Copper Alloys," *Gold Bulletin*, 15 (1982): 69–72

Craddock, Paul T., and Alessandra Giumlia-Mair, "Ḥsmn-Km, Corinthian Bronze, Shakudo: Black-Patinated Bronze in the Ancient Word," in *Metal Plating and Patination: Cultural, Technical and Historical Developments*, ed. Paul T. Craddock and Susan La Niece (Oxford: Butterworth-Heinemann, 1993), 102–27

Craddock, Paul T., *Early Metal Mining and Production* (Edinburgh: Edinburgh University Press, 1995)

Crowfoot, Grace Mary, and Norman de Gary Davies, "The Tunic of Tut'ankhamūn," *Journal of Egyptian Archaeology*, 37 (1941): 113–30

Daiber, Hans, "Democritus in Arabic and Syriac Tradition," in *Proceedings of the First International Congress on Democritus* (Xanthi: International Democritean Foundation, 1984), 251–65

Daumas, François, "Quelques textes de l'Atelier des Orfèvres dans le temple de Dendara," in *Livre du centenaire: 1880–1980* (Cairo: IFAO, 1980), 109–18

De Saint-Denis, Eugène, *Le vocabulaire des animaux marins en latin classique* (Paris: Klincksieck, 1947)

Derchain, Philippe, "L'Atelier des Orfèvres à Dendara et les origines de l'Alchimie," *Chronique d'Égypte*, 65 (1990): 219–42

Diels, Hermann, "Über Epimenides von Kreta," *Sitzungsberichte der Kgl. Pr. Akademie der Wissenschaften zu Berlin* (s.n., 1891), 387–404

Diels, Hermann, *Antike Technik*, second edition (Leipzig: Teubner, 1924)

Dorandi, Tiziano, *Nell'officina dei classici* (Rome: Carocci, 2007)

Dorandi, Tiziano, *Diogenes Laertius, Lives of Eminent Philosophers* (Cambridge: Cambridge University Press, 2013)

Duminil, Marie-Paule, *Hippocrate. Vol. 7: Plaies, Nature des os, Coeur, Anatomie* (Paris: Les Belles Lettres, 1998)

Duval, Rubens, "Notes de lexicographie syriaque et arabe," *Journal asiatique*, n.s. 2 (1893): 290–361

Duval, Rubens, *Lexicon Syriacum auctore Hassano bar Bahlule*, 3 vols. (Paris: Imprimerie Nationale, 1888–1901)

Eichholz, David E., *Theophrastus, De lapidibus, Edited with Introduction, Translation and Commentary* (Oxford: Clarendon Press, 1965)

Ferguson, John, "On the First Editions of the Chemical Writings of Democritus and Synesius," *Proceedings of the Philosophical Society of Glasgow*, 16 (1884–85): 36–46

Ferrini, Maria Fernanda, *Pseudo Aristotele, I Colori. Edizione critica, traduzione e commento* (Pisa: ETS, 1999)

Festugière, Andrè-Jean, *La révélation d'Hermès Trismégiste*, 4 vols. (Paris: Les Belles Lettres, 1944–54)

Festugière, Andrè-Jean, "L'arétalogie isiaque de la Korè Kosmou," in *Mélanges d'archéologie et d'histoire offerts à Charles Picard*, 2 vols. (Paris: Presse universitaire de France, 1949), vol. 1, 376–81

Festugière, Andrè-Jean, "Alchymica," in *Hermétisme et mystique païenne* (Paris: Aubier-Montaigne, 1967), 205–29

Forbes, Robert J., *Studies in Ancient Technology*, second edition, 9 vols. (Leiden: Brill, 1965–72)

Formentin, Maria R., "Domenico Pizzimenti Vibonense: maestro, interprete, copista del sec. XVI," in *Testi medici latini antichi. Le parole della medicina: Lessico e storia*, ed. Maurizio Baldin, Marialuisa Cecere and Daria Crismani (Bologna: Pàtron, 2004), 691–701

Fowden, Garth, *The Egyptian Hermes. A Historical Approach to the Late Pagan Mind* (Princeton, NJ: Princeton University Press, 1993)

Fraser, Kyle A., "Zosimos of Panopolis and the Book of Enoch: Alchemy as Forbidden Knowledge," *Aries*, 4 (2004): 125–47

Fraser, Peter M., *Ptolemaic Alexandria*, 3 vols. (Oxford: Oxford University Press, 1972)

Fück, Johann W., "The Arabic Literature on Alchemy According to An-Nadīm (A.D. 987): A Translation of the Tenth Discourse of the Book of the Catalogue (Al-Fihrist) with Introduction and Commentary," *Ambix*, 4 (1951): 81–144

Gaillard-Seux, Patricia, "Sympathie et antipathie dans l'«Histoire Naturelle» de Pline l'Ancien," in *Rationnel et irrationnel dans la médecine ancienne et médiévale. Aspects historiques, scientifiques et culturels*, ed. Nicoletta Palmieri (Saint-Étienne: Publications de L'Université de Saint-Étienne, 2003), 113–28

Gaillard-Seux, Patricia, "Un pseudo-Démocrite énigmatique: Bolos de Mendès," in *Transmettre les savoirs dans les mondes hellénistique et romain*, ed. Frédéric Le Blay (Rennes: Presse Universitaires de Rennes, 2009), 223–43

Garzya, Antonio, *Opere di Sinesio di Cirene* (Turin: UTET, 1989)

Gazza, Vittorino, "Prescrizioni mediche nei papiri dell'Egitto greco-romano, parte II," *Aegyptus*, 36 (1956): 73–114

Gemelli Marciano, M. Laura, "Le Démocrite technicien. Remarques sur la réception de Démocrite dans la littérature technique," in *Democritus: Science, the Arts, and the Care of the Soul. Proceedings of the International Colloquium on Democritus, Paris, 18–20 September 2003*, ed. Aldo Brancacci and Pierre Marie Morel (Leiden: Brill, 2007), 207–37

Giannini, Alessandro, *Paradoxographorum Graecorum Reliquiae* (Milan: Istituto Editoriale Italiano, 1966)

Giumlia-Mair, Alessandra, and Stephen Quirke, "Black Copper in Bronze Age Egypt," *Revue d'égyptologie*, 48 (1997): 95-108

Griffin, Patricia S., "The Selective Use of Gilding on Egyptian Polychromed Bronzes," in *Gilded Metals. History, Technology and Conservation*, ed. Terry Drayman-Weisser (London: Archetype, 2000), 49–72

Halleux, Robert, "L'orichalque et le laiton," *L'Antiquité Classique*, 42 (1973): 64–81

Halleux, Robert, *Le problème des métaux dans la science antique* (Paris: Les Belles Lettres, 1974)

Halleux, Robert, "L'affinage de l'or, des origines aux premiers alchimistes," *Janus*, 62 (1975): 79–102

Halleux, Robert, "De stagnum «étang» à stagnum «étain», contribution à l'histoire de l'étamage et de l'argenture," *L'Antiquité Classique*, 46 (1977): 557–70

Halleux, Robert, *Les textes alchimiques* (Turnhout: Brepols, 1979)

Halleux, Robert, *Papyrus de Leyde. Papyrus de Stockholm. Fragments de recettes* (Paris: Les Belles Lettres, 1981)

Halleux, Robert, "Nouveaux textes sur la métallurgie antique," in *Mines et fonderies de la Gaule. Université de Toulouse-Le Mirail, 21–22 novembre 1980, table ronde du CNRS* (Paris: éd. du CNRS, 1982), 193–204

Halleux, Robert, "Méthodes d'essai et d'affinage des alliages aurifères dans l'Antiquité et au Moyen Âge," in *L'or monnayé*. Vol. 1: *Purification et altérations de Rome à Byzance*, ed. Cécile Morrisson (Paris: éd. du CNRS, 1985), 39–77

Halleux, Robert, and Paul Meyvaert, "Les origines de la Mappae clavicula," *AHMA*, 62 (1987): 5–58

Hallum, Benjamin C., "The Tome of Images: an Arabic Compilation of Texts by Zosimos of Panopolis and a Source of the *Turba Philosophorum*," *Ambix*, 56 (2009): 76–88

Hammer Jensen, Ingeborg, "Deux papyrus à contenu d'ordre chimique," in *Oversigt over det Kgl. Danske Videnskabernes Selskabs Forhandlinger* (Copenhagen, 1916), 279–302

Hammer Jensen, Ingeborg, *Die älteste Alchymie* (Copenhagen: Hovedkom-missionær, A.F. Høst & Søn, 1921)

Healy, John F., *Pliny the Elder on Science and Technology* (Oxford: Oxford University Press, 1999)

Heinen, Anton M., "The Treatise on Alloys by Menelaos of Alexandria. An Example of an Ancient Greek Text Lost in the Original but Preserved in an Arabic Translation," in *L'eredità classica nelle lingue orientali*, ed. Massimiliano Pavan and Umberto Cozzoli (Florence: Istit. Encicl. Treccani, 1986), 161–70

Hershbell, Jackson P., "Democritus and the Beginnings of Greek Alchemy," *Ambix*, 34 (1987): 5–20

Heseltine, Michael, *Petronius* (Cambridge, MA and London: Loeb, 1913)

Hicks, Robert Drew, *Diogenes Laertius, Lives of Eminent Philosophers*, 2 vols. (London and New York: Loeb, 1925)

Hoefer, Ferdinand, *Histoire de la chimie*, second edition, 2 vols. (Paris: Firmin Didot frères, fils et Cie, 1866–69)

Holmyard, Eric John, *Alchemy* (Harmondsworth: Penguin Books, 1957; reprint, New York, 1990)

Holmyard, Eric John, *The Makers of Chemistry* (Oxford: Oxford University Press, 1931)

Hopkins, Arthur J., "Transmutation by Colour. A Study of Earliest Alchemy," in *Studien zur Geschichte der Chemie: Festgabe Edmund O. Von Lippmann zum siebzigsten Geburtstage dargebracht aus Nah und Fern und im Auftrage der deutschen Gesellschaft für Geschichte der Medizin und der Naturwissenschaften*, ed. Julius Ruska (Berlin: Julius Springer, 1927), 9–14

Hort, Arthur, *Theophrastus, Enquiry into Plants and Minors Works on Odours and Weather Signs*, 2 vols. (London and New York: Loeb, 1916)

Hunt, Leslie B., "The Oldest Metallurgical Handbook. Recipes of a Fourth Century Goldsmith," *Gold Bulletin*, 9 (1976): 24–31

Hunter, Erica C.D., "Beautiful Black Bronzes: Zosimos' Treatises in Cam. Mm.6.29," in *I bronzi antichi: produzione e tecnologia. Atti de XV Congresso internazionale sui bronzi antichi organizzato dall'Università di Udine, sede di Gorizia, Grado-Aquileia, 22-26 maggio 2001*, ed. Alessandra Giumlia-Mair (Montagnac: Monique Mergoil, 2002), 655–9

Irby-Massie, Georgia L., and Paul T. Keyser, *Greek Science of the Hellenistic Era: A Sourcebook* (London and New York: Taylor & Francis Routledge, 2002)

Jacobson, David M., "Corinthian Bronze and the Gold of Alchemists," *Gold Bulletin*, 33 (2000): 60–6

Jackson, John, *Tacitus, Annals 13–16* (Cambridge, MA and London: Loeb, 1937)

Jacobson, David M., and Michael P. Weitzman, "Black Bronze and 'Corinthian Alloy'," *The Classical Quarterly*, 45 (1995): 580–3

Jones, William H.S., *Pliny, Natural History, Books 24–27* (Cambridge, MA and London: Loeb, 1956)

Jones, William H.S., *Pliny, Natural History, Books 28–32* (Cambridge, MA and London: Loeb, 1963)

Keyser, Paul T., "Greco-Roman Alchemy and Coins of Imitation Silver," *American Journal of Numismatics*, Second Series, 7–8 (1995–6): 209–33

Keller, Otto, *Rerum naturalium scriptores Graeci minores*. Vol. 1: *Paradoxographi Antigonus, Apollonius, Phlegon, Anonymus Vaticanus* (Leipzig: Teubner, 1877)

Khanikoff, Nicolas, "Analysis and Extracts of the Book of the Balance of Wisdom, An Arabic Work on the Water-Balance Written by 'Al-Khâzînî in the Twelfth Century," *Journal of the American Oriental Society*, 6 (1858): 1–129

Kingsley, Peter, "From Pythagoras to the *Turba philosophorum*: Egypt and Pythagorean Tradition," *Journal of the Warburg and Courtauld Institutes*, 57 (1994): 1–13

Kingsley, Peter, *Ancient Philosophy, Mystery and Magic. Empedocles and Pythagorean Tradition* (Oxford: Clarendon Press, 1995)

Kissling, Robert Ch., "The ΟΧΗΜΑ - ΠΝΕΥΜΑ of the Neo-Platonists and the *De Insomniis* of Synesius of Cyrene," *American Journal of Philology*, 43 (1922): 318–30

Kopp, Hermann F.M., *Beiträge zur Geschichte der Chemie*, 3 vols. (Braunschweig: F. Vieweg und Sohn, 1869–75)

Kroll, Wilhelm, "Bolos und Democritos," *Hermes*, 69 (1934): 228–32

Lacombrade, Christian, *Synésios de Cyrène. Hellène et Chrétien* (Paris: Les Belles Lettres, 1951)

Lacombrade, Christian, *Synésios de Cyrène. Hymnes* (Paris: Les Belles Lettres, 1978)

Lacombrade, Christian, "Le *Dion* de Synésios de Cyrène et ses quatre sages barbares," *KOINΩNIA*, 12 (1988): 17–26

Lamoureux, Jacques, and Noël Aujoulat, *Synésios de Cyrène. Opuscules I* (Paris: Les Belles Lettres, 2004)

Lagercrantz, Otto, *Papyrus Graecus Holmiensis. Recepte für Silber, Steine und Purpure* (Uppsala: A.B. Akademiscka Bockhandeln, 1913)

Laurenti, Renato, "La questione Bolo-Democrito," in *L'atomo fra scienza e letteratura* (Genoe: Istituto di Filologia Classica e Medievale, 1985), 75–106

Le Déaut, Roger, and Jacques Robert, *Targum du Pentateuque*. Vol. 2: *Exode et Lévitique* (Paris: Éd. du Cerf, 1979)

Lee, Henry D.P., *Aristotle, Meteorologica* (Cambridge, MA and London: Loeb, 1952)

Leemans, Conrad, *Papyri Graeci Musei Antiquarii Publici Lugduni Batavi*, 2 vols. (Leiden: Brill, 1843–85)

Leszl, Walter, "Democritus' Works: from their Titles to their Contents," in *Democritus: Science, the Arts, and the Care of the Soul. Proceedings of the International Colloquium on Democritus, Paris, 18-20 September 2003*, ed. Aldo Brancacci and Pierre Marie Morel (Leiden: Brill, 2007), 11–76

Letrouit, Jean, "Chronologie des alchimistes grecs," in *Alchimie: art, histoire et mythes. Actes du Ier Colloque international de la Société d'Étude de l'Histoire de l'Alchimie*, ed. Didier Kahn and Sylvain Matton (Paris and Milan: S.É.H.A and Arché, 1995), 11–93

Linden, Stanton J., *The Alchemy Reader. From Hermes Trismegistus to Isaac Newton* (Cambridge: Cambridge University Press, 2003)

Lindsay, Jack, *The Origins of Alchemy in Graeco-Roman Egypt* (New York: Barnes & Noble, 1970)

Lippmann, Edmund O. von, *Entstehung und Ausbreitung der Alchemie*, 3 vols. (Berlin: Julius Springer, 1919–54)

Littré, Émile, *Pline l'Ancien. Histoire naturelle*, 2 vols. (Paris: Dubochet, 1848–50)

Long, Herbert S., *Diogenis Laertii Vitae philosophorum*, 2 vols. (Oxford: Oxford Classical Texts, 1964)

Longo, Odone, "La zoologia delle porpore nell'antichità greco-romana," in *La porpora. Realtà ed immaginario di un colore simbolico. Atti del Convegno di studio, Venezia, 24 e 25 ottobre 1996*, ed. Oddone Longo (Venice: Istituto Veneto di scienze lettere ed arti, 1998), 79–90

Lorenzoni, Alberta, "Eustazio: paura 'verde' e oro 'pallido' (Ar. *Pax* 1176, Eup. fr. 253 K.-A., Com. adesp. frr. 390 e 1380A E.)," *Eikasmos*, 5 (1994): 139–63

Lucas, Alfred, and John Richard Harris, *Ancient Egyptian Materials and Industries*, fourth edition (London: E. Arnold, 1962)

Ludwich, Arthur, *Maximi et Ammonis carminum de actionum auspiciis reliquiae. Accedunt anecdota astrologica* (Leipzig: Teubner, 1877)

Marganne, Marie-Hélène, "Le système chromatique dans le Corpus Aristotélicien," *Études Classiques*, 46 (1978): 185–203

Martelli, Matteo, "L'opera alchemica pseudo-democritea: un riesame del testo," *Eikasmos*, 14 (2003): 161–84

Martelli, Matteo, "Divine Water in the Alchemical Writings of Pseudo-Democritus," *Ambix*, 56 (2009): 5–22

Martelli, Matteo, *Pseudo-Democrito, scritti alchemici con il commentario di Sinesio* (Milan and Paris: S.É.H.A and Archè, 2011)

Martelli, Matteo, "Medicina e alchimia. 'Estratti galenici' nel Corpus degli scritti alchemici siriaci di Zosimo," *Galenos*, 4 (2010): 207–28

Martelli, Martelli, "La tradizione tecnico-artigianale e l'influenza orientale: lo Pseudo-Democrito alchimista," in *La scienza antica e la sua tradizione*, ed. Franco Repellini and Gianni Micheli (Milan: Cisalpino, 2011), 175–212

Martelli, Matteo, "Greek Alchemists at Work: 'Alchemical Laboratory' in the Greco-Roman Egypt," *Nuncius*, 26 (2011): 271–311

Matton, Sylvain, "L'influence de l'humanisme sur la tradition alchimique," *Micrologus* 3 (1995): 279–345

Maxfield, Valerie A., "Stone Quarrying in the Eastern Desert with Particular Reference to Mons Claudianus and Mons Porphyrites," in *Economies beyond Agriculture in the Classical Word*, ed. David J. Mattingly and John Salmon (London: Routledge, 2001), 143–170

Mejer, Jirgen, "Demetrius of Magnesia: On Poets and Authors of the Same Name," *Hermes*, 109 (1981): 447–72

Mertens, Michèle, *Un traité gréco-égyptien d'alchimie: la 'Lettre d'Isis à Horus'* (Liège: Mémoire de licence inédit, Université de Liège, 1984)

Mertens, Michèle, "Une scène d'initiation alchimique: la *Lettre d'Isis à Horus*," *Revue de l'histoire des religions*, 205 (1988): 3–23

Mertens, Michèle, "Pourquoi Isis est-elle appelée 'prophetis'?," *Chronique d'Égypte*, 64 (1989): 260–6

Mertens, Michèle, "Sur la trace des anges rebelles dans les traditions ésotériques du début de notre ère jusqu'au XVIIᵉ siècle," in *Anges et démons. Actes du Colloque de Liège et de Louvain-la-Neuve (25–26 novembre 1987)*, ed. Julien Ries and Henri Limet (Louvain-la-Neuve: Centre d'histoire des religions, 1989), 383–98

Mertens, Michèle, *Zosime de Panopolis, Mémoires authentiques* (Paris: Les Belles Lettres, 1995)

Milik, Jazef T., *The Books of Enoch. Aramaic Fragments of Qumrân Cave 4* (Oxford: Clarendon Press, 1976)

Momigliano, Arnaldo, *Alien Wisdom. The Limits of Hellenization* (Cambridge: Cambridge University Press, 1975)

Monat, Pierre, *Firmicus Maternus. Mathesis, livres I–II* (Paris: Les Belles Lettres, 1992)

Mott Gummerre, Richard, *Seneca, Epistles 66–92* (Cambridge, MA and London: Loeb, 1920)

Multhauf, Robert P., *The Origins of Chemistry* (London: Oldbourne, 1966)

Nicholson, Paul T., and Ian Shaw, *Ancient Egyptian Materials and Technology* (Cambridge: Cambridge University Press, 2000)

Nriagu, Jerome O., *Lead and Lead Poisoning in Antiquity* (New York: John Wiley and Sons, 1983)

Oder, Eugen, "Beiträge zur Geschichte der Landwirtschaft bei den Griechen, Teil I," *Rheinisches Museum für Philologie*, 45 (1890): 58–98

Papathanassiou, Maria K., "L'oeuvre alchimique de Stéphanos d'Alexandrie: structure et transformations de la matière, unité et pluralité, l'énigme des philosophes," *Chrysopoeia*, 7 (2000-3): 11-31 (reprint in *L'alchimie et ses racines philosophiques*, ed. Cristina Viano [Paris, 2005], 113–33)

Partington, James R., *A Short History of Chemistry* (New York: St. Martin's Press, 1957)

Partington, James R., *A History of Chemistry,* vol. I/1: *Theoretical Background* (London: Macmillan & Co, 1970)

Pensabene, Patrizio, "Le cave del Mons Claudianus. Conduzione statale, appalti, e distribuzione," *Journal of Roman Archaeology,* 12 (1999): 721–36

Pfister, René, "Teinture et alchimie dans l'Orient hellénistique," *Seminarium Kondakovianum,* 7 (1935): 1–59

Pfister, René, *Textiles de Palmyre,* 3 vols. (Paris: Les éditions d'Art et d'Histoire, 1940)

Plass, Paul, "A Greek Alchemical Formula," *Ambix,* 26 (1982): 69–73

Principe, Lawrence M., *The Secrets of Alchemy* (Chicago and London: University of Chicago Press, 2013)

Quack, Joachim F., "Les Mages Égyptianisés? Remarks on Some Surprising Points in Supposedly Magusean Texts," *Journal of Near Eastern Studies,* 64 (2006): 267–82

Rackham, Harris, *Pliny, Natural History, Books 33–35* (Cambridge, MA and London: Loeb, 1971)

Raïos, Dimitris R., *Recherches sur le «Carmen de ponderibus et mensuribus»* (Ioannina: Panepistēmio Iōanninōn, 1983)

Raïos, Dimitris R., *Archimède, Ménélaos d'Alexandrie et le «Carmen de ponderibus et mensuris»: contributions à l'histoire des sciences*, (Ioannina: Panepistēmio Iōanninōn, 1989)

Raïos, Dimitris R., "L'invention de l'hydroscope et la tradition arabe," *Graeco-Arabica*, 5 (1993): 275–86

Raïos, Dimitris R., "Autour de la paraphrase du «Carmen de ponderibus et mensuris»," in *Science antique, science médiévale (Autour du manuscrit d'Avranches 235). Actes du Colloque international, Mont-Saint-Michel, 4–7 septembre 1998*, ed. Louis Callebat and Olivier Desbordes (Hildesheim, Zürich and New York: Olms and Weidmann, 2000), 297–318

Renna, Enrico, "Ricette per succedanei della porpora in due papiri greci," in *La porpora. Realtà ed immaginario di un colore simbolico. Atti del Convegno di studio, Venezia, 24 e 25 ottobre 1996*, ed. Oddone Longo (Venice: Istituto Veneto di scienze lettere ed arti, 1998), 133–47

Ribichini, Sergio, "Fascino dall'Oriente e prime lezioni di magia," in *La questione delle influenze vicino-orientali sulla religione greca. Stato degli studi e prospettive della ricerca. Atti del Colloquio internazionale di Roma, 20–22 maggio 1999*, ed. Sergio Ribichini, Maria Rocchi and Paolo Xella (Rome: ed. del CNRS, 2001), 103–16

Riess, Ernst, "Nechepsonis et Petosiridis Fragmenta Magica," *Philologus*, suppl. 6 (1892): 325–94

Rolfe, John C., *The Attic Nights of Aulus Gellius*, 3 vols. (London: Loeb, 1927)

Romano, Elisa, "I colori artificiali e le origini della chimica," in *Sciences exactes et sciences appliquées à Alexandrie, Actes du Colloque international qui s'est tenu à Saint-Étienne du 6 au 8 juin 1996*, ed. Gilbert Argoud and Jean-Yves Guillaumin (Saint-Étienne: Publications de L'Université de Saint-Étienne, 1998), 115–26

Romano, Francesco, "Porfirio technologos?," *Siculorum Gymnasium. Rassegna semestrale della Facoltà di Lettere e Filosofia dell'Università di Catania*, 31 (1978): 517–20

Ruska, Julius, *Tabula Smaragdina. Ein Beitrag zur Geschichte der hermetischen Literatur* (Heidelberg: Carl Winter's Universitätsbuchhandlung, 1926)

Ruska, Julius, *Turba philosophorum. Ein Beitrag zur Geschichte der Alchemie* (Berlin: Julius Springer, 1931)

Sachau, Eduard, *Aramäische Papyrus und Ostraka aus einer jüdischen militär-Kolonie zu Elephantine* (Leipzig: J.C. Hinrichs'sche Buchhandlung, 1911)

Schorsch, Deborah, "Precious-Metal Polychromy in Egypt in the Time of Tutankhamun," *Journal of Egyptian Archaeology*, 87 (2001): 55–71

Secret, François, "Notes sur quelques alchimistes italiens de la Renaissance," *Rinascimento*, 23 (1973): 211–7

Sezgin, Fuat, *Geschichte des arabischen Schrifttums*. Vol. 4: *Alchimie-Chemie, Botanik-Agrikultur* (Leiden: Brill, 1971)

Smith, Cyril S., and John C. Hawthorne, "Mappae Clavicula: A Little Key to the World of Medieval Techniques," *Transactions of the American Philosophical Society*, n.s. 64 (1974): 3–128

Steele, Robert R., "The Treatise of Democritus on Things Natural and Mystical," *Chemical News*, 61 (1890): 88–125

Steele, Robert R., "Practical Chemistry in the Twelfth Century. *Rasis de Aluminibus et Salibus* Translated by Gerard of Cremona," *Isis*, 12 (1929): 11–46

Stéphanidès, Michel, "Petites contributions à l'histoire des sciences," *Revue des études grecs*, 31 (1918): 197–206

Stéphanidès, Michel, "Notes sur les textes chymeutiques," *Revue des études grecs*, 35 (1922): 296–320

Susanetti, Davide, *Sinesio di Cirene. I sogni, introduzione, traduzione e commento* (Bari: Adriatica editrice, 1992)

Tannery, Paul, "Études sur les alchimistes grecs. Synésius à Dioscore, " *Revue des études grecs*, 3 (1890): 282–8

Taylor, F. Sherwood, "A Survey of Greek Alchemy," *Journal of Hellenic Studies*, 50 (1930): 109–39

Taylor, F. Sherwood, "The Origins of Greek Alchemy," *Ambix*, 1 (1937): 30–48

Taylor, F. Sherwood, "The Alchemical Works of Stephanos of Alexandria. Translations and Commentary, Part. II," *Ambix*, 2 (1938-40): 38–49

Taylor, F. Sherwood, "The Evolution of the Still," *Annals of Science*, 5 (1945): 185–202

Takacs, Laszlo, "Quicksilver from Cinnabar: the First Documented Mechano-chemical Reaction?," *Journal of Minerals, Metals and Materials Society*, 52 (2000): 12–3

Tarrant, Harold, *Thrasyllian Platonism* (Ithaca, NY and London: Cornell University Press, 1993)

Tourtelle, Étienne, *Histoire philosophique de la médecine, depuis son origine jusqu'au commencement du 18ᵉ siècle*, 2 vols. (Paris: Levrault, 1804)

Traunecker, Claude, "Le Château de l'Or de Thoutmosis III et les magasins nord du temple d'Amon," *CRIPEL*, 11 (1989): 89–111

Ullmann, Manfred. *Die Natur- und Geheimwissenschaften im Islam* (Leiden: Brill 1972)

Verhecken, André, "Relation between age and dyes of 1st millennium AD textiles found in Egypt," in *Methods of Dating Ancient Textiles of the 1ˢᵗ Millennium AD from Egypt and Neighbouring Countries*, ed. Antoine De Moor and Cäcilia Flück (Tielt: Lannoo, 2007), 206–13

Vereno, Ingolf, *Studien zum ältesten alchemistischen Schrifttum. Auf der Grundlage zweier erstmals edierter arabischer Hermetica* (Berlin: Klaus Schwarz Verlag, 1992)

Viano, Cristina, "Gli alchimisti greci e l'acqua divina," *Rendiconti dell'Accademia Nazionale delle Scienze detta dei LX. Parte II: Memorie di Scienze Fisiche e Narurali*, 21 (1997): 61–70

Viano, Cristina, "Les alchimistes gréco-alexandrins et le *Timée* de Platon," in *L'alchimie et ses racines philosophiques. La tradition grecque et la tradition arabe*, ed. Cristina Viano (Paris: Vrin, 2005), 91–107

Viano, Cristina, *La matière des choses. Le livre IV des Météorologiques d'Aristote et son interprétation par Olympiodore* (Paris: Vrin, 2006)

Vittori, Ottavio, "Pliny the Elder on Gilding: A New Interpretation of His Comments," *Gold Bulletin*, 12 (1979): 35-9

Węcowski, Marek, "Pseudo-Democritus, or Bolos of Mendes (n. 263)," in *Brill's New Jacoby*, ed. Ian Worthington (Brill Online, 2013)

Wellmann, Max, *Pedanii Dioscuridis Anazarbei de materia medica libri quinque*, 3 vols. (Berlin: Weidmann, 1906–14)

Wellmann, Max, "Aelius Promotus: *Iatrika physika kai antipathētika*," *Sitzungsberichte der preußischen Akademie der Wissenschaften*, 37 (1908): 772–7

Wellmann, Max, *Die Georgika des Demokritos* (Berlin: "Abhandlungen der Preußischen Akademie der Wissenschaften, Phil.-hist. Klasse" 4, 1921)

Wellmann, Max, "Zu Demokrit," *Hermes*, 61 (1926), 474–5

Wellmann, Max, *Die phsyika des Bolos Demokritos und der Magier Anaxilaos aus Larissa, Teil I* (Berlin: "Abhandlungen der Preußischen Akademie der Wissenschaften. Phil.-hist. Klasse" 7, 1928)

Wilson, C. Anne, "Distilling, Sublimation, and the Four Elements: The Aims and Achievements of the Earliest Greek Chemists," in *Science and Mathematics in Ancient Greek Culture*, ed. Christopher J. Tuplin and Tracey E. Rihll (Oxford: Oxford University Press, 2002), 307–22

Wood, Robert W., "The Purple Gold of Tut'ankhamūn," *Journal of Egyptian Archaeology*, 20 (1934): 62–5

Würschmidt, Joseph, "Die Schrift des Menelaos über die Bestimmung der Zusammensetzung von Legierungen," *Philologus*, 80 (1925): 377–409

Zaccagnini, Carlo, "Patterns of Mobility among Ancient Near Eastern Craftsmen," *Journal of Near Eastern Studies*, 42 (1983): 245–64